Topics in Geometric Group Theory

Chicago Lectures in Mathematics Series
Robert J. Zimmer, series editor
J. P. May, Spencer J. Bloch, Norman R. Lebovitz, and Carlos Kenig, editors

Other *Chicago Lectures in Mathematics* titles available from the
University of Chicago Press:

Simplicial Objects in Algebraic Topology, by J. Peter May (1967)
Fields and Rings, Second Edition, by Irving Kaplansky (1969, 1972)
Lie Algebras and Locally Compact Groups, by Irving Kaplansky (1971)
Several Complex Variables, by Raghavan Narasimhan (1971)
Torsion-Free Modules, by Eben Matlis (1973)
Stable Homotopy and Generalised Homology, by J. F. Adams (1974)
Rings with Involution, by I. N. Herstein (1976)
Theory of Unitary Group Representation, by George V. Mackey (1976)
Commutative Semigroup Rings, by Robert Gilmer (1984)
Infinite-Dimensional Optimization and Convexity, by Ivar Ekeland and
Thomas Turnbull (1983)
Navier-Stokes Equations, by Peter Constantin and Ciprian Foias (1988)
Essential Results of Functional Analysis, by Robert J. Zimmer (1990)
Fuchsian Groups, by Svetlana Katok (1992)
*Unstable Modules over the Steenrod Algebra and Sullivan's Fixed Point
Set Conjecture*, by Lionel Schwartz (1994)
Topological Classification of Stratified Spaces, by Shmuel Weinberger
(1994)
Lectures on Exceptional Lie Groups, by J. F. Adams (1996)
Geometry of Nonpositively Curved Manifolds, by Patrick B. Eberlein
(1996)
*Dimension Theory in Dynamical Systems: Contemporary Views and
Applications*, by Yakov B. Pesin (1997)
*Harmonic Analysis and Partial Differential Equations: Essays in
Honor of Alberto Calderón*, edited by Michael Christ, Carlos Kenig,
and Cora Sadosky (1999)
A Concise Course in Algebraic Topology, by J. P. May (1999)

Pierre de la Harpe

Topics in Geometric Group Theory

The University of Chicago Press

Chicago and London

The University of Chicago Press, Chicago 60637
The University of Chicago Press, Ltd., London
© 2000 by The University of Chicago
All rights reserved. Published 2000
Printed in the United States of America
09 08 07 06 05 04 03   2 3 4 5

ISBN: 0-226-31719-6 (cloth)
ISBN: 0-226-31721-8 (paper)

Library of Congress Cataloging-in-Publication Data

La Harpe, Pierre de.
    Topics in geometric group theory / Pierre de la Harpe.
        p.   cm.—(Chicago lectures in mathematics)
    Includes bibliographic references and index.
    ISBN 0-226-31719-6 (alk. paper)—ISBN 0-226-31721-8 (pbk. : alk. paper)
    1. Geometric group theory. I. Title. II. Series.

QA183.L3   2000
512'.2—dc21

                                        00-020271

CONTENTS

INTRODUCTION

All ornament should be based upon a
geometrical construction
(Proposition 8 of [Jon–1856])

Groups as abstract structures were recognized progressively during the 19th century by mathematicians including Gauss (*Disquisitiones arithmeticae* in 1801), Cauchy, Galois, Cayley, Jordan, Sylow, Frobenius, Klein (*Erlangen program* in 1872), Lie, Poincaré ... ; see e.g. Chapter III in [Dieud–78] and [NeumP–99]. Groups are of course sets given with appropriate "multiplications", and they are often given together with actions on interesting geometric objects. But the fact which we want to stress here is that groups are also interesting geometric objects by themselves — a point of view illustrated in the past by Cayley and Dehn (see Chapter I.5 in [ChaMa–82]), and more recently by Gromov. More precisely, a finitely-generated group can be seen as a metric space (the distance between two points being defined "up to quasi-isometry", in the sense of Section IV.B below), and this gives rise to a very fruitful approach to group theory.

The purpose of this book is to provide an introduction to this point of view, with emphasis on finitely-generated versus finitely-presented groups, on growth of groups, and on examples.

It has developed from a postgraduate course given in Geneva and Lausanne during the winter term 1995/96. Several lectures[1] were given by R. Grigorchuk, and one by D.B.A. Epstein. The lectures about random walks on groups and about amenability have not been written up; among several possible references for these subjects, let us mention a recent book on random walks, [Woess–00], and an article on amenability, [CecGH–99].

The backgrounds of the course participants were far from homogeneous. Similarly, in the present book, the knowledge assumed of the reader varies a great deal from one page to another. As we have chosen to concentrate on examples, we hope that the reader who does not have the background for one part of our exposition can nevertheless read later parts. We also hope that

---

[1]Part of the material exposed by R. Grigorchuk appears in parts of Chapters VI to VIII. D. Epstein lectured on rewriting systems, which do not appear here (see [Epst6–92], in particular Chapter 6, and [Brown–92]).

mature mathematicians, working outside geometric group theory, will find the notes of interest and that they will enjoy the connections between various fields.

However, we definitely assume that the reader will have some knowledge of basic group theory, such as a *small part* of a book like [KarMe–85] or [Rotma–95], as well as some knowledge of elementary algebraic topology (fundamental group and covering spaces), for which two among many good references are the de Rham notes [Rham–69] and a book by Massey [Masse–67].

Most sections are followed by a choice of *exercises,* most of which should be easy for the reader, and a list of *problems and complements;* problems range from slightly more difficult exercises to research problems, and complements often reduce to some comments on the literature. We feel it is crucial for the reader to study many of the exercises (even if it is safe to avoid quite a few others). On the other hand, problems and complements can be used for relaxed bar-conversations after more serious lectures: what the reader will grasp there could well depend strongly on the mood of the day.

An index of research problems appears after the references.

One consequence of the style adopted for these problems and complements is a very long list of references. Even if nobody will imagine that I have studied all of them in detail, my hope is that some readers will find it helpful; others will forgive me for missing important contributions.

My choice of material to illustrate "geometric group theory" is clearly subjective, and there are many important topics which are completely missing, or to say the least under-represented, in this book. Besides random walks and amenability, already alluded to, here are some of the topics that would certainly deserve more attention.

- Small cancellation theory, and the geometry of defining relations [Ol's–89].
- Mostow rigidity ([Mosto–68], [Mosto–73]).
- Margulis super-rigidity and arithmeticity [Margu–91].
- Thurston's geometrization conjecture
  ([Thurs–82], [Scott–83], [NeumW–99]).
- The congruence subgroup problem ([PrasG–91], [Rapin–92], [Rapin–99]).
- A *proof* of Gromov's theorem on polynomial growth and nilpotent groups [Gromo–81] (see VII.29).
- Asymptotic cones; see [Gromo–81] and Chapter 2 in [Gromo–93], as well as [VdDW–84a], [KapLe–95] and [KleLe–98].
- Finite generation of kernels of abelian quotients of finitely-generated groups, and Bieri-Neumann-Strebel invariants [BieNS–87].
- Groups acting on real trees [Morga–92], the work of Rips [Pauli–97], and JSJ-decomposition of groups (see III.15).
- Moduli spaces of representations of a finitely-generated group in some algebraic group (see [CulSh–83] and [BauSh–90]).

The course could not have been given nor the notes written up without the help of many students, colleagues, librarians, and readers of earlier versions. We would like to thank in particular

Slava Grigorchuk

as well as

Goulnara Arzhantseva, Roland Bacher, Laurent Bartholdi, Bachir Bekka, Dietmar Bisch, Martin Bridson, Marc Burger, Tullio Ceccherini-Silberstein, Bernard Dudez, Shalom Eliahou, Benson Farb, Steve Gersten, Etienne Ghys, André Haefliger, Jean-Claude Hausmann, Michel Kervaire, Gilbert Levitt, Alex Lubotzky, Igor Lysenok, Antonio Machì, Volodia Nekrashevych, Alexander Yu. Ol'shanskii, Panos Papasoglu, Christophe Pittet, Gilles Robert, Felice Ronga, Jan Saxl, Yehuda Shalom, Tatiana Smirnova-Nagnibeda, John Steinig, Jacques Thévenaz, Aryeh Vaillant, Alain Valette, and Thierry Vust

for their much appreciated contributions. This incomplete list shows among other things that support of the Swiss National Science Foundation has been vital for this book, and more generally to keep Geneva in the flow of current research.

One of my personal beliefs is that fascination with symmetries and groups is one way of coping with the frustrations of life's limitations: we like to recognize symmetries which allow us to grasp more than what we can see. In this sense, the study of geometric group theory is a part of culture, and reminds me of several things that Georges de Rham practised on many occasions, such as teaching mathematics, reciting Mallarmé, or greeting a friend.

CHAPTER I

GAUSS' CIRCLE PROBLEM AND
PÓLYA'S RANDOM WALKS ON LATTICES

Much of the material in the later chapters is about results of the last 50 years. However, there are related results which are classic; this introductory chapter is about two of them. One goes back to Gauss; the other goes back at least to Pólya (1921), and possibly to the first mathematically inclined drunkard.

I.A. The circle problem

Consider a group Γ and a function $\sigma : \Gamma \to \mathbb{R}_+$ which measures in some sense the "size" of elements of Γ.

What can be said about $\quad \sharp \left\{ \gamma \in \Gamma \mid \sigma(\gamma) \leq t \right\} \quad$ for large $t \in \mathbb{R}_+$?

This theme, concerning growth, has many variations. Here we illustrate it with a classical result due to Gauss about the *circle problem*, for which $\Gamma = \mathbb{Z}^2$ and $\sigma(a,b) = a^2 + b^2$. (See [Chand–68].) More on this in our Chapters VI and VII (in particular VI.42).

More precisely, set

$$R(t) \ = \ \sharp \left\{ (a,b) \in \mathbb{Z}^2 \mid a^2 + b^2 \leq t \right\}$$

for all $t \geq 0$. For example,

$$
\begin{aligned}
R(0) &= 1 & R(1) &= 5 \\
R(10) &= 37 & R(100) &= 317 \\
R(1000) &= 3149 & R(10\,000) &= 31\,417.
\end{aligned}
$$

The problem is to understand the asymptotic behaviour of $R(t)$ when t is large.

1. Theorem (Gauss). *We have* $R(t) - \pi t = O(\sqrt{t})$.

Proof. To each lattice point $(a,b) \in \mathbb{Z}^2$, we associate the unit square of the Euclidean plane with (a,b) as its "south-west corner". If $a^2 + b^2 \leq t$, the square corresponding to (a,b) is inside the disc of radius $\sqrt{t} + \sqrt{2}$; hence

$$R(t) \ \leq \ \pi \left(\sqrt{t} + \sqrt{2} \right)^2$$

for all $t \geq 0$. If the square corresponding to (a, b) touches the disc of radius $\sqrt{t} - \sqrt{2}$, then $a^2 + b^2 \leq t$; hence

$$R(t) \geq \pi \left(\sqrt{t} - \sqrt{2} \right)^2$$

for all $t \geq 0$. Thus

$$|R(t) - \pi t| \leq 2\pi \left(1 + \sqrt{2t} \right)$$

for all $t \geq 0$. $\square$

2. Later work. Many mathematicians have worked to improve the error term in Gauss' result. For example, it is known that

$$R(t) - \pi t = O(t^\alpha)$$

for the following values of α :

$$\alpha = \frac{1}{3} \qquad \sim 0.3333 \qquad \text{(Sierpiński, 1906)}$$

$$\alpha = \frac{37}{112} \qquad \sim 0.3304 \qquad \text{(Van der Corput, 1923)}$$

$$\alpha = \frac{15}{46} \qquad \sim 0.3261 \qquad \text{(Titchmarch, 1934)}$$

$$\alpha = \frac{13}{40} + \epsilon \qquad \sim 0.3250 \qquad \text{(Hua, 1941)}$$

$$\alpha = \frac{12}{37} \qquad \sim 0.3243 \qquad \text{(Chen, 1963)}$$

$$\alpha = \frac{35}{108} + \epsilon \qquad \sim 0.3241 \qquad \text{(Kolesnik, 1982)}$$

$$\alpha = \frac{7}{22} + \epsilon \qquad \sim 0.3182 \qquad \text{(Iwaniec and Mozzochi, 1988)}$$

$$\alpha = \frac{23}{73} + \epsilon \qquad \sim 0.3151 \qquad \text{(Huxley, 1993)}$$

to name but a few (where "$+\epsilon$" means as usual "for all $\epsilon > 0$"). Moreover,

$$\text{if} \qquad R(t) - \pi t = O(t^\alpha) \qquad \text{then} \qquad \alpha > \frac{1}{4}$$

(due independently to Landau, 1912, and Hardy, 1915). Apparently, specialists believe that $R(t) - \pi t = O(t^{1/4+\epsilon})$ for all $\epsilon > 0$. More on this in the following references.

S.W. Graham and G. Kolesnik: *Van der Corput's method of exponential sums,* Cambridge Univ. Press, 1991.

E. Grosswald: *Representation of integers by sums of squares,* Springer, 1985; see pages 20–22.

G.H. Hardy and E.M. Wright: *An introduction to the theory of numbers*, 5th ed., Oxford Univ. Press 1979; see page 272.

M.N. Huxley: *Exponential sums and lattice points II*, Proc. London Math. Soc. **66** (1993) 279–301 and **68** (1994) 264.

I.A. Ivic: *The Riemann zeta function*, J. Wiley, 1985; see in particular pages 372, 375, 384.

B. Lichtin: *Geometric features of lattice point problems*, in *Singularity theory*, D.T. Lê, K. Saito and B. Teissier Editors, World Scientific, 1995; see pages 370–443. This has an exposition of $R(t) - \pi t = O\left(t^{1/3+\epsilon}\right)$.

W. Sierpiński: *Elementary theory of numbers*, 2nd ed., North Holland, 1988; see pages 383 ff.

Problems analogous to the circle problem have also been considered in $\mathbb{R}^d$ for $d \geq 3$, with ellipsoids instead of balls, and with other plane domains; see, e.g., the papers by Lichtin and Huxley cited above.

There are also natural problems in hyperbolic spaces (rather than in $\mathbb{R}^d$), about the asymptotic behaviour of numbers of lattice points in large balls. Work on these problems goes back at least to Delsarte [Delsa–42]. See § III.3 in [BekMa] for an exposition of particular cases of results of Margulis and Eskin-McMullen; see [Patte–87] for an exposition of more precise results of Patterson and Sullivan.

I.B. Pólya's recurrence theorem

Consider the *simple random walk* on the lattice $\mathbb{Z}$ of integers. In loose terms, a walker is at the origin at time 0 and moves one step to the left or one step to the right, equiprobably, after each unit of time.

The question of *recurrence* for $\mathbb{Z}$ is to ask whether the walker has a 100 % chance of visiting the origin again infinitely many times. (In the original paper [Pólya–21], Pólya's question was slightly different; see Section 5.3 in [DoySn–84].)

The answer is yes, as we show, following Section 7.2 of [DoySn–84]; other arguments are sketched in Exercise 5.

The number of all paths of length $2n$ is 2^{2n}; the number of those ending at the origin is $\binom{2n}{n}$ because they involve a choice of n steps to the right among their $2n$ steps. Hence the probability u_{2n} of being at the origin after $2n$ steps is

$$u_{2n} = \frac{1}{2^{2n}} \binom{2n}{n}$$

(observe that $u_{2n+1} = 0$). Using Stirling's formula $k! \sim k^k e^{-k} \sqrt{2\pi k}$, we have

$$u_{2n} \sim \frac{1}{2^{2n}} \frac{(2n)^{2n} e^{-2n} \sqrt{2\pi 2n}}{n^{2n} e^{-2n} 2\pi n} = \frac{1}{\sqrt{\pi n}}$$

(where $\sim$ indicates that the quotient of the left-hand side by the right-hand side converges to 1 when $n \to \infty$). As $\sum_{n=1}^{\infty} \frac{1}{\sqrt{n}} = \infty$, we have

$$\sum_{k=1}^{\infty} u_k = \sum_{n=1}^{\infty} u_{2n} = \infty.$$

This means that the simple random walk on $\mathbb{Z}$ just considered is *recurrent*, namely the random walker has probability 1 of returning to his starting point infinitely often.

From the elementary theory of infinite products, and as $u_{2n} > 0$ for all $n \geq 1$, the series $\sum_{n=1}^{\infty} u_{2n}$ is divergent if and only if the infinite product $\prod_{n=1}^{\infty} (1 - u_{2n})$ is not absolutely convergent. (Recall that an infinite product of real numbers of the form $\prod_{n=1}^{\infty} (1 - a_n)$, with $0 \leq a_n < 1$ for all $n \geq 1$, diverges to zero if and only if the series $\sum_{n=1}^{\infty} a_n$ diverges; see e.g. Theorem 1.17 of Chapter VII in [SakZy–65].) In other words, the random walk is recurrent if and only if the probability for the walker of *never* re-visiting his starting point is zero.

Consider the analogous problem on the lattice $\mathbb{Z}^2$ of the Euclidean plane. The walk is again recurrent.

Let us show this. The number of all paths of length $2n$ is now 4^{2n}. Among these, the number of paths that return to the origin after k steps North, k steps South, $n - k$ steps East and $n - k$ steps West is the multinomial coefficient

$$\binom{2n}{k \ , \ k \ , \ n-k \ , \ n-k} = \frac{(2n)!}{k! \, k! \, (n-k)! \, (n-k)!} \cdot$$

Hence the probability of being at the origin after $2n$ steps is

$$u_{2n} = \frac{1}{4^{2n}} \sum_{k=0}^{n} \frac{(2n)!}{k! \, k! \, (n-k)! \, (n-k)!} = \frac{1}{4^{2n}} \frac{(2n)!}{n! \, n!} \sum_{k=0}^{n} \binom{n}{k} \binom{n}{n-k} \cdot$$

Now, comparing the coefficient of x^n in the two sides of the identity $(1+x)^{2n} = (1+x)^n (1+x)^n$, one obtains $\binom{2n}{n} = \sum_{k=0}^{n} \binom{n}{k} \binom{n}{n-k}$. Thus

(*) $$u_{2n} = \frac{1}{4^{2n}} \frac{(2n)!}{n! \, n!} \binom{2n}{n} = \left\{ \frac{1}{2^{2n}} \binom{2n}{n} \right\}^2 \cdot$$

Using Stirling's formula, we have

$$u_{2n} \sim \frac{1}{\pi n} \quad \text{and} \quad \sum_{k=1}^{\infty} u_k = \sum_{n=1}^{\infty} u_{2n} = \infty$$

as above, so that the simple random walk on $\mathbb{Z}^2$ is indeed recurrent.

The situation is different in one more dimension: the simple random walk on $\mathbb{Z}^3$ is *transient* (this means precisely "non-recurrent").

As before, we have

$$u_{2n} = \frac{1}{6^{2n}} \sum_{\substack{j,k \geq 0 \\ j+k \leq n}} \frac{(2n)!}{j! \, j! \, k! \, k! \, (n-j-k)! \, (n-j-k)!}$$

$$= \frac{1}{2^{2n}} \binom{2n}{n} \sum_{\substack{j,k \geq 0 \\ j+k \leq n}} \left(\frac{1}{3^n} \frac{n!}{j! \, k! \, (n-j-k)!} \right)^2 .$$

For all $j, k \geq 0$ with $j + k \leq n$, we have

(**) $$\frac{n!}{j! \, k! \, (n-j-k)!} \leq \frac{n!}{\left[\frac{n}{3}\right]! \, \left[\frac{n}{3}\right]! \, \left[\frac{n}{3}\right]!}$$

(Exercise 6). Consequently

$$u_{2n} \leq \frac{1}{2^{2n}} \binom{2n}{n} \frac{1}{3^n} \frac{n!}{\left(\left[\frac{n}{3}\right]!\right)^3} \sum_{\substack{j,k \geq 0 \\ j+k \leq n}} \frac{1}{3^n} \frac{n!}{j! \, k! \, (n-j-k)!} .$$

As the last sum is 1, we have

$$u_{2n} \leq \frac{1}{2^{2n}} \binom{2n}{n} \frac{1}{3^n} \frac{n!}{\left(\left[\frac{n}{3}\right]!\right)^3} .$$

Stirling's fomula shows that

$$\frac{1}{2^{2n}} \binom{2n}{n} \frac{1}{3^n} \frac{n!}{\left(\left[\frac{n}{3}\right]!\right)^3} = \frac{1}{2^{2n}3^n} \frac{(2n)!}{n! \, \left(\left[\frac{n}{3}\right]!\right)^3} \sim \frac{\sqrt{2}}{\left(\sqrt{\frac{2\pi}{3}}\right)^3 n^{\frac{3}{2}}}$$

and we have finally

$$\sum_{k=1}^{\infty} u_k = \sum_{n=1}^{\infty} u_{2n} \leq K \sum_{n=1}^{\infty} \frac{1}{n^{\frac{3}{2}}} < \infty$$

for an appropriate constant K.

3. Theorem (Pólya, 1921). *The simple random walk on $\mathbb{Z}^d$ is*

$$\begin{cases} \text{recurrent if } d = 1 \text{ or } d = 2 \\ \text{transient if } d \geq 3. \end{cases}$$

Proof. The above proof for $d = 3$ carries over (with only minor changes) to the case $d \geq 3$. $\square$

Feller adds that, in dimension 3, the probability of returning to the initial position is about 0.35, and the expected number of returns is consequently

$$0.65 \sum_{k \geq 1} k(0.35)^k \; = \; 0.65 \frac{0.35}{(1-0.35)^2} \; = \; \frac{0.35}{0.65} \; = \; 0.53$$

(Section 7 of Chapter 14 in [Felle–50]).

Recall that a *probability measure* on a group Γ is a function $p : \Gamma \to [0,1]$ such that $\sum_{\gamma \in \Gamma} p(\gamma) = 1$; such a probability is *symmetric* if $p(\gamma^{-1}) = p(\gamma)$ for all $\gamma \in \Gamma$. A probability p on Γ defines a *left-invariant random walk,* with the probability that a walker at some point $\gamma_1 \in \Gamma$ will go in one step to a point γ_2 being $p(\gamma_1^{-1}\gamma_2)$. Random walks of this kind constitute an important subject of "geometric group theory"; one of the foundational works is [Keste–59]; see also the reviews in § 7 of [MohWo–89] and in [Woess–94].

The following result is due to Varopoulos (see [VarSC–92], in particular the end of Chapter VI, as well as the exposition in [Ancon–90]). It is considerably deeper than the theorem of Pólya, and depends among other things on Gromov's Theorem VII.29.

4. Theorem (Varopoulos, 1986). *Let* Γ *be a finitely-generated group and let*

$$p \; : \; \Gamma \; \longrightarrow \; [0,1]$$

be a symmetric probability measure on Γ *with finite support which generates* Γ. *If the random walk defined by* Γ *and* p *is recurrent, then*

> *either* Γ *is a finite group,*
>
> *or* Γ *has a subgroup of finite index isomorphic to* $\mathbb{Z}$,
>
> *or* Γ *has a subgroup of finite index isomorphic to* $\mathbb{Z}^2$.

Exercises for I.B

5. Exercise. Fill in the details of the following arguments, showing that the simple random walk on $\mathbb{Z}$ is recurrent.

(i) Let $\pi(k)$ denote the probability that a random walker starting at some point $k \in \mathbb{Z}$ will sooner or later be at the origin; in particular $\pi(0) = 1$. As

$$\pi(k) \; = \; \frac{1}{2}\pi(k-1) + \frac{1}{2}\pi(k+1)$$

the graph of the function π is a straight line. As $\pi(k) \geq 0$ for any $k \in \mathbb{Z}$, the equation of this line is $\pi(k) = 1$. (This appears on page 1 of [DynYu–67].)

(ii) Let π denote the probability that a random walker starting at the origin will later return to the origin. As

$$\pi = \frac{1}{2} + \frac{1}{4}\pi + \frac{1}{8}\pi^2 + \ldots = \frac{1}{2-\pi}$$

we have $\pi = 1$. (This comes from the beginning of Chapter 3 in [Hughe–95].)

6. Exercise. (i) Explain Formula (*) above, showing that the probability u_{2n} for $\mathbb{Z}^2$ is the square of the probability u_{2n} for $\mathbb{Z}$.

[Hint. The walk in $\mathbb{Z}^2$ is the "product" of a first walk with steps $\pm\left(\frac{1}{2},\frac{1}{2}\right)$ and of a second walk with steps $\pm\left(\frac{1}{2},-\frac{1}{2}\right)$. If a broader hint is necessary, see [DoySn–84], Section 7.6.]

(ii) Check that $a!b! \geq (a+1)!(b-1)!$ for positive integers a, b such that $a < b$. Deduce from it the inequality (**) used above, shortly before Theorem 3.

7. Exercise. A walker on $\mathbb{N}$ is at[1] 0 at time 0 and at 1 at time 1. If he is at k at time $n \geq 1$, then either $k = 0$ and he stays there at time $n + 1$, or $k \geq 1$ and he moves equiprobably one step to the left or one step to the right. Denote by P_k^n the probability that the walker is at k at time n, and set $P_k(z) = \sum_{n=0}^{\infty} P_k^n z^n$ (a formal power series, which is called the *generating function* of the sequence $(P_k^n)_{n\geq 0}$).

(i) Compute P_k^n for $n \leq 10$.

(ii) Check that

$$P_0(z) = 1 + z\Big[P_0(z) - 1\Big] + \frac{z}{2}P_1(z),$$

$$P_1(z) = z\Big[1 + \frac{1}{2}P_2(z)\Big],$$

$$P_k(z) = z\Big[\frac{1}{2}P_{k-1}(z) + \frac{1}{2}P_{k+1}(z)\Big] \qquad \text{for all} \qquad k \geq 2.$$

(iii) Deduce from (ii) that

$$P_k(z) = 2\left(\frac{1 - \sqrt{1-z^2}}{z}\right)^k$$

for all $k \geq 1$ and that

$$P_0(z) = 1 + \frac{1 - \sqrt{1-z^2}}{1-z}.$$

Check that the first terms of the Taylor expansion of this function at the origin agree with the numbers found in (i).

---

[1] In this book, $\mathbb{N}$ denotes the set of natural numbers, *including* 0.

(iv) For any convergent power series $f(z) = \sum_{n \geq 0} f_n z^n$ about the origin, check that

$$(1 - z)f'(z) - f(z) = \sum_{n \geq 1} n(f_n - f_{n-1})z^{n-1}.$$

In particular, for $f = P_0$,

$$(1 - z)P_0'(z) - P_0(z) = \frac{z}{\sqrt{1 - z^2}} - 1 = \sum_{n \geq 1} n(P_0^n - P_0^{n-1})z^{n-1}$$

$$= -1 + z + \frac{1}{2}z^3 + \frac{1 \cdot 3}{2 \cdot 4}z^5 + \frac{1 \cdot 3 \cdot 5}{2 \cdot 4 \cdot 6}z^7 + \ldots + \frac{1}{2^{2k}}\binom{2k}{k}z^{2k+1} + \ldots .$$

Check that these values of $n(P_0^n - P_0^{n-1})$ agree with the values found in (i).

(v) Observe that $P_0^n \geq P_0^{n-1}$ for all $n \geq 0$ and show that the *expected duration*

$$\sum_{n=1}^{\infty} n\left(P_0^n - P_0^{n-1}\right)$$

of return to the origin is infinite.

[Hint. One may use the following consequence of Abel's theorem, from the theory of functions of one complex variable: if $f(z) = \sum_{n \geq 0} f_n z^n$ is a holomorphic function in the open unit disc and if the series is convergent at the point 1, then

$$\lim_{r \longrightarrow 1-} f(r) = \sum_{n \geq 0} f_n .$$

See § III.2 in [SakZy–65].]

(vi) For a direct estimate of the expected duration of return to the origin, see also Section 3 of Chapter 14 in [Felle–50].

8. Exercise. Consider a connected regular graph, given together with a root, and the corresponding simple random walk: the random walker starts from the root and, at each step, walks with equal probability to one of the neighbouring steps. For each integer $k \geq 1$, denote by

N_k the number of all paths of length k starting from the root,
U_k the number of all paths of length k starting from and ending at the root,
V_k the number of all paths of length k starting from the root, ending at the root, and not visiting the root in between.

Set $u_k = U_k/N_k$ and $f_k = V_k/N_k$. One defines the *Green's function*

$$G(z) = 1 + \sum_{k=1}^{\infty} u_k z^k$$

and introduces the series

$$F(z) = \sum_{k=1}^{\infty} f_k z^k.$$

(i) Check that $G(z) = \frac{1}{1-F(z)}$.

(ii) For the lattice $\mathbb{Z}$, verify that

$$G(z) = 1 + \sum_{n=1}^{\infty} \frac{1}{2^{2n}} \binom{2n}{n} z^{2n} = \frac{1}{\sqrt{1-z^2}}.$$

(iii) If $P_0(z)$ is as in Exercise 7 and if $F(z) = \sum_{k=1}^{\infty} f_k z^k$ refers to the lattice $\mathbb{Z}$, check that

$$P_0(z) = 1 + \sum_{k=1}^{\infty} \left(\sum_{n=1}^{k} f_n \right) z^k = 1 + \frac{1}{1-z} \sum_{n=1}^{\infty} f_n z^n = 1 + \frac{1}{1-z} \sum_{n=1}^{\infty} f_{2n} z^{2n}$$

and deduce from this formula that

$$f_{2n} = \frac{1}{2^{2n-1}} \frac{1}{n} \binom{2n-2}{n-1}$$

for all $n \geq 1$. [For a standard combinatorial trick proving the identity for f_{2n}, see [Felle–50], Section 2 of Chapter III.]

Observe that the formula for $P_0(z)$ shows that

$$F(z) = 1 - \sqrt{1-z^2},$$

in agreement with (i) and (ii).

(iv) For the lattice $\mathbb{Z}^2$, check that

$$G(z) = 1 + \sum_{n=1}^{\infty} \frac{1}{4^{2n}} \binom{2n}{n}^2 z^{2n} = F\left(\frac{1}{2}, \frac{1}{2}, 1; z^2 \right),$$

where $F(a, b, c; z)$ is the traditional notation for the hypergeometric series, as in [Ahlfo–53].

For $a = b = \frac{1}{2}$ and $c = 1$ as above, this is non-algebraic. This follows from apparently well-known arguments on the growth of coefficients or of transcendence theory (end of Section 2 in [Stanl–80], Sections 3.4 and 4 in [Flajo–87]); this follows also from a computation of the monodromy group of the function $F\left(\frac{1}{2}, \frac{1}{2}, 1; z \right)$, as in Section E.1 of [GriHa–97].

9. Exercise. Consider now the simple random walk on the regular tree of degree $d \geq 2$. With the notation of the previous exercise, show successively that

$$f_{2n} = \frac{1}{d^{2n}} d(d-1)^{n-1} \frac{1}{n} \binom{2n-2}{n-1}$$

$$F(z) = \sum_{n=1}^{\infty} f_{2n} z^{2n} = \frac{d}{2(d-1)} \left[1 - \sqrt{1 - 4\frac{d-1}{d^2} z^2} \right]$$

$$G(z) = \frac{2(d-1)}{d-2 + \sqrt{d^2 - 4(d-1)z^2}}$$

(in case d is even, the formula for G appears as Formula (4.8) in [Keste–59]).

[Sketch for f_{2n}. Let V_{2n}^N denote the number of paths of length $2n$ on $\mathbb{N}$ which start from 0, end at 0, and do not visit 0 in between. It follows from Step (iii) in Exercise 8 that

$$V_{2n}^N = \frac{1}{n}\binom{2n-2}{n-1}.$$

The number V_{2n} of Exercise 8 for the regular tree of degree d is

$$V_{2n} = d(d-1)^n V_{2n}^N = d(d-1)^{n-1}\frac{1}{n}\binom{2n-2}{n-1}$$

because there are d choices for the first step away from the origin, $d-1$ choices for each other step away, and just one choice for each step back. This provides the answer for $f_{2n} = d^{-2n}V_{2n}$.]

Problems and complements for I.B

10. Problem. Check the details of the proof of Theorem 3 for $d \geq 3$.

11. Green's functions for other graphs and rationality. Let $G_X(z)$ be the Green's function of the simple random walk on a connected graph X. (For the definition of the Green's function, see Exercise 8 when X is regular, and [Woess–00] in general.) If X is finite, it is easy to see that $G_X(z)$ is a rational function. For regular connected graphs which are almost vertex-transitive, it is known that the converse holds; see Theorem 3.10 in [Barth–99], as well as [Kouks–98]. See also Complement VII.37 below.

Recall that a graph is *almost vertex-transitive* if its automorphism group acting on its vertex set has finitely many orbits. For example, Cayley graphs (see Section IV.A below) are vertex-transitive, and the barycentric subdivision of a vertex-transitive graph of finite degree is almost vertex-transitive.

12. Self-avoiding random walks. Much work has been done on problems of self-avoiding walks. Consider an integer $d \geq 2$ and the lattice $\mathbb{Z}^d$. A *self-avoiding walk* of length N in $\mathbb{Z}^d$ is a sequence $(\omega(0),\omega(1),\ldots,\omega(N))$ of points in $\mathbb{Z}^d$ such that $\omega(j) \neq \omega(k)$ for $j,k \in \{0,1,\ldots,N\}, j \neq k$, and such that $\omega(j)$ is a neighbour of $\omega(j-1)$ for $j \in \{1,\ldots,N\}$. Let $c_{d,N}$ denote the number of these with $\omega(0)$ the origin and let

$$\langle |\omega(N)|^2\rangle = \frac{1}{c_{d,N}}\sum_{\substack{\omega \text{ of length } N \\ \omega(0)=0}} \|\omega(N)\|^2$$

denote the *mean-square displacement*, where $\|\omega(N)\|$ denotes a Euclidean norm. It is an easy exercise to show that

$$d \leq \limsup_{N\to\infty}(c_{d,N})^{\frac{1}{N}} \leq 2d-1$$

for all $d \geq 2$.

Two basic problems are to understand $c_{d,N}$ and $\langle |\omega(N)|^2 \rangle$ for large N. One of the first results in this area is that the limit

$$\mu_d = \lim_{N \to \infty} (c_{d,N})^{\frac{1}{N}}$$

exists (Hammersley and Morton, 1954); it is called the *connective constant* for dimension d. The exact value of μ_d is unknown, even if $d = 2$ or $d = 3$. For the numbers of self-avoiding walks and for the mean-square displacement, there are *conjectures* according to which

$$c_{d,N} \sim A\mu^N N^{\gamma-1} \qquad \text{if} \quad d \neq 4$$

$$c_{4,N} \sim A\mu^N (\log N)^{\frac{1}{4}} \qquad \text{if} \quad d = 4$$

$$\langle |\omega(N)|^2 \rangle \sim DN^{2\nu} \qquad \text{if} \quad d \neq 4$$

$$\langle |\omega(N)|^2 \rangle \sim DN (\log N)^{\frac{1}{4}} \qquad \text{if} \quad d = 4$$

where A, D, μ, γ, ν are constants, depending on the dimension d, with

$$\gamma = \frac{43}{52} \qquad \text{and} \quad \nu = \frac{3}{4} \qquad \text{if} \quad d = 2$$

$$\gamma = 1.162\ldots \quad \text{and} \quad \nu = 0.59\ldots \quad \text{if} \quad d = 3$$

$$\gamma = 1 \qquad \text{and} \quad \nu = \frac{1}{2} \qquad \text{if} \quad d \geq 5$$

(see Chapter 1 of [MadSl–93]).

Research in this field is strongly motivated by various applications, such as models for long-chain polymers in a dilute solution (in chemical physics), spin models for ferromagnetism (in statistical physics), or percolation theory. See [MadSl–93] and Chapter 7 in [Hughe–95].

CHAPTER II

FREE PRODUCTS AND FREE GROUPS

This chapter is built around the so-called "Table-Tennis Lemma", due essentially to F. Klein, and some of its applications, exposed in Section II.B.

II.A. Free products of groups

A *monoid* is a set given together with a multiplication which is associative and which has a unit. Given a set A, the *free monoid* on A, also called the *set of words* on A, is the set $W(A)$ of finite sequences of elements of A; the product is given by juxtaposition and the unit is the empty word. We write a typical element of $W(A)$ as $w = a_1a_2...a_n$, where each a_j is in A; the integer n is then the *length* of w; it is 0 if w is the empty word. We identify A with the subset of $W(A)$ of words of length one.

Let $(\Gamma_\iota)_{\iota \in I}$ be a family of groups. Set $A = \coprod_{\iota \in I} \Gamma_\iota$ (disjoint union). Let $\sim$ be the equivalence relation on $W(A)$ generated by

$$we_\iota w' \sim ww' \qquad \text{where } e_\iota \text{ is the unit in } \Gamma_\iota \text{ for some } \iota \in I$$

$$wabw' \sim wcw' \qquad \text{whenever } a, b, c \in \Gamma_\iota \text{ with } c = ab \text{ for some } \iota \in I$$

for all $w, w' \in W(A)$. It is easy to check that the quotient $W(A)/\sim$ is naturally a monoid *which is indeed a group;* it is called the *free product* of the groups Γ_ι ($\iota \in I$) and is denoted by

$$*_{\iota \in I}\Gamma_\iota .$$

The inverse of the class represented by the word $w = a_1a_2...a_n$ is the class represented by $a_n^{-1}a_{n-1}^{-1}...a_1^{-1}$.

For A as above, a word $w = a_1a_2...a_n \in W(A)$, with $a_j \in \Gamma_{\iota_j}$ say, is said to be *reduced* if $\iota_{j+1} \neq \iota_j$ ($1 \leq j \leq n-1$) and if a_j is not the unit element of Γ_{ι_j} ($1 \leq j \leq n$). For example, if $I = \{1,2\}$, if $\Gamma_1 = \{e_1, a\}$ is a group of order 2, and if $\Gamma_2 = \{e_2, x, y\}$ is a group of order 3, then $axaya$ is reduced, but ae_2, xxa, and xy are not.

1. Proposition. *Let $(\Gamma_\iota)_{\iota \in I}$ be a family of groups. Let $A = \coprod_{\iota \in I} \Gamma_\iota$ and $*_{\iota \in I} \Gamma_\iota = W(A)/\sim$ be as above.*

Then any element in the free product $*_{\iota \in I} \Gamma\iota$ is represented by a unique reduced word in $W(A)$.

Proof [VandW–48]. *Existence.* Consider an integer $n \geq 0$, a reduced word $w = a_1 a_2 ... a_n$ in $W(A)$, an element $a \in A$ and the word $aw \in W(A)$. Set

$$
\mathcal{R}(aw) \;=\; \begin{cases} w & \text{if } a \text{ is the unit element } e_\iota \in \Gamma_\iota \text{ for some } \iota \in I, \\ aa_1 a_2 ... a_n & \text{if } a \in \Gamma_\iota \,,\ a \neq e_\iota \text{ for some } \iota \in I \,,\ \iota \neq \kappa \\ ba_2 ... a_n & \text{if } a \in \Gamma_\kappa \text{ and } aa_1 = b \neq e_\kappa \\ a_2 ... a_n & \text{if } a \in \Gamma_\kappa \text{ and } a_1 = a^{-1} \in \Gamma_\kappa \end{cases}
$$

where $\kappa \in I$ is that index for which $a_1 \in \Gamma_\kappa$.

Then $\mathcal{R}(aw)$ is again a reduced word, and $\mathcal{R}(aw) \sim aw$. It follows that any word $w \in W(A)$ is equivalent to some reduced word (by induction on the length of w).

Uniqueness. For each $a \in A$, let $T(a)$ be the mapping $w \mapsto \mathcal{R}(aw)$ from the set of reduced words to itself. For each word $w = b_1 b_2 ... b_n \in W(A)$, reduced or not, set $T(w) = T(b_1)T(b_2)...T(b_n)$. For $a, b, c \in \Gamma_\iota$ with $ab = c$, one has $T(a)T(b) = T(c)$; also $T(e_\iota) = id$ for all $\iota \in I$. It follows that $T(w_1) = T(w_2)$ whenever $w_1, w_2 \in W(A)$ are equivalent.

For each *reduced* word w, observe that $T(w)w_0 = w$ (where w_0 denotes the empty word).

Now let w be some word in $W(A)$ and let w_1, w_2 be two reduced words equivalent to w. As $w_1 \sim w_2$, we have $T(w_1) = T(w_2)$, so that

$$
w_1 \;=\; T(w_1)w_0 \;=\; T(w_2)w_0 \;=\; w_2
$$

and this concludes the proof. $\square$

2. Corollary. *Let $(\Gamma_\iota)_{\iota \in I}$ and let $\Gamma = *_{\iota \in I}\Gamma_\iota$ be as above. For each $\iota_0 \in I$, the canonical homomorphism*

$$
\Gamma_{\iota_0} \;\longrightarrow\; \Gamma
$$

is injective.

Proof. The image in Γ of any $a \in \Gamma_{\iota_0} \setminus \{e_{\iota_0}\}$ is represented by a one-letter word, and in particular by a reduced word which is not equivalent to the empty word. $\square$

3. Definitions. Let X be a set. The *free group over X* is the free product of a family of copies of the group $\mathbb{Z}$ indexed by X; it is denoted by $F(X)$. We identify each $x \in X$ with the generator "+1" in the copy of $\mathbb{Z}$ corresponding to x, so that X is viewed as a subset of the free product; thus $F(X)$ can be identified with the set of reduced words in $X \cup X^{-1}$, the product in the group corresponding to the juxtaposition of words followed by appropriate multiplications and reductions (i.e., by deletion of any pair of consecutive letters of the form xx^{-1} or $x^{-1}x$).

The cardinality of X is called the *rank* of the free group $F(X)$.

Let Γ be a group. A *free subset* of Γ is a subset X of Γ such that the inclusion $X \hookrightarrow F(X)$ extends to an isomorphism of the subgroup of Γ generated by X onto the free group $F(X)$.

4. Basic examples. *The fundamental group of a wedge of k circles is a free group of rank k. More generally, fundamental groups of*

> *connected graphs,*
>
> *connected surfaces with non-empty boundaries,*
>
> *connected surfaces which are not compact*

are free groups.

For these very classical consequences of the so-called *Seifert – van Kampen theorem*, see e.g. [Masse–67], respectively Theorem VI.5.1, Exercise VI.5.2, and Exercise VI.5.5.

5. Universal property. *Let Γ be a group, let $(\Gamma_\iota)_{\iota \in I}$ be a family of groups and let $(h_\iota : \Gamma_\iota \to \Gamma)_{\iota \in I}$ be a family of homomorphisms.*

Then there exists a unique homomorphism $h : *_{\iota \in I}\Gamma_\iota \to \Gamma$ such that the diagram

$$\begin{array}{ccc} \Gamma_{\iota_0} & & \\ \downarrow & \searrow^{h_{\iota_0}} & \\ *_{\iota \in I}\,\Gamma_\iota & \xrightarrow{h} & \Gamma \end{array}$$

is commutative for each $\iota_0 \in I$.

In particular, if Γ is a group, if X is a set, and if $\phi : X \to \Gamma$ is a mapping, there exists a unique homomorphism $\Phi : F(X) \to \Gamma$ such that

$$\Phi(x) \;=\; \phi(x)$$

for all $x \in X$.

Proof. For a reduced word $w = a_1 a_2 \ldots a_n \in *_{\iota \in I}\Gamma_\iota$, with $a_j \in \Gamma_{\iota_j} \setminus \{e_{\iota_j}\}$ for $j \in \{1, \ldots, n\}$ and $\iota_{j-1} \neq \iota_j$ for $j \in \{2, \ldots, n\}$, one sets

$$h(w) \;=\; h_{\iota_1}(a_1)h_{\iota_2}(a_2)\ldots h_{\iota_n}(a_n) \in \Gamma$$

and this defines h uniquely in terms of the h_ι 's. $\square$

Let X, Y be two sets. It follows from the universal property of free groups that any mapping $X \to Y$ extends canonically to a group homomorphism $F(X) \to F(Y)$. In particular, if X and Y are equipotent, any bijection $X \to Y$ extends to a group isomorphism $F(X) \to F(Y)$.

6. Corollary. *Any group is a quotient of a free group.*

Proof. A group Γ is, for example, a quotient of the free group on the under-lying set of Γ. $\square$

Exercises for II.A

7. Exercise. State and prove a universal property for free monoids.

8. Exercise. Observe that a submonoid P in a free monoid $W(A)$ on an alphabet A need not be free. [For example, if $A = \{a, b\}$, the submonoid of $W(A)$ generated by $\{a, ab, ba\}$ is not free because $a(ba) = (ab)a$.]
Compare with Complement 17 below.

9. Exercise. In the free monoid on a set of finite cardinality k, count for each $n \geq 1$ the number σ_n of words of length exactly n. Show that the formal power series $\sum_{n \geq 0} \sigma_n z^n$ is the Taylor series at the origin of a rational function, by writing this function explicitely.
[More on this in Chapters VI and VII; see in particular Exercise VI.12, as well as Proposition VII.1.]
For an interesting reference on monoids, see [Lotha–83].

10. Exercise. Show that two free groups of finite or countable infinite rank are isomorphic if and only if they have the same rank.
[Hint. In a free group of rank r, check that there are $2^r - 1$ subgroups of index 2. See also Complement 21.]

11. Exercise. Let X be a decent topological space with fundamental group Γ ("decent" means for example "arcwise connected and locally arcwise connected"). Recall that there is a natural bijection between isomorphism classes of connected coverings of X and conjugacy classes of subgroups of Γ (see Section V.6 in [Masse–67]).
In case X is a wedge of two circles, check that there are precisely three non-isomorphic classes of coverings of degree 2, corresponding to the three subgroups of index 2 in the fundamental group of X, namely in a free group of rank two.

12. Exercise. Let γ be an element of finite order in a free product $*_{\iota \in I} \Gamma_\iota$; assume that $\gamma \neq 1$. Show that there exists one and only one $\kappa \in I$ such that γ is conjugate to an element of Γ_κ.

13. Exercise. Let $\Gamma = *_{\iota \in I} \Gamma_\iota$ be a free product of $|I| \geq 2$ groups, none of the Γ_ι's being reduced to one element. Show that the centre of Γ is reduced to one element.

14. Exercise. Show that the quotient of a free product $*_{\iota \in I} \Gamma_\iota$ of *abelian* groups by its commutator subgroup is isomorphic to the direct sum $\bigoplus_{\iota \in I} \Gamma_\iota$. (See also Complement 16.)

15. Exercise. Let $\Gamma = *_{i=1}^{p} C_i$ and $\Gamma' = *_{j=1}^{q} C_j'$ be two free products of cyclic groups. If Γ and Γ' are isomorphic, show that $p = q$ and that the factors C_i are the same as the factors C_j' (up to re-indexing).

[Hint. For the infinite cyclic groups, use the previous exercise. For the finite cyclic groups, consider the conjugacy classes of maximal finite cyclic subgroups of Γ and Γ'.]

For Exercises 12 to 15, see if necessary the solutions in § 2 of [Rham–69].

Problems and complements for II.A

16. Commutators in a free product. Let A, B be two groups and let Γ be their free product. Let X be the set of commutators $(a, b) = a^{-1}b^{-1}ab$, with $a \in A \setminus \{1\}$ and $b \in B \setminus \{1\}$, and let F be the kernel of the canonical homomorphism

$$A * B \longrightarrow A \times B$$

from the free product onto the direct product.

Show that X is free and generates F, so that F is isomorphic to $F(X)$.

In particular, if A and B are abelian groups, this shows that the *commutator subgroup* generated by commutators in $A * B$ is a free group over X.

[For hints, see Exercise 32, page I.149 of [Bourb–70]. For a complete solution, see Proposition 4 of N° I.1.3 in [Serr–77], which is based on Problems 11 and 24 of Section 4.1 in [MagKS–66]; alternatively, see [Lyndo–73].

For the special case of the modular group $\Gamma = (\mathbb{Z}/2\mathbb{Z}) * (\mathbb{Z}/3\mathbb{Z})$, with Γ modulo its commutator group isomorphic to $\mathbb{Z}/6\mathbb{Z}$, a geometric proof involving the Poincaré theorem of Section V.B appears at Proposition VII.5.2 of [Ivers–92].]

17. The Nielsen-Schreier theorem on subgroups of free groups. *Any subgroup of a free group is a free group.*

This is due to J. Nielsen (1921) for finitely-generated groups and to O. Schreier (1927) in general. For the very simple "topological proof", in terms of fundamental groups of graphs and of coverings, see e.g. [Masse–67], Theorem VI.7.2. For "algebraic proofs", see e.g. [KarMe–85], Theorem 14.3.5, or [Johns–90].

Historically, it seems that free groups go back to W. Dyck (1882), but were baptized by the topologist Nielsen (1924); see [Bourb–70], I, page 165, "note historique". According to a remark at the end of Section 4.1 in [MagKS–66], free products as explicit and special objects of investigation were introduced by Schreier in 1927.

18. Free groups as groups acting freely on trees. *A group is free if and only if it can act freely by automorphisms on a tree.*

This statement clearly implies the Nielsen-Schreier theorem of the previous item. It can be found in Reidemeister's book [Reide–32, Section 4.20] or in the more recent book of Serre [Serr–77, § I.3]; see also [Biggs–89].

There are other free actions by *homeomorphisms,* for example of $\mathbb{Z}^d$ on $\mathbb{R}$ for any $d \geq 1$; in this context, recall the following result of Hölder (1901):

    if a group acts freely on the real line or on the circle by homeomorphisms,
    it is abelian

(see Section VIII.3.1 in [HecHi–83] or Theorem 6.10 in [Ghys–99]). It follows that groups which act freely on $\mathbb{R}$ by homeomorphisms are precisely the subgroups of $\mathbb{R}$.

The theory of Bass and Serre of *graphs of groups* provides a good understanding of more general actions of groups on trees by automorphisms (= by isometries for the combinatorial distance), actions for which each group element may have fixed points but such that no point is fixed by the whole group; see [Serr–77] and [Bass–93]. For example, one has the following characterization (Theorem 8.4 in [Bass–93]):

    A group has a free subgroup of finite index
    if and only if it can act properly on a tree.

In this case, "properly" is equivalent to "when the isotropy group of any vertex is finite".

Finitely-generated groups acting freely by isometries on *real trees* have been characterized by E. Rips as free products of free abelian groups and of surface groups; see [GabLP–94] and [BesF–95]. A real tree is an arcwise connected metric space in which every arc is homeomorphic to an interval of $\mathbb{R}$, or equivalently a geodesic metric space which is 0-hyperbolic in the sense of Section V.D. Ordinary trees are clearly real trees; other examples of real trees occur in the theory of codimension 1 foliations with a transverse measure, or as Gromov limits of hyperbolic metric spaces.

19. Free groups are groups of cohomological dimension one. There is a *cohomological characterization* of free groups due to Stallings [Stall–68] and Swan [Swan–69]. One defines cohomology groups $H^k(\Gamma, M)$ associated to a group Γ, a Γ-module M and an integer $k \geq 0$. The *cohomological dimension* of Γ is the supremum (possibly ∞) of the integers k for which there exists a Γ-module M with $H^k(\Gamma, M) \neq \{0\}$.

From the precise definitions, it is easy to check that a group has cohomological dimension zero if and only if it is reduced to one element. Also, a free group (not reduced to one element) has cohomological dimension one; this follows for example from the fact that there exists a CW-complex of dimension one (a wedge of circles) with fundamental group the given free group and with universal cover contractible (a tree). The theorem of Stallings and Swan shows that, conversely,

    a group of cohomological dimension one is necessarily free.

More on the cohomological dimension of groups in [Serr–71].

20. Ranks of subgroups of finite index in free groups. Let $k \geq 2$ be an integer, let F_k be a free group of rank k and let $n \geq 2$ be an integer. Show that there exist subgroups of F_k of index n which are free of rank

$$l = 1 + n(k-1).$$

In particular,

a free group of rank 2 contains subgroups which are free of rank l for any $l \geq 2$.

Conversely, any subgroup of finite index, say n, in F_k is again free of finite rank, say l, and n, k, l satisfy the formula above.

[Hint: the Euler characteristic of a connected covering of degree n of a bouquet of k circles is $n - nk$.]

21. M. Hall's formula for the number of subgroups of given index in a free group of finite rank. For integers $r \geq 1$ and $n \geq 1$, let $N_{n,r}$ denote the number of subgroups of index n in the free group of rank r. M. Hall's inductive formula [HallM–49] states that

$$N_{n,r} = n(n!)^{r-1} - \sum_{j=1}^{n-1} ((n-j)!)^{r-1} N_{j,r}$$

so that, for example, $N_{3,r} = 3 \cdot 6^{r-1} - 3 \cdot 2^{r-1} + 1$ (compare with Exercise 10).

For a given r, this implies that the radius of convergence of the formal power series $\sum_{n \geq 0} N_{n,r} z^n$ is zero. In particular, the growth of $(N_{n,r})_{n \geq 0}$ is *superexponential*.

There are other interesting groups Γ for which the numbers

$$N_{n,\Gamma} = \text{number of subgroups of } \Gamma \text{ of index } n$$

have been studied with some detail (see [GruSS–88], [Lubo–95a], [Lubo–95c]). It is a conjecture of [Lubo–95e] that, for any lattice Γ in $SO(n,1)$, $n \geq 3$, the growth of $(N_{n,\Gamma})_{n \geq 1}$ is superexponential.

The following five steps provide a proof of M. Hall's formula.

(i) For a group Γ, there is a natural bijection between

the set of subgroups Γ_0 of Γ of index n, together with a numbering of the classes in Γ/Γ_0 with number 1 on the class $\{\Gamma_0\} = \Gamma_0/\Gamma_0$

and

the set of transitive actions of Γ on $X_n = \{1, ..., n\}$.

Thus $(n-1)! N_{n,r}$ is the number of *transitive* actions of a free group of rank r, say F_r, on the set $X_n = \{1, ..., n\}$.

(ii) The total number of actions of F_r on X_n is $(n!)^r$.

(iii) The number of actions of F_r on X_n with the orbit of 1 of cardinality j is

$$(n-1)(n-2) \; \ldots \; (n-j+1)N_{j,r}\big((n-j)!\big)^r.$$

(iv) The three previous steps imply that

$$(n!)^r = (n-1)!N_{n,r} + \sum_{j=1}^{n-1} \frac{(n-1)!}{(n-j)!}N_{j,r}\big((n-j)!\big)^r$$

and Hall's formula follows.

(v) For $r \geq 2$, one has $N_{j,r} \leq j\big(j!\big)^{r-1}$ for all $j \geq 1$ and consequently

$$N_{n,r} \geq (n!)^{r-1}\left\{ n - \sum_{j=1}^{n-1} \frac{j}{\big[\binom{n}{j}\big]^{r-1}} \right\} \geq n!$$

so that the radius of convergence of the series $\sum_{n\geq 0} N_{n,r}z^n$ is zero.

(vi) For what is known about the number of subgroups of some given index in the *modular group*

$$PSL(2,\mathbb{Z}) \approx \mathbb{Z}/2\mathbb{Z} * \mathbb{Z}/3\mathbb{Z},$$

see [Jones–86]. For surface groups, see [Medny–79].

22. The Kurosh subgroup theorem. *In a free product* $\Gamma = \Gamma_1 * \Gamma_2$, *any subgroup is itself a free product of a free group and of subgroups of conjugates in Γ of Γ_1 and Γ_2.*

The original paper goes back to 1934. Appendix C of Volume II of [Kuros–56] comments on six proofs; we recommend the expostion in Chapter 7 of [Masse–67] and in [ScoWa–79]. Proofs of more general results appear in n° I.5.5 of [Serr–77] and in Theorem 5 of [Morga–92].

For example, a finitely-generated subgroup of $\mathbb{Z}/2\mathbb{Z} * \mathbb{Z}/3\mathbb{Z}$ is isomorphic to a free product of s copies of $\mathbb{Z}$, a copies of $\mathbb{Z}/2\mathbb{Z}$, and b copies of $\mathbb{Z}/3\mathbb{Z}$, for some triple of integers $s,a,b \geq 0$, and any such triple occurs. In connexion with (vi) of the previous complement, let us mention that any such triple occurs for some subgroup *of finite index* with four exceptions, which are $\mathbb{Z}$, $\mathbb{Z}/2\mathbb{Z}$, $\mathbb{Z}/3\mathbb{Z}$, and $\mathbb{Z}/2\mathbb{Z} * \mathbb{Z}/2\mathbb{Z}$ (Corollary 5.2 in [Kulka–83]).

23. Free groups, random walks, operator algebras, unrestricted free products, and free topological groups. For the importance of free groups in the theory of random walks and in that of operator algebras, see respectively [Ledra] and [Harpe–95].

For an infinite family of groups $(\Gamma_\iota)_{\iota \in I}$, the free product $*_{\iota \in I}\Gamma_\iota$ is the *direct* limit of the family of free products $*_{\iota \in F}\Gamma_\iota$, where F runs over the finite subsets of I, with respect to the natural inclusions $*_{\iota \in F}\Gamma_\iota \longrightarrow *_{\iota \in G}\Gamma_\iota$ whenever $F \subset G$. The *unrestricted free product* is the *inverse* limit of the same family, with respect to the natural epimorphisms $*_{\iota \in G}\Gamma_\iota \longrightarrow *_{\iota \in F}\Gamma_\iota$ whenever $F \subset G$, and is naturally a *topological group*. Subgroups of unrestricted free products occur as fundamental groups of topological spaces such as Hawaiian earrings. See [Higma–52], various papers by K. Eda including [Eda–92], and Item III.6 below.

For *free topological groups,* see [Marko–45] and [Graev–48].

II.B. The Table-Tennis Lemma (Klein's criterion) and examples of free products

The following criterion was used on many occasions by F. Klein, though the formulation below is more recent. See for example the study of Schottky groups by Klein (reference in Item 26 below); for recent references, see among others [Tits–72], Proposition 12.2 of Chapter III in [LynSc–77], [Harpe–83], and [Hausm–81]. This same result is also called the "Ping-Pong Lemma" and the "Schottky Lemma".

24. The Table-Tennis Lemma. *Let G be a group acting on a set X, let Γ_1, Γ_2 be two subgroups of G, and let Γ be the subgroup of G generated by Γ_1 and Γ_2; assume that Γ_1 contains at least 3 elements and Γ_2 at least 2 elements.*

Assume that there exist two non-empty subsets X_1, X_2 in X, with X_2 not included in X_1, such that

$$\gamma(X_2) \subset X_1 \qquad \text{for all} \qquad \gamma \in \Gamma_1 \,,\, \gamma \neq 1\,,$$
$$\gamma(X_1) \subset X_2 \qquad \text{for all} \qquad \gamma \in \Gamma_2 \,,\, \gamma \neq 1\,.$$

Then Γ is isomorphic to the free product $\Gamma_1 * \Gamma_2$.

Proof. Let w be a non-empty reduced word spelled with letters from the disjoint union of $\Gamma_1 \setminus \{1\}$ and $\Gamma_2 \setminus \{1\}$. We have to show that the element of Γ defined by w (again written w) is not the identity.

If $w = a_1 b_1 a_2 b_2 ... a_k$ with $a_1, ..., a_k \in \Gamma_1 \setminus \{1\}$ and $b_1, ..., b_{k-1} \in \Gamma_2 \setminus \{1\}$, then

$$w(X_2) = a_1 b_1 ... a_{k-1} b_{k-1} a_k(X_2) \subset a_1 b_1 ... a_{k-1} b_{k-1}(X_1)$$
$$\subset a_1 b_1 ... a_{k-1}(X_2) \subset ... \subset a_1(X_2) \subset X_1.$$

As $X_2 \not\subset X_1$ this implies that $w \neq 1$.

If $w = b_1 a_2 b_2 a_2 ... b_k$, choose $a \in \Gamma_1 \setminus \{1\}$; the previous argument shows that $awa^{-1} \neq 1$, so that $w \neq 1$. If $w = a_1 b_1 ... a_k b_k$, choose $a \in \Gamma_1 \setminus \{1, a_1^{-1}\}$ and argue similarly with awa^{-1}. If $w = b_1 a_2 b_2 ... a_k$, choose $a \in \Gamma_1 \setminus \{1, a_k\}$ and argue with awa^{-1}. $\square$

There are more precise results: see e.g. [PayVa–91], which has applications to groups acting on trees and their free subgroups. There is an analogous result for free monoids (Proposition VII.2).

25. Example (free subgroup of $SL(2,\mathbb{Z})$). *The two matrices* $\begin{pmatrix} 1 & 2 \\ 0 & 1 \end{pmatrix}$ *and* $\begin{pmatrix} 1 & 0 \\ 2 & 1 \end{pmatrix}$ *generate a subgroup of* $SL(2,\mathbb{Z})$ *which is free of rank two.*

Proof. Let

$$\Gamma_1 = \left\{ \begin{pmatrix} 1 & 2n \\ 0 & 1 \end{pmatrix} \in SL(2,\mathbb{Z}) \,\middle|\, n \in \mathbb{Z} \right\}$$

and

$$\Gamma_2 = \left\{ \begin{pmatrix} 1 & 0 \\ 2n & 1 \end{pmatrix} \in SL(2,\mathbb{Z}) \,\middle|\, n \in \mathbb{Z} \right\}$$

be the infinite cyclic subgroups of $SL(2,\mathbb{Z})$ generated respectively by the two matrices $\begin{pmatrix} 1 & 2 \\ 0 & 1 \end{pmatrix}$ and $\begin{pmatrix} 1 & 0 \\ 2 & 1 \end{pmatrix}$. The group $SL(2,\mathbb{Z})$ acts linearly on $\mathbb{R}^2$ in the usual way. One can set

$$X_1 = \left\{ \begin{pmatrix} x \\ y \end{pmatrix} \in \mathbb{R}^2 \,\middle|\, |x| > |y| \right\}$$

and

$$X_2 = \left\{ \begin{pmatrix} x \\ y \end{pmatrix} \in \mathbb{R}^2 \,\middle|\, |x| < |y| \right\}$$

to apply the Table-Tennis Lemma. $\square$

[A variant of this proof appears in [LynSc–77], Proposition III.12.3, where the result is attributed to Sanov (1947). One can of course speculate that Poincaré knew the result (see Section V.B below).]

Observe that the two matrices $\begin{pmatrix} 1 & k \\ 0 & 1 \end{pmatrix}$ and $\begin{pmatrix} 1 & 0 \\ k & 1 \end{pmatrix}$ generate a free subgroup of rank two in $SL(2,\mathbb{Z})$ for any $k \geq 2$, for the same reasons, but not for $k = 1$, because

$$\begin{pmatrix} 1 & -1 \\ 0 & 1 \end{pmatrix} \begin{pmatrix} 1 & 0 \\ 1 & 1 \end{pmatrix} \begin{pmatrix} 1 & -1 \\ 0 & 1 \end{pmatrix} = \begin{pmatrix} 0 & 1 \\ -1 & 0 \end{pmatrix}$$

is of finite order. It is well-known that $\begin{pmatrix} 1 & 1 \\ 0 & 1 \end{pmatrix}$ and $\begin{pmatrix} 1 & 0 \\ 1 & 1 \end{pmatrix}$ generate $SL(2,\mathbb{Z})$.

There is a large number of papers on the following question: for which $\lambda \in \mathbb{C}$ do the matrices $\begin{pmatrix} 1 & 2 \\ 0 & 1 \end{pmatrix}$ and $\begin{pmatrix} 1 & 0 \\ \lambda & 1 \end{pmatrix}$ generate a free subgroup in $SL(2,\mathbb{C})$? As the conjugates of these matrices by $\begin{pmatrix} x & 0 \\ 0 & x^{-1} \end{pmatrix}$ are $\begin{pmatrix} 1 & 2x^2 \\ 0 & 1 \end{pmatrix}$ and $\begin{pmatrix} 1 & 0 \\ \lambda x^{-2} & 1 \end{pmatrix}$, it is essentially the same question to ask for which $z \in \mathbb{C}$ the matrices $\begin{pmatrix} 1 & z \\ 0 & 1 \end{pmatrix}$ and $\begin{pmatrix} 1 & 0 \\ z & 1 \end{pmatrix}$ generate a free subgroup in $SL(2,\mathbb{C})$. See

Exercise 33; see also Appendix III in [GooHJ–89] and the list of references in [Bambe]. A related question is: for which $\lambda \in \mathbb{R}$ [respectively $\lambda \in \mathbb{C}$] do the two matrices above generate a *discrete* subgroup of $SL(2, \mathbb{R})$? [respectively $SL(2, \mathbb{C})$?] See [Gilma–95]; the question makes sense for Lie groups of higher ranks [Oh–98].

Let us quote a result from [Satô–99] about the semi-direct product $\mathbb{Z}^2 \rtimes SL(2, \mathbb{Z})$ acting on $\mathbb{R}^2$ by affine transformations in the usual way: the subgroup Γ_0 generated by the two transformations

$$\begin{pmatrix} x \\ y \end{pmatrix} \longmapsto \begin{pmatrix} 7 & 3 \\ 9 & 4 \end{pmatrix} \begin{pmatrix} x \\ y \end{pmatrix} + \begin{pmatrix} 1 \\ -1 \end{pmatrix} \text{ and } \begin{pmatrix} x \\ y \end{pmatrix} \longmapsto \begin{pmatrix} 94 & 39 \\ 147 & 61 \end{pmatrix} \begin{pmatrix} x \\ y \end{pmatrix} + \begin{pmatrix} 3 \\ 2 \end{pmatrix}$$

is free of rank 2 *and acts freely* on $\mathbb{Z}^2$; for each $\gamma \in \Gamma_0, \gamma \neq 1$, the fixed-point set $\{v \in \mathbb{R}^2 \mid \gamma v = v\}$ is reduced to the point $\begin{pmatrix} 2/3 \\ -5/3 \end{pmatrix} \in \mathbb{R}^2$.

26. Schottky groups. Recall that the group $PSL(2, \mathbb{C})$ acts by fractional linear transformations on the Riemann sphere $\hat{\mathbb{C}}$. More precisely, if $\begin{bmatrix} a & b \\ c & d \end{bmatrix}$ denotes the class in the group $PSL(2, \mathbb{C})$ of a matrix $\begin{pmatrix} a & b \\ c & d \end{pmatrix} \in SL(2, \mathbb{C})$, then

$$\begin{bmatrix} a & b \\ c & d \end{bmatrix} z = \frac{az + b}{cz + d}$$

for all $z \in \hat{\mathbb{C}} = \mathbb{C} \cup \{\infty\}$.

By restriction, one obtains the standard action of $PSL(2, \mathbb{R})$ on the *real projective line* $\hat{\mathbb{R}} = \mathbb{R} \cup \{\infty\}$, as well as on the *Poincaré half-plane*

$$\mathcal{P} = \{z \in \mathbb{C} \mid \Im(z) > 0\}.$$

Example. *Consider an integer $g \geq 1$ and $2g$ closed discs $D_1, \ldots, D_{2g}$ in $\hat{\mathbb{C}}$ with disjoint interiors. For each $k \in \{1, \ldots, g\}$, choose an element $\gamma_k \in PSL(2, \mathbb{C})$ which maps the exterior of D_{2k-1} onto the interior of D_{2k}.*

Then the subgroup Γ of $PSL(2, \mathbb{C})$ generated by $\gamma_1, \ldots, \gamma_g$ is free on these elements.

Proof. This follows from the straightforward generalization of the Table-Tennis Lemma to g subgroups of $PSL(2, \mathbb{C})$ respectively generated by $\gamma_1, \ldots, \gamma_g$, with $X_k = D_{2k-1} \cup D_{2k}$ for $k \in \{1, \ldots, g\}$. □

An early appearance of Schottky groups is in [Klein–83, page 200]. More on Schottky groups in [Lehne–64, pages 119 and 400] and [Magnu–74, Chapter IV].

There is an analogous notion of Schottky groups acting on locally finite trees. Any finitely-generated torsion-free discrete subgroup of the automorphism group of a tree is such a Schottky group (Section 1 of [Lubo–91]).

27. Higher dimensional Schottky groups. Consider the projective space $\mathbb{P}^3_{\mathbb{C}}$ with homogeneous coordinates $[z_0 : z_1 : z_2 : z_3]$ and with its standard *Hausdorff* topology (*not* the Zariski topology). The relevant group of automorphisms (or of biholomorphic transformations) is the quotient $PGL(4, \mathbb{C})$ of $GL(4, \mathbb{C})$ by its centre, which acts naturally on the projective space. We consider also the two projective lines σ, of equations $z_2 = z_3 = 0$, and ρ, of equations $z_0 = z_1 = 0$.

Given real numbers λ_j $(0 \leq j \leq 3)$ with $0 < \lambda_0 < \lambda_1 < \lambda_2 < \lambda_3$, there is a transformation

$$\gamma : [z_0 : z_1 : z_2 : z_3] \longmapsto [\lambda_0 z_0 : \lambda_1 z_1 : \lambda_2 z_2 : \lambda_3 z_3]$$

of $\mathbb{P}^3_{\mathbb{C}}$ for which σ and ρ are two invariant lines containing the 4 fixed points. For a real number $\epsilon > 0$, the set

$$S_\epsilon = \left\{ [z_0 : z_1 : z_2 : z_3] \in \mathbb{P}^3_{\mathbb{C}} \;\middle|\; \frac{|z_0|^2 + |z_1|^2}{|z_2|^2 + |z_3|^2} > \frac{1}{\epsilon} \right\}$$

is a neighbourhood of σ and the set

$$R_\epsilon = \left\{ [z_0 : z_1 : z_2 : z_3] \in \mathbb{P}^3_{\mathbb{C}} \;\middle|\; \frac{|z_0|^2 + |z_1|^2}{|z_2|^2 + |z_3|^2} < \epsilon \right\}$$

is a neighbourhood of ρ. If $\epsilon < 1$ then $\overline{S_\epsilon} \cap \overline{R_\epsilon} = \emptyset$. If

$$\frac{\lambda_1}{\lambda_2} < \epsilon < 1$$

then

$$\gamma \left(\mathbb{P}^3_{\mathbb{C}} \setminus S_\epsilon \right) \subset R_\epsilon \qquad \text{and} \qquad \gamma^{-1} \left(\mathbb{P}^3_{\mathbb{C}} \setminus R_\epsilon \right) \subset S_\epsilon.$$

The set $\mathbb{P}^3_{\mathbb{C}} \setminus (S_\epsilon \cup R_\epsilon)$ contains a fundamental domain for the free action of the cyclic group $(\gamma)^{\mathbb{Z}}$ on $\mathbb{P}^3_{\mathbb{C}} \setminus (\sigma \cup \rho)$.

The letters "sigma" and "rho" stand for "source" and "range" respectively. The dynamics of γ on $\mathbb{P}^3_{\mathbb{C}}$ is in some sense similar to that of a hyperbolic element in $PSL(2, \mathbb{C})$ on $\mathbb{P}^1_{\mathbb{C}} = \mathbb{C} \cup \{\infty\}$. (For "hyperbolic", see Exercise 32.)

For any pair (σ', ρ') of disjoint lines with disjoint neighbourhoods S', R' in $\mathbb{P}^3_{\mathbb{C}}$, and for ϵ small enough, there exists an automorphism $x \in PGL(4, \mathbb{C})$ such that

$$\begin{aligned} x(\sigma) &= \sigma' & x(S_\epsilon) &\subset S', \\ x(\rho) &= \rho' & x(R_\epsilon) &\subset R'. \end{aligned}$$

Indeed, as $PGL(4, \mathbb{C})$ acts transitively on the set of pairs of disjoint lines, there exists $x_1 \in PGL(4, \mathbb{C})$ such that $x_1(\sigma) = \sigma'$ and $x_1(\rho) = \rho'$. One may then set $x = x_1 x_2$ for $x_2 = [diag(a, a, b, b)]$ with $a, b > 0$ and ab^{-1} large enough. Consequently the transformation $\gamma' = x\gamma x^{-1}$ has two invariant lines, which are σ' (its "source") and ρ' (its "range"), and

$$(\gamma')^n \left(\mathbb{P}^3_{\mathbb{C}} \setminus S' \right) \subset R'$$

$$(\gamma')^{-n} \left(\mathbb{P}^3_{\mathbb{C}} \setminus R' \right) \subset S'$$

for all integers $n \geq 1$.

Example. *Let* $\sigma_1, \ldots, \sigma_g, \rho_1, \ldots, \rho_g$ *be* $2g$ *pairwise disjoint lines in* $\mathbb{P}^3_{\mathbb{C}}$, *with pairwise disjoint neighbourhoods* $S_1, \ldots, S_g, R_1, \ldots, R_g$ *respectively. For each* $k \in \{1, \ldots, g\}$, *choose an element* $\gamma_k \in PGL(4, \mathbb{C})$ *with "source"* σ_k *and "range"* ρ_k, *such that* $\gamma_k \left(\mathbb{P}^3_{\mathbb{C}} \setminus S_k \right) \subset R_k$.
Then the subgroup Γ *of* $PGL(4, \mathbb{C})$ *generated by* $\gamma_1, \ldots, \gamma_g$ *is free on these elements.*

Proof: analogous to that of Example 26. □

The same construction provides "Schottky groups" in $PGL(n, \mathbb{C})$ for every $n \geq 2$, and indeed in any so-called reductive group: [Benoi–96], [Benoi–97]; see also [Nori–96].

28. Example (Hecke groups). *Consider an integer* $q \geq 3$, *the number* $\lambda = 2 \cos(\frac{\pi}{q})$, *the elements*

$$a_\lambda = \begin{bmatrix} 1 & \lambda \\ 0 & 1 \end{bmatrix} \quad and \quad j = \begin{bmatrix} 0 & 1 \\ -1 & 0 \end{bmatrix}$$

in $G = PSL(2, \mathbb{R})$ *and the subgroup* Γ_λ *of* G *generated by them.*
Then

$$\Gamma_\lambda \approx (\mathbb{Z}/q\mathbb{Z}) * (\mathbb{Z}/2\mathbb{Z}) .$$

In particular, for the modular group:

$$PSL(2, \mathbb{Z}) \approx (\mathbb{Z}/3\mathbb{Z}) * (\mathbb{Z}/2\mathbb{Z})$$

(case $q = 3$*).*

Proof. We have

$$a_\lambda(z) = z + \lambda \quad and \quad j(z) = -\frac{1}{z}$$

for all $z \in \mathcal{P} \cup \hat{\mathbb{R}}$, with $\mathcal{P}$ and $\hat{\mathbb{R}}$ as in Example 26.
The matrix $\begin{pmatrix} 1 & \lambda \\ 0 & 1 \end{pmatrix} \begin{pmatrix} 0 & 1 \\ -1 & 0 \end{pmatrix} = \begin{pmatrix} -\lambda & 1 \\ -1 & 0 \end{pmatrix}$ has trace $-\lambda$ and determinant 1. Thus its eigenvalues are $-exp\left(\pm i\frac{\pi}{q}\right)$. It follows that

$$b_\lambda \doteq a_\lambda j = \begin{bmatrix} -\lambda & 1 \\ -1 & 0 \end{bmatrix} : z \longmapsto -\frac{1}{z} + \lambda$$

generates a subgroup $\Gamma_{(1)}$ of G which is cyclic of order q. More precisely, as the conjugate

$$\begin{bmatrix} -exp\left(-i\frac{\pi}{q}\right) & exp\left(i\frac{\pi}{q}\right) \\ -1 & 1 \end{bmatrix}^{-1} \begin{bmatrix} -\lambda & 1 \\ -1 & 0 \end{bmatrix} \begin{bmatrix} -exp\left(-i\frac{\pi}{q}\right) & exp\left(i\frac{\pi}{q}\right) \\ -1 & 1 \end{bmatrix}$$

$$= \begin{bmatrix} -exp\left(-i\frac{\pi}{q}\right) & 0 \\ 0 & -exp\left(i\frac{\pi}{q}\right) \end{bmatrix}$$

of b_λ is a rotation of $\hat{\mathbb{C}}$ of angle $-\frac{2\pi}{q}$ with fixed point set $\{0, \infty\}$, the element b_λ itself is a hyperbolic rotation of $\mathcal{P}$ of angle $-\frac{2\pi}{q}$ with fixed point $exp\left(i\frac{\pi}{q}\right)$.

We denote by $\Gamma_{(2)}$ the subgroup $\{1, j\}$ of G. It is straightforward that j is a hyperbolic half-turn with fixed point i.

Consider the action of G on $\hat{\mathbb{R}}$; set

$$X_1 =]0, \infty[\cup \{\infty\} \qquad \text{and} \qquad X_2 =]-\infty, 0].$$

It is clear that $j(X_1) = X_2$. We claim that $b_\lambda^k(X_2) \subset X_1$ for all $k \in \{1, 2, ..., q-1\}$.

Indeed, b_λ maps the point 0 [respectively $-\epsilon$ (where $\epsilon > 0$), and ∞] in $\hat{\mathbb{R}}$ to the point ∞ [respectively $\frac{1}{\epsilon} + \lambda$, and λ]; thus b_λ maps X_2 onto the interval $]\lambda, \infty[\cup \{\infty\}$ of $\hat{\mathbb{R}}$, and $b_\lambda^2, ..., b_\lambda^{q-1}$ map X_2 onto subintervals of $]0, \lambda[$.

Consequently, the Table-Tennis Lemma applies, so that Γ_λ is the free product of $\Gamma_{(1)} \approx \mathbb{Z}/q\mathbb{Z}$ and $\Gamma_{(2)} \approx \mathbb{Z}/2\mathbb{Z}$.

If $q = 3$, the group generated by a_λ and j is the modular group $PSL(2, \mathbb{Z})$; we leave this as an exercise. [Hint for one possible solution: use the Euclidean division algorithm in $\mathbb{Z}$; or refer to § VII.1 of [Serr–70a].] □

There is more on these groups Γ_λ in Appendix III of [GooHJ–89]; see also Example V.44 below.

The seminal paper on Hecke groups is [Hecke–36]. The modular group is one of the most important groups in mathematics, with at least one book devoted to it [Ranki–69].

For lattices in a semi-simple group G (say with trivial centre and no compact factors), there is a notion of *arithmeticity*. (This is delicate; see [Zimme–84] or [Margu–91] in general, [Katok–92] for $G = PSL(2, \mathbb{R})$, and [ElsGM–98] and [NeumW–99] for $G = PSL(2, \mathbb{C})$.) It is a deep result of Margulis that, in "most" semi-simple groups (for example in those of so-called real rank at least 2), irreducible lattices are *necessarily* arithmetic.

Let Γ_1, Γ_2 be two subgroups of a group G. Say that Γ_1 and Γ_2 are *commensurable in G* if there exists $g \in G$ such that $\Gamma_1 \cap g\Gamma_2 g^{-1}$ is of finite index in both Γ_1 and $g\Gamma_2 g^{-1}$. (This is *not* the same notion as in Definition IV.27 ! For example, it follows from Exercise 33 below that there are discrete subgroups of $PSL(2, \mathbb{R})$ which are isomorphic and which are not commensurable *in $PSL(2, \mathbb{R})$*.) *A lattice Γ in $G = PSL(2, \mathbb{R})$ such that $\Gamma \backslash G$ is not compact is arithmetic if and only if it is commensurable in G with $PSL(2, \mathbb{Z})$*; see [Katok–92].

For $q \geq 3$ and $\lambda = 2\cos(\frac{\pi}{q})$, the Hecke group Γ_λ is arithmetic in $PSL(2, \mathbb{R})$ if and only if $q \in \{3, 4, 6\}$; see Example 5.5.H in [Katok–92], and Exercise 35 below.

29. Example (automorphisms of the line). *Let f be the piecewise linear homeomorphism of the interval $[0, 1]$ defined by*

$$f(t) = \begin{cases} 4t & \text{if} \quad t \in \left[0, \frac{1}{5}\right] \\ \frac{4}{5} + \frac{1}{4}\left(t - \frac{1}{5}\right) & \text{if} \quad t \in \left[\frac{1}{5}, 1\right]. \end{cases}$$

Let γ_1 be the piecewise linear homeomorphism of the real line defined by

$$\gamma_1(t) = [t] + f(\{t\}) \quad \text{for all} \quad t \in \mathbb{R}$$

where $[t]$ denotes the integral part of a real number t, and $\{t\} = t - [t] \in [0, 1[$ its fractional part. Set also

$$\gamma_2 = T\gamma_1 T^{-1}$$

where T denotes the translation $t \longmapsto t - \frac{1}{2}$.

Then γ_1, γ_2 generate a free group of rank 2 in the group $Homeo_+(\mathbb{R})$ of all orientation-preserving piecewise linear homeomorphisms of the real line.

Observe that one may view γ_1 and γ_2 in the group of all order-preserving isomorphisms of $\mathbb{R}$, or in the group of all order-preserving isomorphisms of $\mathbb{Q}$, instead of in $Homeo_+(\mathbb{R})$.

The present example is the simplest one from [Benne–97].

Proof. For any $n \in \mathbb{Z}$ let f^n denote the nth iterate of f. It is important to observe that

$$f^n\left(\left[\frac{1}{5}, 1\right]\right) \subset \left[\frac{4}{5}, 1\right] \quad \text{for} \quad n \geq 1,$$

$$f^n\left(\left[0, \frac{4}{5}\right]\right) \subset \left[0, \frac{1}{5}\right] \quad \text{for} \quad n \leq -1.$$

Set now

$$X_1 = \bigcup_{k \in \mathbb{Z}}\left[k - \frac{1}{5}, k + \frac{1}{5}\right] \quad \text{and} \quad X_2 = \bigcup_{k \in \mathbb{Z}}\left[k + \frac{1}{2} - \frac{1}{5}, k + \frac{1}{2} + \frac{1}{5}\right] = T(X_1).$$

The previous observation implies that

$$\gamma_1^n(X_2) \subset X_1$$

for all $n \in \mathbb{Z} \setminus \{0\}$. This in turn implies that

$$\gamma_2^n(X_1) = T\gamma_1^n T^{-1}(T(X_2)) \subset T(X_1) = X_2$$

for all $n \in \mathbb{Z} \setminus \{0\}$.

The statement of Example 29 now follows from the Table-Tennis Lemma. $\square$

For other free subgroups of $Homeo_+(\mathbb{R})$, see Complement 40 below.

Exercises for II.B

30. Exercise. Let Γ be a group of transformations of the real line generated by two half-turns with distinct centres. Show that Γ is isomorphic to the *infinite dihedral group* $(\mathbb{Z}/2\mathbb{Z}) * (\mathbb{Z}/2\mathbb{Z})$.

31. Exercise. Let $a = \begin{bmatrix} 1 & 2 \\ 0 & 1 \end{bmatrix}$ and j be as in Example 28 (Hecke groups, case $q = \infty$) and let $s_1, ..., s_p \in \mathbb{Z} \setminus \{0\}$, with $p \geq 2$. Check that

$$a^{s_1} j a^{s_2} j ... j a^{s_p} \infty \neq \infty.$$

Deduce from this that, for two distinct integers $k, l \geq 0$, the subgroup of $PSL(2, \mathbb{R})$ generated by $a^k j a^k$ and $a^l j a^l$ is free of rank two.

[This is the method used by von Neumann in [vonNe–29, pages 633–634] to show that $PSL(2, \mathbb{R})$ contains non-abelian free subgroups.]

32. Exercise on parabolic and hyperbolic elements. In the group $G = PSL(2, \mathbb{R})$, an element $g = \begin{bmatrix} a & b \\ c & d \end{bmatrix} \neq 1$ is *parabolic* if it has a unique fixed point in $\hat{\mathbb{R}}$ (equivalently, if $|a + d| = 2$, equivalently if it is conjugate in G to a translation of $\mathbb{R}$), *hyperbolic* if it has two distinct fixed points in $\hat{\mathbb{R}}$ (equivalently, if $|a + d| > 2$, equivalently if it is conjugate in G to a dilation $x \mapsto \lambda x$ of $\mathbb{R}$), and *elliptic* if it has a fixed point in the Poincaré half-plane $\mathcal{P}$ (equivalently, if $|a + d| < 2$, equivalently if it is conjugate in G to a rotation $\begin{bmatrix} \cos\theta & \sin\theta \\ -\sin\theta & \cos\theta \end{bmatrix}$).

For the equivalences, see for example Section I.2 in [Magnu–74].

Let Γ be a subgroup of G generated by two elements γ_1, γ_2. Show that Γ contains a free group of rank two in each of the following cases.

(i) γ_1, γ_2 are parabolic with distinct fixed points in $\hat{\mathbb{R}}$,
(ii) γ_1 is parabolic with fixed point $x \in \hat{\mathbb{R}}$ and $\gamma_2(x) \neq x$,
(iii) γ_1, γ_2 are hyperbolic, respectively with fixed points α_1, ω_1 and α_2, ω_2, and $\{\alpha_1, \omega_1\} \cap \{\alpha_2, \omega_2\} = \emptyset$.

33. Exercise on free subgroups of $PSL(2, \mathbb{C})$. For each $z \in \mathbb{C}$, let Γ_z be the subgroup of $PSL(2, \mathbb{C})$ generated by $\begin{bmatrix} 1 & z \\ 0 & 1 \end{bmatrix}$ and $\begin{bmatrix} 1 & 0 \\ z & 1 \end{bmatrix}$.

(i) Show that $\begin{pmatrix} 1 & z \\ 0 & 1 \end{pmatrix}$ and $\begin{pmatrix} 1 & 0 \\ z & 1 \end{pmatrix}$ generate a free subgroup of rank two in $SL(2, \mathbb{C})$ if $|z| \geq 2$. [See Example 25.]

(ii) Deduce from (i) that the subgroup Γ_z of $PSL(2, \mathbb{C})$ is free of rank 2 if $|z| \geq 2$.

(iii) Deduce from (ii) that Γ_z is free of rank 2 if z is transcendental. [Use a wild automorphism of $\mathbb{C}$.]

(iv) For $z \in \mathbb{R}$ with $|z| > 2$, show that Γ_z and Γ_2 are not conjugate inside $PSL(2,\mathbb{R})$. [The volume of the quotient of the Poincaré half-plane $\mathcal{P}$ (see Example 26) by Γ_z is infinite, and the volume of $\mathcal{P}/\Gamma_2$ is finite.]

(v) Anticipating on Poincaré's theorem on fundamental polygons of Fuchsian groups and Section V.B, show that Γ_2 is the fundamental group of an open surface obtained from a 2-sphere by removing three points.

[Consider in the Poincaré half-plane the fundamental "polygon" P limited by the two half-lines of equation $\Re(z) = \pm 1$ and the two half-circles of radius $\frac{1}{2}$ centred at $\pm\frac{1}{2}$. Then $\begin{bmatrix} 1 & 2 \\ 0 & 1 \end{bmatrix}$ applies the left half-line in the boundary of P onto its right half-line and $\begin{bmatrix} 1 & 0 \\ 2 & 1 \end{bmatrix}$ applies the right half-circle of the boundary of P onto its left half-circle. It is easy to check that the *cusp conditions* of Item V.39 are satisfied.]

(vi) There exists also a pair of transformations in $PSL(2,\mathbb{R})$ generating a free group Γ'_2 of rank 2 which is the fundamental group of an open surface obtained from a 2-torus by removing one point.

[With the notation of (v), consider appropriate elements of $PSL(2,\mathbb{R})$ mapping the left half-line of the boundary of P onto its right half-circle and its right half-line onto its left half-circle, and the group Γ'_2 generated by these two elements.]

(vii) Observe that the subgroups Γ_2 and Γ'_2 of (v) and (vi) above are two discrete subgroups of $PSL(2,\mathbb{R})$ which are both free of rank 2 and for which the quotients $\mathcal{P}/\Gamma_2$ and $\mathcal{P}/\Gamma'_2$ have the same volume. Now the surfaces $\mathcal{P}/\Gamma_2$ and $\mathcal{P}/\Gamma'_2$ are not homeomorphic: any compact subset K of $\mathcal{P}/\Gamma'_2$ is contained in a larger compact subset L such that $(\mathcal{P}/\Gamma'_2) \setminus L$ is connected (i.e., $\mathcal{P}/\Gamma'_2$ has one end, in the sense of Example IV.25.vi), whereas this does not hold for $\mathcal{P}/\Gamma_2$ (which has three ends). It follows that Γ_2 and Γ'_2 are not conjugate inside $PSL(2,\mathbb{R})$. This is a first way of showing that "rigidity in the sense of Mostow does not hold for $PSL(2,\mathbb{R})$" (see [Mosto–73], as well as [GroPa–91]).

(viii) By Teichmüller theory, hyperbolic structures on an orientable closed surface Σ_g of genus $g \geq 2$ are parametrized by the quotient of a disc of dimension $6g - 6$ by a countable equivalence relation. (See e.g. [ImaTa–72] and [Schmu–99]; see also the end of Example V.46.) In particular there are uncountably many embeddings of the fundamental group of Σ_g in $PSL(2,\mathbb{R})$, all as discrete subgroups with a finite volume given by the Gauss-Bonnet formula, which are pairwise not conjugate. This is a second way of showing that "rigidity in the sense of Mostow does not hold for $PSL(2,\mathbb{R})$".

34. Exercise on some subgroups of Hecke groups. With the same notation as in Example 28 (Hecke groups), set

$$a'_\lambda = j a_\lambda^{-1} j = \begin{bmatrix} 1 & 0 \\ \lambda & 1 \end{bmatrix}$$

and let Γ'_λ be the subgroup of Γ_λ generated by $\{a_\lambda, a'_\lambda\}$.

If q is odd, check that $\Gamma'_\lambda = \Gamma_\lambda$. [Hint: $a_\lambda (a'_\lambda)^{-1} = b_\lambda^2$.]

If $q = 2q'$ is even, check that Γ'_λ is of index two in Γ_λ. [Hint: Γ'_λ is the kernel of the homomorphism $\Gamma_\lambda \to \mathbb{Z}/2\mathbb{Z}$, which counts the parity of the number of occurrences of j in reduced words in $\{b_\lambda, j\}$.] It is known that $\Gamma'_\lambda \approx \mathbb{Z} * \mathbb{Z}/q'\mathbb{Z}$. (This is Theorem 7 in [LynUl–69], but it is also an easy consequence of the theorem of Poincaré of Section V.B.)

35. Exercise on arithmeticity. The purpose of this exercise is to show that the Hecke group

$$\Gamma_{2\cos\frac{\pi}{q}} = \left\langle \begin{bmatrix} 1 & 2\cos\frac{\pi}{q} \\ 0 & 1 \end{bmatrix}, \begin{bmatrix} 0 & 1 \\ -1 & 0 \end{bmatrix} \right\rangle$$

is arithmetic in $G = PSL(2, \mathbb{R})$ if $q = 4$ or $q = 6$ but not if $q = 5$. [We are grateful to Paul Schmutz Schaller for his indications. For "arithmetic", see the end of Example 28.]

(i) Let n be an integer, $n \geq 2$. Let $\Delta'(n)$ denote the subgroup of G consisting of elements of the form $\begin{bmatrix} a & b\sqrt{n} \\ c\sqrt{n} & d \end{bmatrix}$, with $a, b, c, d \in \mathbb{Z}$ and $ad - nbc = 1$. Let $\Delta(n)$ denote the subgroup of G consisting of $\Delta'(n)$ and of elements of the form $\begin{bmatrix} a\sqrt{n} & b \\ c & d\sqrt{n} \end{bmatrix}$, with $a, b, c, d \in \mathbb{Z}$ and $nad - bc = 1$.

Check that $\Delta'(2)$ is a subgroup of index 2 in $\Delta(2)$, and that $\Gamma_{\sqrt{2}} \subset \Delta(2)$; similarly $\Delta'(3)$ is a subgroup of index 2 in $\Delta(3)$, and $\Gamma_{\sqrt{3}} \subset \Delta(3)$.

(ii) For $n \geq 2$ as above, set

$$\Gamma_0(n) \doteq \left\{ \begin{bmatrix} a & b \\ c & d \end{bmatrix} \in PSL(2, \mathbb{Z}) \,\Big|\, c \equiv 0 \pmod{n} \right\}.$$

For any prime p, show that $\Gamma_0(p)$ is of index $p + 1$ in $PSL(2, \mathbb{Z})$. [Hint: think of a transitive action of $PSL(2, \mathbb{Z})$ on the projective line over the prime field of order p.]

More generally, the index of $\Gamma_0(n)$ in $PSL(2, \mathbb{Z})$ is known to be $n \prod \left(1 + \frac{1}{p}\right)$, where the product is taken over all primes p which divide n.

(iii) For all $n \geq 2$, show that

$$\begin{bmatrix} \omega & 0 \\ 0 & \omega^{-1} \end{bmatrix} \Delta'(n) \begin{bmatrix} \omega^{-1} & 0 \\ 0 & \omega \end{bmatrix} = \Gamma_0(n)$$

for an appropriate $\omega \in \mathbb{R}$.

(iv) It follows from (i) to (iii) that $\Gamma_{\sqrt{2}}$ has a subgroup $\Gamma_{\sqrt{2}} \cap \Delta'(2)$ of index 2 which is conjugate in G to a subgroup of $\Gamma_0(2)$, and thus to a subgroup of index a multiple of 3 in $PSL(2, \mathbb{Z})$.

If $\mathcal{P}$ denotes the Poincaré half-plane (as in Example 26), there is a natural notion of volume for quotients of $\mathcal{P}$, for which

$$vol\left(\Gamma_{\sqrt{2}} \setminus \mathcal{P}\right) = \frac{\pi}{2} \quad \text{and} \quad vol\left(PSL(2,\mathbb{Z}) \setminus \mathcal{P}\right) = \frac{\pi}{3}.$$

Show that this implies that $\Gamma_{\sqrt{2}} \cap \Delta'(2)$ is conjugate to the whole of $\Gamma_0(2)$; in particular, $\Gamma_{\sqrt{2}}$ and $PSL(2,\mathbb{Z})$ are commensurable in G.

(v) Check as in (iv) that $\Gamma_{\sqrt{3}}$ has a subgroup of index 2 which is conjugate in G to $\Gamma_0(3)$, so that $\Gamma_{\sqrt{3}}$ and $PSL(2,\mathbb{Z})$ are commensurable in G.

(vi) If γ is the element

$$\gamma \doteq \begin{bmatrix} 1 & 2\cos\frac{\pi}{5} \\ 0 & 1 \end{bmatrix} \begin{bmatrix} 0 & 1 \\ -1 & 0 \end{bmatrix} \in \Gamma_{2\cos\frac{\pi}{5}},$$

show that

$$|trace\,(\gamma^n)| \notin \mathbb{Q}$$

for all $n \geq 1$. (We denote by $|trace|$ the function on $PSL(2,\mathbb{R})$ defined via the usual trace of $SL(2,\mathbb{R})$; of course, without the absolute value, the "trace of an element in $PSL(2,\mathbb{R})$" would make no sense.)

Deduce from this that $\Gamma_{2\cos\frac{\pi}{5}}$ and $PSL(2,\mathbb{Z})$ are not commensurable in G. [Hint: $2\cos(\frac{\pi}{5}) = \frac{1}{2}(1 + \sqrt{5})$.]

Problems and complements for II.B

36. Free groups generated by hyperbolic automorphisms of trees. (See also [Hausm–81].) Let X be a tree; we denote by X^0 its vertex set and by $d: X^0 \times X^0 \longrightarrow \mathbb{N}$ the usual combinatorial distance function. An automorphism γ of X has an *inversion* if there exist two vertices in X which are joined by an edge and which are exchanged by γ. An automorphism γ of X is *hyperbolic* if it has no inversion and if its *amplitude*

$$a_\gamma = \min_{x \in X^0} d(x, \gamma x)$$

is strictly positive; it follows that

$$\left\{ x \in X^0 \mid d(x, \gamma x) = a_\gamma \right\}$$

is the vertex set L_γ^0 of a subgraph L_γ of X which is a geodesic line in X, called the *axis* of γ, and that γ induces on L_γ a translation of amplitude a_γ. If γ is hyperbolic, observe that γ^n is also hyperbolic for any $n \in \mathbb{Z}, n \neq 0$, with same axis as γ and with amplitude $|n|$ times larger. An automorphism γ of X is *elliptic* if there exists a vertex of X fixed by γ; it can be shown that an automorphism without inversion is either hyperbolic or elliptic. (See n° I.6.4 in [Serr–77].)

Let k be an integer, $k \geq 0$, and let Γ be a group of automorphisms of X without inversion. Say Γ is k-*acylindrical* if the fixed-point set of any element $\gamma \neq 1$ in Γ has diameter bounded by k. In particular, saying that Γ is 0-acylindrical is the same as saying that Γ acts freely on the edge set of X. (This notion of k-acylindrical action, which appears in [Sela–97a], was explained to me by T. Delzant, to whom I am grateful. See Proposition VII.18 below for an application, and [Delz–99] for more general notions.)

Consider two hyperbolic automorphisms γ_1, γ_2 of X, respectively with amplitudes a_1, a_2 and with axis L_1, L_2. We denote by $\langle \gamma_1, \gamma_2 \rangle$ the group of automorphisms of X generated by γ_1 and γ_2.

(i) Assume that $L_1 \cap L_2$ is either empty or reduced to one vertex. Show that $\langle \gamma_1, \gamma_2 \rangle$ is free on γ_1 and γ_2.

[Hint. Let $y_1 \in L_1^0$ and $y_2 \in L_2^0$ be the end-vertices of the shortest geodesic segment connecting L_1 to L_2. (One has $y_1 = y_2$ in case $L_1 \cap L_2$ consists of one vertex.) Define X_1^0 to be the subset of X^0 of those vertices x for which the geodesic segment from x to y_1 contains one of $\gamma_1(y_1)$, $\gamma_1^{-1}(y_1)$, and define X_2^0 similarly. Apply the Table-Tennis Lemma to the subgroups $\langle \gamma_1 \rangle$, $\langle \gamma_2 \rangle$, of $Aut(X)$ and the subsets X_1^0, X_2^0 of the vertex set of X.]

(ii) Assume that $L_1 \cap L_2$ has finite length $l \geq 1$, and let $m_1, m_2 \in \mathbb{N}$ be such that $m_1 a_1 > l$, $m_2 a_2 > l$. Show that the group $\langle \gamma_1^{m_1}, \gamma_2^{m_2} \rangle$ is free on $\gamma_1^{m_1}$ and $\gamma_2^{m_2}$.

(iii) Assume that $a_1 = a_2 = a \geq 1$, that $\gamma_2 \neq \gamma_1^{\pm 1}$, and that $\langle \gamma_1, \gamma_2 \rangle$ is k-acylindrical for some $k \geq 0$. Check that $L_1 \cap L_2$ is either empty or a segment of length at most $a + k$.

[Hint: observe first that, if γ_1, γ_2 are hyperbolic elements of amplitude a in a k-acylindrical group of automorphisms of X and if $L_1 \cap L_2$ contains a segment of length $a + k + 1$, then $\gamma_2 = \gamma_1^{\pm 1}$.]

37. Free groups of rotations. It can be shown by completely elementary computations that the matrices

$$g = \begin{pmatrix} 0 & 1 & 0 \\ 1 & 0 & 0 \\ 0 & 0 & -1 \end{pmatrix} \quad \text{and} \quad h = \begin{pmatrix} 1 & 0 & 0 \\ 0 & -\frac{1}{2} & \frac{\sqrt{3}}{2} \\ 0 & -\frac{\sqrt{3}}{2} & -\frac{1}{2} \end{pmatrix}$$

generate in $SO(3)$ a subgroup isomorphic to $(\mathbb{Z}/2\mathbb{Z}) * (\mathbb{Z}/3\mathbb{Z})$.

See [Osofs–78], or [Harpe–83], or [Wagon–85, Theorem 2.1].

It follows that $SO(3)$ contains non-abelian free groups of rank n for any $n \geq 2$ (see Problems 16 and 20).

To show that $SO(3)$ contains free subgroups, there is another method (longer but quite instructive) which uses the arithmetic of Hurwitz integral quaternions. See Proposition 2.1.7 in [Lubo–94].

38. Free groups of rational matrices. There are free subgroups of rank 2 in $SO(3) \cap SL(3, \mathbb{Q})$. There are such subgroups for which any element distinct from the identity has no fixed point on the rational unit sphere $\mathbb{S}^2 \cap \mathbb{Q}^3$ [Satô–95], and similarly for the sphere $(\sqrt{q}\mathbb{S}^2) \cap \mathbb{Q}^3$ for any rational number $q > 0$ [Satô–98].

39. Highly transitive actions of free groups. There exists an action of the free group F_2 of rank two on an infinite countable set X which is *highly transitive*, i.e., which satisfies the following property: given any integer $n \geq 1$ and two labelled subsets of size n in X, say $\{x_1, \ldots, x_n\}$ and $\{y_1, \ldots, y_n\}$, there exists $\gamma \in F_2$ such that $\gamma(x_j) = y_j$ for $j \in \{1, \ldots, n\}$. See Section 9.3 of [DixMo–96].

40. Research problem. Find an action of an appropriate group on an appropriate set and use the Table-Tennis Lemma to show that the transformations

$$t \longmapsto t + 1 \qquad \text{and} \qquad t \longmapsto t^3$$

generate a free group of rank 2. These transformations can be seen

either as homeomorphisms of $\mathbb{R}$,
or as germs of homeomorphisms at ∞ in $\mathbb{R}$.

For the existing proof (using algebra à la Galois), see [White–88], [AdeGM–91], and [CohGl–97]. Another proof would possibly cover other cases, such as one germ of homeomorphisms at ∞ "tangent to the identity" (as $t \longmapsto t + 1$) and another such germ with appropriate "attracting" properties (as $t \longmapsto t^3$).

41. Dense free subgroups of topological groups. For a topological group G, let $D_n(G)$ [respectively $F_n(G)$] denote the subset of G^n of n-uples $(g_1, \ldots, g_n)$ generating a subgroup of G which is dense [respectively free of rank n].

Observe that $F_2(PSL(2, \mathbb{R}))$ and $F_2(SO(3))$ are non-empty by Exercise 28 and Complement 37. It follows from *Lie theory* that $F_n(G)$ is non-empty for any connected real Lie group which is not solvable; under the same hypothesis on G, Epstein has pointed out that the complement of $F_n(G)$ in G^n has measure zero [Epste–71].

Standard arguments in Lie theory show also that, if G is a *semi-simple connected real Lie group,* then there exist two elements in G generating a subgroup which is both dense and free of rank 2; see [Bourb–75], Chapter VIII, § 3, Exercise 10.

If G is *a metrizable connected compact group,* it is a classical result [SchUl–35] that $D_2(G)$ is residual in $G \times G$ (residual subset = countable intersection of open dense subsets). If G is *moreover non-abelian,* then $F_n(G)$ is residual in G^n for all $n \geq 2$; see [Mycie–73] for a stronger result.

If G is a metrizable connected *locally compact* group which is not solvable, $F_n(G)$ is residual in G^n for all $n \geq 2$ by results going back to [Kuran–51]; see again [Mycie–73].

Let Γ be a subgroup of $SO(3)$ which is dense and free of rank 2. Arnold and Krylov have shown that there is a phenomenon of equidistribution, in the following sense [ArnKr–63]. Choose two generators s, t of Γ. Let $\mathcal{D}$ be a region of the two-dimensional sphere which is bounded by a piecewise smooth curve. For an integer $k \geq 0$, observe that there are 2^k elements in $SO(3)$ which can be written as words in s and t (not using the inverses), and let $p_k(\mathcal{D})$ denote the number of these which map the North Pole into $\mathcal{D}$. Then the numbers $2^{-k}p_k(\mathcal{D})$ have a limit when $k \to \infty$, which is the ratio between the area of $\mathcal{D}$ and the area of the sphere. (There are more recent results due to Lubotzky, Phillips, and Sarnak; see Chapter 9, "Distributing points on the sphere", in [Lubo–94], as well as [Sarna–90].)

Similar results hold for homogenous spaces of motion groups $\mathbb{R}^n \rtimes O(n)$ and more generally of Lie groups G containing a nilpotent closed invariant subgroup N such that G/N is compact, by results of Kazhdan [Kazhd–65] and Guivarc'h [Guiva–76].

It is an open problem to know whether this carries over to other cases, such as appropriate subgroups of the group $PSL(2, \mathbb{R})$ acting by isometries on the hyperbolic plane. I have been informed that R. Grigorchuk and Yu.B. Vorobets have partial results for this case (unpublished).

If, moreover, G is an algebraic group such that the quotient of G by its solvable radical is not compact, it is also known that G contains free subgroups which are discrete for the Hausdorff topology and Zariski-dense. Appropriate formulations of this fact carry over to algebraic groups over other locally compact fields [Wink–97a].

For Zariski-dense free subgroups of algebraic groups defined over arbitrary fields, see Theorems 3 and 4 in [Tits–72].

42. Tits' alternative [Tits–72]. *Let Γ be a subgroup of $GL(N, \mathbb{K})$, for some integer $N \geq 1$ and some field $\mathbb{K}$ of characteristic zero. Then either Γ has a non-abelian free subgroup or Γ possesses a soluble subgroup of finite index.*

There is an integer function $n \longmapsto \lambda(n)$ such that, in case Γ has no free subgroup of finite index, Γ possesses a normal soluble subgroup of index at most $\lambda(n)$ [Wang–81].

As stated above, the alternative is not true for a field of finite characteristic, as shown by the example of the full linear group over an infinite algebraic extension of a finite field. However an appropriate formulation carries over to fields of characteristic > 0 (Theorem 2 in [Tits–72]).

A class $\mathcal{C}$ of groups is said *to satisfy Tits' alternative* if any group in $\mathcal{C}$ either has a non-abelian free subgroup or possesses a soluble subgroup of finite index. Besides the class of linear groups in characteristic zero (as above), other classes are known to satisfy Tits' alternative:

- finitely-generated linear groups in finite characteristic [Tits–72],
- one-relator groups (Theorem 3 of [KarSo–71], see also [CecGr–97]),
- subgroups of mapping class groups of surfaces ([Ivano–84] and [McCar–85]),
- fundamental groups of appropriate 3-manifolds [Parr–92a],

- subgroups of Gromov hyperbolic groups
  (see 8.2.F in [Gromo–87] or Theorem 37 of Chapter 8 in [GhyHp–90]),
- free products studied in [FinLR–88],
- subgroups of the group of polynomial automorphisms of $\mathbb{C}^2$
  ([Lamy–98] and [Lamy]),
- groups of outer automorphisms of free groups of finite rank (see [BesFH–a]
  and [BesFH–b], as well as the discussion in Section 10 of [BasLu–94]).

The class of subgroups of automorphisms of locally finite trees does *not* satisfy
Tits' alternative (see [GupS–83b], as well as [PayVa–91]).

43. An open problem on free subgroups. A group Γ is said to have
Property $P_{\text{naï}}$ if, for any finite subset F of $\Gamma \setminus \{1\}$, there exists $y_0 \in \Gamma$ of
infinite order such that, for each x in F, the canonical epimorphism from the
free product $\langle x \rangle * \langle y_0 \rangle$ onto the subgroup $\langle x, y_0 \rangle$ of Γ generated by x and y_0 is
an isomorphism. (The subscript "naï" refers to "naïve", and distinguishes $P_{\text{naï}}$
from other properties defined in [BekCH–94].)

It is known and easy to show that the reduced C*-algebra of a group with
property $P_{\text{naï}}$ is simple with unique trace. Known examples of groups having
this property include non-abelian free groups, $PSL(2, \mathbb{Z})$, and more generally
Zariski-dense subgroups of connected simple Lie groups with $\mathbb{R}$-rank 1 and
trivial centre [BekCH–94].

Does $PSL(n, \mathbb{Z})$ have property $P_{\text{naï}}$ for $n \geq 3$?

There are "many" groups which have Property $P_{\text{naï}}$, as shown by the following "asymptotic Freiheitssatz", of independent interest [ArzOl–96] (see also
[CheSc–98]); this uses notions exposed in Chapter V below.

Consider two integers $n \geq 2$ and $m \geq 0$. For each $t \geq 0$, let $N_{m,n,t}$ denote
the number of presentations of groups of the form

$$(*) \qquad \qquad \Gamma = \langle s_1, \ldots, s_n \mid r_1, \ldots, r_m \rangle$$

where the r_i are cyclically reduced words in $\left\{ s_1, s_1^{-1}, \ldots, s_n, s_n^{-1} \right\}$ of length $\leq t$.
Let $N_{m,n,t}^f$ denote the number of such presentations for which, moreover, any
subgroup of Γ generated by at most $n - 1$ elements is free. It is then a theorem
that, for fixed m and n, the ratio $\frac{N_{m,n,t}^f}{N_{m,n,t}}$ approaches 1 exponentially fast when
$t \to \infty$.

In particular for fixed m and $n \geq 3$, and for t large enough, the proportion
of presentations (*) providing groups with Property $P_{\text{naï}}$ is almost 100 %.

44. Homomorphisms of free groups onto free groups. Let k, l be
integers such that $k \geq l$. Let F_k denote the free group on k generators $s_1, \ldots, s_k$
and let F_l denote the free group on l generators $t_1, \ldots, t_l$; denote by $\pi_{k,l}$:
$F_k \longrightarrow F_l$ the homomorphism mapping s_j to t_j for $j \in \{1, \ldots, l\}$ and to 1
for $j \in \{l + 1, \ldots, k\}$. The theory of "Nielsen reduction" for sets of generators
of subgroups of free groups provides the following result (see Theorem 3.3 in
[MagKS–66], page 132).

Proposition. *If $k \geq l$, any homomorphism π from F_k onto F_l can be written as $\pi = \pi_{k,l}\alpha$ for some automorphism α of F_k, with $\pi_{k,l}$ as above.*

Concerning homomorphisms of (not necessarily free) groups onto free groups, there are open problems with a low-dimensional topology flavour in [Stall–95].

45. Open problem. For an integer $g \geq 1$, consider the homomorphism

$$\pi'_g : F_{2g} \longrightarrow F_g \times F_g$$

mapping the first g [respectively the last g] generators of F_{2g} onto the g generators of the first factor [respectively of the second factor] in $F_g \times F_g$.

Is is true that any homomorphism π from F_{2g} onto $F_g \times F_g$ can be written as $\pi = \pi'_g\alpha$ for some automorphism α of F_{2g} ?

The question has an affirmative answer for $g = 1$ (easy exercise). If the answer were affirmative for all $g \geq 2$, it would imply the truth of a conjecture of Andrews and Curtis on presentation of the trivial group with the same number of generators and relations; see [Wrigh–75], page 205 in [GriKu–93], [BurMa–93], [HogMe–93], and Problem 5.2 of [Kirby–97].

46. Kervaire's conjecture. Let Γ be a group, let $\tilde{\Gamma}$ be the free product of Γ and of an infinite cyclic group, let r be an element of $\tilde{\Gamma}$ and let N be the smallest normal subgroup of $\tilde{\Gamma}$ containing r. If the quotient $\tilde{\Gamma}/N$ is reduced to one element, does it follow that Γ is reduced to one element ?

The problem arose from higher-dimensional knot theory [Kerva–65]. It was first expressed in private conversation with Gilbert Baumslag in winter 1963/64 [Kervaire, private communication]; record of this appears in [MagKS–66], page 403. Though progress has been achieved by A. Klyachko, the problem is still open (see [Klyac–93] and [FenRo–96]). A discussion appears in Problem 5.7 of [Kirby–97].

47. Combinatorial group formulation of the Poincaré conjecture. For an integer $g \geq 1$, consider the surface group

$$\Gamma_g = \langle a_1, b_1, \ldots, a_g, b_g \mid [a_1, b_1] \ldots [a_g, b_g] = 1 \rangle$$

(we anticipate here on presentations of groups, as in Chapter V below). For homomorphisms from Γ_g onto free groups, the following two results are due to H. Zieschang. (See the comment which follows Proposition I.6.4 in [LynSc–77], as well as [GriKZ–92].)

(i) There exists a homomorphism from Γ_g *onto* F_k if and only if $k \leq g$. [To show the inequality, use the fact that the homomorphism $\Gamma_g \longrightarrow F_k$ has a section and the fact that, since the cup product in the cohomology $H^1(F_k, \mathbb{Q})$ is zero, $H^1(\Gamma_g, \mathbb{Q})$ must have a subspace of dimension k on which the cup product is zero.]

(ii) For $k \leq g$ and two homomorphisms ϕ, ϕ' from Γ_g onto F_k, there exists an automorphism α of Γ_g such that $\phi' = \phi\alpha$.

Consider now the homomorphism

$$\pi_g : \Gamma_g \longrightarrow F_g \times F_g$$

mapping $a_1, \ldots, a_g$ [respectively $b_1, \ldots, b_g$] onto the g generators of the first factor [respectively of the second factor] in $F_g \times F_g$. (A Heegard splitting of genus g of a closed 3-manifold gives such a pair of homomorphisms into free groups of rank g.)

Is it true that any homomorphism π from Γ_g onto $F_g \times F_g$ can be written as $\pi = \pi_g \alpha$ for some automorphism α of Γ_g ?

It is an exercise to show that the answer is "yes" for $g = 1$. It is known that this question has an affirmative answer for all $g \geq 1$ if and only if the 3-dimensional Poincaré conjecture is true. See [Stall–66], [Jaco–69], and Theorem 14.6 in [Hempe–76].

There are other open questions of low-dimensional topology which have a group-theoretic formulation, for example *Whitehead's conjecture* ([Bogle–93], [BogDy–95]): if Y is a sub-complex of a 2-complex X and if the universal cover of X is contractible, then the universal cover of Y is conjecturally also contractible. But at least one of this conjecture and another famous conjecture, the "Eilenberg-Ganea conjecture", does *not* hold [BesBr–97].

48. Groups without non-abelian free subgroups. It may be true *and non-obvious* that some given group does NOT contain any non-abelian free group.

This is for example the case of the group $PL_+([0,1])$ of piecewise linear orientation-preserving homeomorphisms of the real line, by Theorem 3.1 in [BriSq–85] (see also Corollary 4.9 in [CanFP–96] and Theorem 4.6 in [Ghys–99]). However, it is also known that this group satisfies no laws: for any non-empty reduced word w in the free group on two letters, there exist $g, h \in PL_+([0,1])$ such that $w(g, h) \neq 1$ [BriSq–85]. The same results hold for the Thompson group which is a subgroup of $PL_+([0,1])$.

For linear groups, Tits' alternative of Complement 42 provides a criterion for the existence of non-abelian free subgroups.

CHAPTER III

FINITELY-GENERATED GROUPS

After a first section with simple examples, we concentrate in this chapter on two-generator groups, and give in particular an exposition of B.H. Neumann's argument (1937) showing that there are uncountably many such groups.

III.A. Finitely-generated and infinitely-generated groups

A group Γ is *finitely generated,* or of *finite type,* if there is a finite subset $S \subset \Gamma$ such that, for any $\gamma \in \Gamma$, there exists a finite sequence $s_1, s_2, ..., s_n$ of elements in $S \cup S^{-1}$ with $\gamma = s_1 s_2 ... s_n$. Equivalently, a group Γ is finitely generated if it is a quotient of a free group of finite rank.

Observe that a finitely-generated group is necessarily countable. Finitely-generated groups are easier to understand than countable groups in general, both in terms of their structure (think of abelian groups!) and in terms of their actions (examples are given in Complements 20 and 25).

The present Section III.A is essentially a description of groups, both finitely generated and not. Some examples require knowledge outside the scope of these notes, but they can also be skipped by the reader without making it impossible to read later sections.

1. Example. *The group $\mathbb{Z}^n$ is finitely generated for any integer n.*

2. Example. *The group $GL(n, \mathbb{Z})$ is finitely generated for any integer n. More precisely, the four matrices*

$$
s_1 = \begin{pmatrix} 0 & 0 & 0 & \ldots & 0 & 1 \\ 1 & 0 & 0 & \ldots & 0 & 0 \\ 0 & 1 & 0 & \ldots & 0 & 0 \\ \ldots & \ldots & \ldots & \ldots & \ldots & \ldots \\ 0 & 0 & 0 & \ldots & 0 & 0 \\ 0 & 0 & 0 & \ldots & 1 & 0 \end{pmatrix}
\qquad
s_2 = \begin{pmatrix} 0 & 1 & 0 & \ldots & 0 & 0 \\ 1 & 0 & 0 & \ldots & 0 & 0 \\ 0 & 0 & 1 & \ldots & 0 & 0 \\ \ldots & \ldots & \ldots & \ldots & \ldots & \ldots \\ 0 & 0 & 0 & \ldots & 1 & 0 \\ 0 & 0 & 0 & \ldots & 0 & 1 \end{pmatrix}
$$

$$
s_3 = \begin{pmatrix} 1 & 1 & 0 & \ldots & 0 & 0 \\ 0 & 1 & 0 & \ldots & 0 & 0 \\ 0 & 0 & 1 & \ldots & 0 & 0 \\ \ldots & \ldots & \ldots & \ldots & \ldots & \ldots \\ 0 & 0 & 0 & \ldots & 1 & 0 \\ 0 & 0 & 0 & \ldots & 0 & 1 \end{pmatrix}
\qquad
s_4 = \begin{pmatrix} -1 & 0 & 0 & \ldots & 0 & 0 \\ 0 & 1 & 0 & \ldots & 0 & 0 \\ 0 & 0 & 1 & \ldots & 0 & 0 \\ \ldots & \ldots & \ldots & \ldots & \ldots & \ldots \\ 0 & 0 & 0 & \ldots & 1 & 0 \\ 0 & 0 & 0 & \ldots & 0 & 1 \end{pmatrix}
$$

generate $GL(n, \mathbb{Z})$ for any integer $n \geq 2$.

Left [respectively right] multiplication by s_1 permutes rows [respectively columns] cyclically, left [respectively right] multiplication by s_2 permutes the first 2 rows [respectively the first 2 columns], s_3 differs from the unit matrix by the single entry $(s_3)_{1,2} = 1$ and s_4 differs from the unit matrix by the single entry $(s_4)_{1,1} = -1$.

Sketch of proof. Finite products of the s_j 's include permutation matrices (this uses s_1, s_2) and "elementary matrices" differing from the unit matrix in one off-diagonal entry only (this uses also s_3).

Now using the Euclidean algorithm, one can express any $\gamma \in GL(n, \mathbb{Z})$ as a product of elementary matrices and of s_4. This is a particular case of reduction to Hermite's normal form of n-by-n matrices with coefficients in a Euclidean ring.

For details, see e.g. Section 22 in [MacDu–46]. See also Example V.9. $\square$

Hua and Reiner have shown that s_1, s_3, s_4 suffice, by writing down a (rather long) formula for s_2 in terms of s_1, s_3, s_4. Trott has shown that s_1, s_3 generate $GL(n, \mathbb{Z})$ when n is even and $SL(n, \mathbb{Z})$ when n is odd. (References in § 7.2 of [CoxMo–57].) It is known that $SL(n, \mathbb{Z})$ can be generated by two elements for all $n \geq 2$; see Proposition 5 in [Wilso–76].

There is another proof in terms of Siegel domains in $GL(n, \mathbb{R})$; see [Borel–69], in particular Corollary 4.8.

Comments. (i) Let R be the ring of integers of a number field and let $n \geq 3$ be an integer. Then $SL(n, R)$ is finitely generated; this goes back to 1895 and to A. Hurwitz (quoted on page 96 of [BasMS–67]), but classical methods are considerably more difficult to apply for this case than for the previous example of $R = \mathbb{Z}$. Since Kazhdan's work [Kazhd–67], the usual way of showing this uses the so-called "Property (T)"; see Complement V.20.

The fact that $SL(n, R)$ is finitely generated is relevant to K-theory. (It implies that "the group $SK_1(R)$ is finite"; see the discussion in page 81 of [Rosen–94].)

(ii) It is not always easy to decide whether a given countable group is finitely generated or not. See (ii) in Complement 16, the discussion in Example V.7, or the open problem stated at the end of Exercise V.18.

(iii) It is also elementary to show that, for any $n \geq 3$, there is a number ν_n such that any matrix in $SL(n, \mathbb{Z})$ can be written as a product of at most ν_n elementary matrices [AdiMe–92]; in other words $SL(n, \mathbb{Z})$ is *boundedly elementary generated;* analogous results for other rings require more machinery ([CarKe–83], [Tauge–91]). Bounded generation is an important ingredient in Shalom's approach to Kazhdan Property (T); see [Shalo–a] and [Shalo–b].

3.   Example. *Let F be a free group of finite rank, let N be a normal subgroup of F distinct from $\{1\}$, and let Q denote the quotient group F/N.*

Then N is finitely generated if and only if N is of finite index, namely if and only if Q is a finite group.

Proof. As F is the fundamental group of a bouquet of circles B, the normal subgroup N corresponds to a covering $\pi : X \longrightarrow B$; the covering space X is a connected graph and the fibers of π are in bijection with Q.

If Q is finite, X is a finite graph and $N \approx \Pi_1(X)$ (the fundamental group of X) is finitely generated.

If Q is infinite, X is a connected infinite graph which is not a tree (because of the assumption $N \neq \{1\}$), on which N acts properly discontinuously. Thus X has infinitely many disjoint closed paths, and consequently its first homology group $H_1(X, \mathbb{Z})$ is not finitely generated. A fortiori $N \approx \Pi_1(X)$ is not finitely generated. $\square$

For other conditions equivalent to Q being finite, see [Grune–78]. For growth functions of generating sets of Q when Q is infinite, see Complement VI.41.

Chapter V is devoted to finitely-presented groups Γ, appearing in short exact sequences of the form $1 \longrightarrow F' \longrightarrow F \longrightarrow \Gamma \longrightarrow 1$ where F is a free group of finite rank and F' a normal subgroup generated *as a normal subgroup* by a finite subset of F; if Γ is infinite, Example 3 shows that F' is *never* finitely generated *as a group*.

Here are two generalizations of Example 3.

(i) Let F be as in Example 3 and let R be a subgroup of F of infinite index. Assume that R contains a normal subgroup of F not reduced to $\{1\}$. Then R is *not* finitely generated, by Theorem 2.10 in Section 2.4 of [MagKS–66].

(ii) Let $\Gamma = A * B$ be a free product of two non-trivial groups and let Δ be a finitely-generated normal subgroup of Γ. Then either Δ is reduced to one element or Δ has finite index in Γ (Theorem 3.11 in [ScoWa–79]).

4. Examples of groups which are *not* finitely generated.

Let X be an infinite set and let $Sym_f(X)$ be the group of permutations of X of finite support, namely of permutations σ for which $\{x \in X | \sigma(x) \neq x\}$ is finite. For any finite subset Y of X, the symmetric group $Sym(Y)$ of Y is naturally a subgroup of $Sym_f(X)$. Any finitely-generated subgroup of $Sym_f(X)$ is contained in $Sym(Y)$ for an appropriate finite subset Y of X; in particular, $Sym_f(X)$ is not finitely generated.

Similarly, any locally finite infinite group is not finitely generated. It is known that there are uncountably many pairwise non-isomorphic countable groups which are infinite and locally finite; see [KegWe–73]. (Recall that a group is *locally finite* if all its finitely-generated subgroups are finite; the terminology is well established, even if it is not consistent with that for "locally compact group"!)

The additive group of $\mathbb{Q}$ is not finitely generated. All its finitely-generated subgroups distinct from $\{0\}$ are isomorphic to $\mathbb{Z}$.

More generally, the additive group of any infinite field is not finitely generated. (If the field is infinite dimensional over its prime field, this should be clear to the reader. Otherwise, the additive group of the field contains that of $\mathbb{Q}$, and the conclusion follows from the case of $\mathbb{Q}$, because a subgroup of a finitely-generated *abelian* group is necessarily finitely generated.)

The multiplicative group $\mathbb{Q}^*$ of $\mathbb{Q}$ is not finitely generated. More precisely, any finitely-generated subgroup of $\mathbb{Q}^*$ is contained in the free abelian group $\langle S \rangle$ of those rationals of the form $\frac{m}{n}$ with m and n products of primes in some *finite* set S of prime numbers.

The same argument shows that the multiplicative group of a pure transcendental extension $\mathbb{F}_q(T)$ of a finite field is not finitely generated. (The only knowledge required is that the polynomial ring $\mathbb{F}_q[T]$ contains infinitely many irreducibles. This follows for example from the fact that there exists a finite field of order q^f for any integer $f \geq 1$. Alternatively, this follows from known exact formulas giving the number $N_q(f)$ of monic irreducible polynomials of degree f in $\mathbb{F}_q[T]$; these formulas show in particular that $N_q(f) \geq 1$ for all $f \geq 1$; see Theorem 3.25 in [LidNi–83].)

If $\mathbb{K}$ is an infinite algebraic extension of a finite field, then $\mathbb{K}^*$ is an infinite locally finite group, and as such it is not a finitely-generated group.

It follows that $\mathbb{K}^*$ is not finitely generated for any infinite field. (The proof again uses the fact that a subgroup of a finitely generated abelian group is itself finitely generated, and the three previous particular cases.)

If $n \geq 2$, the group $SL(n, \mathbb{Q})$ is not finitely generated. All its finitely-generated subgroups are contained in subgroups of the form $SL\left(n, \mathbb{Z}\left[\frac{1}{N}\right]\right)$, for some integer $N \geq 1$ (Exercise 12).

More generally, for any infinite field $\mathbb{K}$ of characteristic 0, the group $SL(n, \mathbb{K})$ is not finitely generated. (This requires some tools — for example the simplicity of $SL(n, \mathbb{K})$ and a result of Mal'cev [Mal'c–40] according to which finitely-generated subgroups of $GL(n, \mathbb{C})$ are "residually finite"— see Complements 18 and 20 below.)

Let q be a prime power, let $\mathbb{F}_q$ be the finite field of order q, and let $\mathbb{F}_q[T]$ denote the ring of polynomials in one variable with coefficients in $\mathbb{F}_q$. It is known that the group $SL(2, \mathbb{F}_q[T])$ is not finitely generated (this is due to Nagao — see page 121 of [Serr–70b]). Similarly, let G be a connected simple Lie group of rank one over the local field $\mathbb{F}_q(T)$ and let Γ be a lattice in G such that G/Γ is *not* compact; then Γ is not finitely generated (see [Lubo–89], [Lubo–91]).

However $SL(n, \mathbb{F}_q[T])$ *is* finitely generated whenever $n \geq 3$, because these groups have the "Property (T) of Kazhdan". (See [Kazhd–67] and [HarVa–89], as well as Item V.20.iv below.)

For $n \geq 2$, let $\Gamma_n \doteq F_n/[[F_n, F_n], [F_n, F_n]]$ denote the free metabelian group on n generators (where F_n denotes the free group of rank n). Then $Aut(\Gamma_n)$ is finitely generated if and only if $n \neq 3$ (see [Bachm–85], [BacMo–85]).

5. Examples of non finitely-generated subgroups of finitely-generated groups.

A first example is provided by the group

$$\Gamma = \left\{ \begin{pmatrix} a & b \\ 0 & 1 \end{pmatrix} \in GL(2, \mathbb{R}) \,\Big|\, a = 2^n \text{ and } b = \frac{p}{2^q} \text{ for some } n, p, q \in \mathbb{Z} \right\}$$

which is generated by the two matrices $\begin{pmatrix} 2 & 0 \\ 0 & 1 \end{pmatrix}$ and $\begin{pmatrix} 1 & 1 \\ 0 & 1 \end{pmatrix}$, and the sub-group

$$\Gamma_0 = \left\{ \begin{pmatrix} 1 & b \\ 0 & 1 \end{pmatrix} \in GL(2, \mathbb{R}) \,\Big|\, b \in \mathbb{Z}\left[\frac{1}{2}\right] \right\}$$

which is not finitely generated. (This Γ is the group $\Gamma_{1,2}$ of Complement 21.)

A second example is $\Gamma_0 \subset \Gamma = \Gamma_0 \rtimes \mathbb{Z}$ where $\Gamma_0 = \bigoplus_{j \in \mathbb{Z}} \Gamma_{(j)}$ is a direct sum of copies of $\mathbb{Z}$ indexed by $\mathbb{Z}$, where $\mathbb{Z}$ acts on Γ_0 by shifts, and where $\rtimes$ indicates the corresponding semi-direct product. (The group Γ is known as the wreath product $\mathbb{Z} \wr \mathbb{Z}$; see VIII.3 below.)

A third example is given by a free group of rank 2 and a subgroup which is free of infinite rank. (Think either of an infinite connected covering of a wedge of two circles, or of the group generated by $\{a^n b a^{-n}\}_{n \geq 1}$ in the free group on $\{a, b\}$.) The kernel F in Problem II.16 for $A \times B$ infinite provides an analogous example.

There exists an example, due to Remeslennikov, of a finitely-presented group whose centre is not finitely generated [Remes–74]. It is an open problem to know whether any countable abelian group is a subgroup of the centre of some finitely-presented group (Problem 5.47 of the Kourovka notebook [Kouro–95]). It is also unknown whether the centre of Γ is always finitely generated, for Γ the fundamental group of a finite *aspherical* complex. (This is Part (A) of Mess' Problem 5.11 in [Kirby–97]; Part (D) asks among other questions whether such a group can be of intermediate growth, in the sense of Chapter VI below.)

In a finitely-generated group which is *virtually*[1] *polycyclic,* namely which contains a normal polycyclic[2] group of finite index, it is easy to show that any subgroup is again finitely generated. (This follows essentially from the same property for finite groups and for finitely-generated abelian groups; see Theorem 9.16 in [MacDo–88] for the nilpotent case or Remark 4.2 in [Raghu–72].)

Here is a partial converse: a finitely-generated solvable group which is *not* virtually polycyclic does contain subgroups which are not finitely generated.

---

[1] More generally, a group Γ is said to have *virtually* some Property (P) if Γ contains a subgroup of finite index which does have Property (P).

[2] A group Γ is *polycyclic* if there exists a nested sequence of subgroups

$$\Gamma = \Gamma_0 > \Gamma_1 > \ldots > \Gamma_m = \{1\}$$

such that Γ_i is a normal subgroup of Γ_{i-1} and Γ_{i-1}/Γ_i is cyclic for each $i \in \{1, \ldots, m\}$. By theorems of Mal'cev and L. Auslander, a group is polycyclic if and only if it is isomorphic to a solvable subgroup of $GL(n, \mathbb{Z})$, for some $n \geq 1$ (see Chapters 2 and 5 in [Segal–83]).

(This follows essentially from the fact that an abelian group is polycyclic if and only if it is finitely generated. See Proposition 4 of § 1.A in [Segal–83].) But examples are known of finitely-generated infinite groups which are simple (in particular not virtually solvable, and even less polycyclic), and in which any subgroup is finitely generated, indeed in which any proper subgroup is infinite cyclic; one example appears as Theorem 28.3 in [Ol's–89]. Ol'shanskii has also shown that any non-elementary torsion-free hyperbolic group has a non-abelian torsion-free infinite quotient in which any proper subgroup is cyclic (Corollary 1 of [Ol's–93]).

A group is said to be *slender* if all its subgroups are finitely generated [DunSa–99].

See also Complement 16 below.

6. Fundamental groups. Fundamental groups of finite CW-complexes and of compact manifolds are finitely generated (and indeed finitely presented: see Example V.7 below). But more complicated topological spaces may have "large" fundamental groups.

For example, for any integer $n \geq 2$, Antoine (1921) has constructed a nested sequence $(N_k)_{k \geq 1}$ of compact subsets of $\mathbb{R}^3$ with the following properties: N_k is the union of $(2n)^k$ solid tori linked with one another, the so-called *Antoine necklace* $A = \cap_{k=1}^{\infty} N_k$ is a Cantor set, and the fundamental group $\pi_1(\mathbb{R}^3 \setminus A)$ is not finitely-generated (see § 6 in [Rham–69]). Here is another example: the fundamental group of the Hawaiian earring is not countable, and a fortiori not finitely generated; it not free, for example because its abelianization contains an element $a \neq 1$ which is a kth power for infinitely many integers k (see [Smit–92] and [CanCo]).

The following general result has been conjectured by Mycielski and proved by Shelah (see [Pawli–98]). The fundamental group of a compact metric space which is path-connected and locally path-connected is either finitely generated or of the cardinality of the continuum; in particular such a group cannot be isomorphic to $\mathbb{Q}$.

It is apparently unknown whether any finitely-generated group can be realized as the fundamental group of a compact metric space which is path-connected and locally path-connected.

Exercises for III.A

7. Exercise. (i) Let Γ be the semi-direct product of $Sym_f(\mathbb{Z})$ by the translation $n \mapsto n + 1$ of $\mathbb{Z}$ (see the first example in Item 4 above for the notation). Check that Γ is finitely generated, though $Sym_f(\mathbb{Z})$ is not.

(ii) Define two permutations ϕ and ψ of the set X consisting of the disjoint union of $\mathbb{Z}$ and of a point $*$ by

$$\phi(x) = x + 1 \text{ for } x \in \mathbb{Z} \quad \text{and} \quad \phi(*) = *$$
$$\psi(0) = * , \quad \psi(*) = 0 \quad \text{and} \quad \psi(x) = x \text{ if } x \notin \{0, *\} .$$

Check that $Sym_f(\mathbb{Z})$ is a subgroup of the group generated by ϕ and ψ (this appears in [Ulam–74], page 677).

8. Exercise. Let Γ be a finitely-generated group.

(i) Show that any subgroup of finite index in Γ is finitely generated.
[Indication. One standard proof uses the so-called Reidemeister-Schreier method, which indeed shows much more; see Corollary 2.7.1 in [MagKS–66]. See also Corollary IV.24.i.]

(ii) Let S be an infinite generating subset of Γ. Check that there exists a finite subset of S which generates Γ.
[See also Proposition V.2 and Exercise V.11.]

9. Exercise. (i) For the free group F_2 of rank two, show that there exists an infinite tower $\Gamma_1 < \Gamma_2 < \ldots < \Gamma_n < \ldots < F_2$ of subgroups of F_2 such that $\Gamma_n \neq \Gamma_{n+1}$ for all $n \geq 1$.

(ii) Let Γ be a finitely-generated group and let $\Gamma_1 < \Gamma_2 < \ldots < \Gamma_n < \ldots < \Gamma$ be an infinite tower such that $\Gamma = \cup_{n \geq 1} \Gamma_n$. Show that $\Gamma = \Gamma_n$ for n large enough.

10. Exercise. Let $p_1, \ldots, p_n$ be distinct prime numbers and let N be their product; set $q_j = \frac{N}{p_j}$ for each $j \in \{1, \ldots, n\}$ and $S = \{q_1, \ldots, q_n\}$. Check that S generates $\mathbb{Z}$ and that no proper subset of S does.

11. Exercise. Show that, for any $n \geq 2$, there exist two finite groups F_1, F_2 and a homomorphism of the free product $F_1 * F_2$ onto $GL(n, \mathbb{Z})$.
[Hint. With the notation of Example 2, take for F_1 the subgroup of $GL(n, \mathbb{Z})$ generated by $\{s_1, s_2, s_4\}$ and for F_2 the cyclic subgroup generated by the product $s_4 s_2 s_3$.]
See also Complement 39 below.

12. Exercise. Show that, for $n \geq 2$, any finitely-generated subgroup of $SL(n, \mathbb{Q})$ is contained in $SL\left(n, \mathbb{Z}\left[\frac{1}{N}\right]\right)$ for some integer $N \geq 1$.
[Hint: take for N the least common multiple of the denominators of the matrix entries of the generators and of their inverses.]

Problems and complements for III.A

13. Indecomposable groups and Grushko's theorem. A group Γ is *indecomposable* if, whenever it can be written as a free product $\Gamma_1 * \Gamma_2$, then Γ_1 or Γ_2 is reduced to one element. Here are a few examples of indecomposable groups:

• finite groups,
• simple groups (see problems 13 and 23 in Section 4.1 of [MagKS–66]),

- groups with non-trivial centres, and, in particular, abelian groups (Exercise II.13),
- non-trivial direct products (see Problem 22 of Section 4.1 of [MagKS–66]),
- finitely-generated groups with one end (see Complement 15), and, in particular, fundamental groups of orientable closed surfaces of genus $g \geq 1$ and the groups $PSL(n, \mathbb{Z})$ for $n \geq 3$.

For other examples, see other problems of Section 4.1 of [MagKS–66].

A theorem of Grushko implies that any finitely-generated group is a free product of a finite family of indecomposable groups, and that the factors (not reduced to one element!) in such a decomposition are well defined up to order and isomorphism. This *does not* carry over to groups which are not finitely generated. See Theorem 3.5 and the example on page 163 in [ScoWa–79].

(As far as we know, Grushko's 1940 paper, which is in Russian, does not quote a 1929 theorem of Kneser, saying that a closed 3-manifold M is in a unique way a connected sum of irreducible 3-manifolds, so that the fundamental group $\Pi_1(M)$ is the free product of the fundamental groups of the factors [Miln–62]. But one of the main steps of Kneser's proof "contains obscurely something like Grushko's Theorem" [Stall–71].)

The theorem of Grushko implies also that $r(\Gamma_1 * \Gamma_2) = r(\Gamma_1) + r(\Gamma_2)$, where $r(\Gamma)$ denotes the minimal number of generators of a group Γ. Anticipating on presentations (see Exercise V.14), we also quote an analogous result of Delzant [Delz–96b]; for a finitely-presented group Γ, let $T(\Gamma)$ denote the minimal number of relations in a presentation of Γ all of whose relations have length two or three; then $T(\Gamma_1 * \Gamma_2) = T(\Gamma_1) + T(\Gamma_2)$.

There are other kinds of indecomposability results. For example, consider on the one hand the fundamental group Γ_g of an orientable closed surfaces of genus $g \geq 2$ (which is indecomposable, as noted above), and on the other hand the smallest class $\mathcal{C}$ of groups which

(i) contains all amenable groups (in particular all finite groups and all solvable groups),

(ii) is closed under free products (i.e. $\Gamma, \Delta \in \mathcal{C} \Longrightarrow \Gamma * \Delta \in \mathcal{C}$),

(iii) contains any group containing a subgroup of finite index in the class.

It is then a theorem that Γ_g is not in $\mathcal{C}$ [HirTh–75].

14. Free products with amalgamation and HNN-extensions. For these constructions, see e.g. § I.1 in [Serr–77], or Chapter 11 in [Rotma–95], or Chapter 1 in [ScoWa–79]. For their historical importance, in particular with respect to the *word problem for groups,* see also [Still–82].

Free products are particular cases of amalgamated free products, going back to [Schre–27]. In the simplest case, one has three groups A, Γ_1, Γ_2 and two inclusions $A \hookrightarrow \Gamma_1$, $A \hookrightarrow \Gamma_2$. The corresponding *free product of Γ_1 and Γ_2 with amalgamation over A* is denoted by $\Gamma_1 *_A \Gamma_2$, and one may identify Γ_1, Γ_2 with subgroups of it. One has the following *universal property*: for a group Γ and two homomorphisms $h_i : \Gamma_i \to \Gamma$ $(i = 1, 2)$ such that $h_1(a) = h_2(a)$ for all $a \in A$, there exists a unique homomorphism $h : \Gamma_1 *_A \Gamma_2 \to \Gamma$ such that $h(\gamma_i) = h_i(\gamma_i)$ for all $\gamma_i \in \Gamma_i$ $(i = 1, 2)$.

The group $\Gamma_1 *_A \Gamma_2$ is finitely generated as soon as Γ_1 and Γ_2 are.

A theorem of Seifert and van Kampen shows that the fundamental group of a complex obtained by glueing two complexes over a common non-empty connected subcomplex (with the above property of injectivity for fundamental groups) is a free product with amalgamation; for example, the fundamental group Γ_g of an orientable closed surface of genus $g \geq 2$ is the free product of a free group of rank $2g - 2$ (the fundamental group of a surface of genus $g - 1$ with a small disc removed) and a free group of rank 2, with amalgamation over an appropriate cyclic subgroup. There are other standard examples, in some sense more algebraic. One is

$$SL(2, \mathbb{Z}) \approx (\mathbb{Z}/6\mathbb{Z}) *_{\mathbb{Z}/2\mathbb{Z}} (\mathbb{Z}/4\mathbb{Z})$$

with $s = \begin{pmatrix} 0 & -1 \\ 1 & 1 \end{pmatrix}$ generating the cyclic group of order 6 and $f = \begin{pmatrix} 0 & 1 \\ -1 & 0 \end{pmatrix}$ the cyclic group of order 4. Observe that the image of fs^4f^{-1} in $PSL(2, \mathbb{Z})$ is $\begin{bmatrix} -1 & 1 \\ -1 & 0 \end{bmatrix}$, and compare with Example II.28. Another one is

$$SL\left(2, \mathbb{Z}\left[\frac{1}{p}\right]\right) \approx SL(2, \mathbb{Z}) *_{\Gamma_0(p)} SL(2, \mathbb{Z})$$

where

$$\Gamma_0(p) = \left\{ \begin{pmatrix} a & b \\ c & d \end{pmatrix} \in SL(2, \mathbb{Z}) \mid c \equiv 0 \pmod{p} \right\}.$$

See n^{os} I.4.2 and II.1.4 in [Serr–77]. Some examples can be "very strange", as one for which $\Gamma \approx \Gamma *_C \Gamma$ with C infinite cyclic [DunJo–99].

Another construction of the utmost importance is the *HNN-extension*: given a group Γ', two subgroups A, B of Γ', and an isomorphism $\theta : A \to B$, it provides a group Γ, an injection $\Gamma' \hookrightarrow \Gamma$, and an element $t \in \Gamma$ such that $\theta(a) = t^{-1}at$ for all $a \in A$ [HigNN–49]. The group Γ is generated by the image of Γ' and by t; a usual notation is $\Gamma = \langle \Gamma', t \mid t^{-1}At = B \rangle$, and another one is $\Gamma = \Gamma' *_A$.

The fundamental group of a complex obtained from another complex by "adding a handle" can be written naturally as an HNN-extension. For example, for $g \geq 2$, the surface group Γ_g is an HNN-extension, with Γ' a free group on $2g - 1$ generators (the fundamental group of an orientable closed surface of genus $g - 1$ with two small discs removed) and with A, B infinite cyclic subgroups of Γ'. See also the Baumslag-Solitar groups of Complement 21 below (with $\Gamma' = \langle b \rangle = \mathbb{Z}$ and with θ mapping the subgroup of $\mathbb{Z}$ generated by b^p to that generated by b^q).

Free products with amalgamation and HNN-extensions are the two basic examples in the theory of *graphs of groups* (due to Bass and Serre, [Serr–77],

[Bass–93]) and the "higher dimensional theory" of *complex of groups* (see [Stall–91], [Haefl–91], and [FloPa–97]).

15. Accessibility. A group Γ is said to *split over a finite group* if it can be written in one of the forms:

$\Gamma = \Gamma_1 *_A \Gamma_2$ with A a *finite* proper subgroup of both Γ_1 and Γ_2,

$\Gamma = \langle\, \Gamma', t \,|\, t^{-1}At = B \,\rangle$ with A, B *finite* subgroups of Γ'.

For example, $SL(2, \mathbb{Z}) = (\mathbb{Z}/6\mathbb{Z}) *_{(\mathbb{Z}/2\mathbb{Z})} (\mathbb{Z}/4\mathbb{Z})$ splits over a finite group. Observe that $\mathbb{Z} = \{1\}*_{\{1\}}$ splits over a finite group.

A fundamental theorem of Stallings (see [ScoWa–79]) shows that a finitely generated group splits over a finite group if and only if it has *at least 2 ends*, namely if and only if there exists a compact subset K of a Cayley graph C of Γ such that $C \setminus K$ has more than one non-compact connected component; it is easy to check that this does not depend on the finite generating set of Γ chosen to define C.

(In the same way that Grushko's theorem relates to Kneser's theorem on connected sums of 3-manifolds, Stallings' theorem is related to work of Dehn and Papakyriakopoulos on spheres in 3-manifolds. See [Stall–71].)

Suppose that Γ splits over a finite group, and is thus isomorphic to some $\Gamma_1 *_A \Gamma_2$ or $\langle\, \Gamma', t \,|\, t^{-1}At = B \,\rangle$ as above. If one of the factors (Γ_1, Γ_2, or Γ') again has at least two ends, it again splits over a finite group, so that the process can be iterated. The original group Γ is said to be *accessible* if this process necessarily terminates. It is a theorem of Dunwoody that finitely-presented groups are accessible [Dunwo–85]. (This does *not* carry over to all finitely-generated groups [Dunwo–93].)

For finitely-generated groups which are *torsion-free*, accessibility follows easily from Grushko's theorem of Complement 13 (see the beginning of Chapter 7 in [ScoWa–79]).

There are generalized notions of accessibility for decompositions of the form $\Gamma_1 *_A \Gamma_2$ or $\langle\, \Gamma', t \,|\, t^{-1}At = B \,\rangle$ with A in various classes of "small" groups, such as the class of cyclic groups, or the class of groups without non-abelian free subgroups; see, e.g., [BesF–91a], [BesF–91b], [Rips–95], [RipSe–97], and [Delz–99]. Much of this work has a strong flavour of low-dimensional topology, such as actions of groups on real trees (and results explained in the summer of 1991 by E. Rips [Pauli–97]), and decompositions of 3-manifolds (results from the late 70's of Jaco-Shalen and Johanssen, for which [NeumW–99] is a useful introduction).

Accessibility makes also sense for graphs: see [ThoWo–93], and references there.

16. Subgroups of finitely-generated groups. (i) There is no obstacle to embedding a countable group in a finitely-generated group, because of the following fact:

every countable group can be embedded in a group with 2 generators

[HigNN–49]. (See also Theorem 11.71 and Corollary 11.80 in [Rotma–95], as well as Complement V.26 below.) Here is a sketch of the argument, using Complement 14 above.

Let Γ be a countable group. If Γ is finite, Γ is a subgroup of a finite symmetric group, and the latter can be generated by two elements.

If Γ is infinite, let $(\gamma_n)_{n \geq 0}$ be an enumeration of its elements such that $\gamma_0 = 1$. Consider

a free group of rank 2, denoted by F_2, generated by two elements x, y,
the free product $\Gamma * F_2$,
the subgroup A of $\Gamma * F_2$ generated by $S \doteq \{x^{-n}yx^n\}_{n \geq 0}$,
the subgroup B of $\Gamma * F_2$ generated by $T \doteq \{\gamma_n y^{-n}xy^n\}_{n \geq 0}$.

One checks that S and T are both free subsets in $\Gamma * F_2$, and consequently that there exists an isomorphism $\theta : A \to B$ such that $\theta(x^{-n}yx^n) = \gamma_n y^{-n}xy^n$ for all $n \geq 0$.

By 14 above, there exists an overgroup $\tilde{\Delta}$ of $\Gamma * F_2$ and an element $t \in \tilde{\Delta}$ such that $t^{-1}at = \theta(a)$ for all $a \in A$. Let Δ be the subgroup of $\tilde{\Delta}$ generated by y and t. Then Δ contains $x = t^{-1}yt$, thus also S and $T = \theta(S)$, and finally γ_n for all $n \geq 0$. In particular Δ contains Γ.

(ii) Given a specific subgroup Γ of some finitely-generated group Δ, it may be a difficult task to decide whether Γ is finitely generated or not. For example, it is not trivial to show that the Torelli group $T(g)$ is finitely generated if and only if $g \geq 3$ (see [Johns–83] and [McCuM–86]). The Torelli group $T(g)$ is the kernel of the homomorphism from $Mod(g)$ to $Sp(g, \mathbb{Z})$, where the domain is the mapping class group of a closed orientable surface of genus g, and where the symplectic group $Sp(g, \mathbb{Z})$ is the group of automorphisms of the first homology group of the surface which preserve the cup product.

17. Research problem on embeddings of $\mathbb{Q}$. Find a "natural and explicit" embedding of $\mathbb{Q}$ in a finitely generated group.

Anticipating Chapter V, we can also record the well-known problem of finding a natural and explicit embedding of $\mathbb{Q}$ in a finitely-*presented* group. Such a group exists, by a theorem of Higman. (See [Higma–61], as well as Theorem 12.28 in [Rotma–95].)

Another formulation of the same problem is: find a "natural and explicit" simplicial complex X which covers a *finite* complex such that the fundamental group of X is $\mathbb{Q}$.

18. Residually finite groups. A group Γ is *residually finite* if, for any $\gamma \in \Gamma, \gamma \neq 1$, there exists a finite group F and a homomorphism $\phi : \Gamma \to F$ such that $\phi(\gamma) \neq 1$.

(i) Observe that a subgroup of a residually finite group is again residually finite.

(ii) Show that $SL(n, \mathbb{Z})$ is residually finite for all $n \geq 1$.
[Hint: the reduction $\mathbb{Z} \to \mathbb{Z}/q\mathbb{Z}$ induces $SL(n, \mathbb{Z}) \to SL(n, \mathbb{Z}/q\mathbb{Z})$, for each $q \geq 2$.]

Deduce from (i) and (ii) that free groups are residually finite.

(iii) It is known that a finitely-presented group Γ is residually finite if and only if the following holds: there exists a probability space (X, μ) on which Γ acts by non-singular invertible transformations, and with the following property:

for any $\epsilon > 0$, for any integer $k \geq 1$, and
    for any sequence $\gamma_1, \ldots, \gamma_k$ of elements in Γ,
there exist invertible transformations $g_1, \ldots, g_k$
    in the "full group" of the action
such that $g_1, \ldots, g_k$ generate a finite group acting freely on X and
    such that $\mu\left(\{x \in X \mid \gamma_j(x) = g_j(x)\}\right) \geq 1 - \epsilon$ for $j = 1, \ldots, k$.

This is one from several related results in [Stepi–83], [Stepi–84] and [Stepi–96].

(iv) Show that a group Γ is residually finite if and only if there exist a Hausdorff topological space M and a faithful action by homeomorphisms of Γ on M which is *chaotic,* i.e. which has the following two properties [Cair5–95]:

• the union of all finite orbits is dense in M,
• given two non-empty open subsets U, V of M, there exists $\gamma \in \Gamma$ such that $\gamma(U) \cap V \neq \emptyset$ (i.e. the action is "topologically transitive").

Check that $SL(n, \mathbb{Z})$ has such a chaotic action on the n-torus.

See [Egoro–99] for another characterization of residual finiteness in terms of an appropriate topological action on a compactum.

(v) A group Γ is *locally approximable by finite groups,* or is a *finite embedding group,* if, for every finite subset E of Γ, there exist a finite group F and an injection $\psi : E \longrightarrow F$ such that $\psi(\gamma_1 \gamma_2) = \psi(\gamma_1)\psi(\gamma_2)$ for all $\gamma_1, \gamma_2 \in E$ with $\gamma_1 \gamma_2 \in E$. Show that any residually finite group is locally approximable by finite groups.

It is known that a finitely-presented group which is locally approximable by finite groups is residually finite, and that this *does not* carry over to finitely-generated groups; see [Stepi–96] and [Stroj–99].

(vi) If a group Γ has a subgroup of finite index which is residually finite, show that Γ itself is residually finite.

(vii) Let $1 \longrightarrow \Gamma' \longrightarrow \Gamma \longrightarrow \Gamma'' \longrightarrow 1$ be a short exact sequence of groups with Γ', Γ'' residually finite and with Γ' finitely generated. If the centre of Γ' is reduced to 1, then Γ is residually finite. Also, if the sequence splits, so that Γ is a semi-direct product $\Gamma' \rtimes \Gamma''$, then Γ is residually finite. (Theorem 7 of Chapter III in [Mille–71].)

(viii) Let Γ be a finitely-generated group which is residually finite. It is a result of Baumslag that the group of automorphisms of Γ is also residually finite. See Section 2 of [BasLu–83].

(ix) We mention here two more classes of residually finite groups: finitely-generated linear groups (see Complement 20) and fundamental groups of compact 3-manifolds for which Thurston's geometrization conjecture holds, in particular fundamental groups of Haken manifolds [Thurs–82].

(x) There exist short exact sequences $1 \longrightarrow C \longrightarrow \Gamma \longrightarrow \Gamma'' \longrightarrow 1$ with C finite and central in Γ, with Γ'' residually finite and with Γ *not* residually finite.

There are examples with Γ'' a cocompact lattice in $Spin(2, 2m+1)$ for $m \geq 1$, as it follows from the main theorem in [Raghu–84]. (The method uses ideas of Deligne [Delig–78]; A. Rapinchuk has informed me that Raghunathan's argument needs a result contained in [PraRa–96].) There are examples of a different kind with Γ'' a 2-generated simple group of exponent p, where p is a large prime, by Theorems 31.8 and 21.8 in [Ol's–89]. There are finally examples with C of order two, Γ, Γ'' finitely presented, Γ'' with centre reduced to $\{1\}$, and Γ, Γ'' without any proper subgroup of finite index (Ol'shanskii, private communication).

(xi) If Γ, Γ'' are as in the examples of (x), deduce from (vi) that there cannot exist any group which is isomorphic to a subgroup of finite index in both Γ and Γ''. (In the terminology of Definition IV.27, Γ and Γ'' are commensurable up to finite kernels but are not commensurable.)

(xii) The following finitely-generated groups are *not* residually finite: non-Hopfian groups and in particular appropriate Baumslag-Solitar groups (see Complement III.21), infinite simple groups (see Complement V.26), and infinite p-groups of finite exponent (see Remark VIII.33).

(xiii) It is not known whether a Gromov-hyperbolic group is necessarily residually finite. See the discussion in [IvaOl–96].

19. Hopfian groups and a theorem of Mal'cev. A group is *Hopfian* if any homomorphism $\Gamma \to \Gamma$ which is onto is necessarily a bijection. The following is a result of Mal'cev [Mal'c–40]:

a finitely-generated residually finite group is Hopfian.

Sketch of one proof. Let N be a normal subgroup of Γ such that Γ/N and Γ are isomorphic. Observe that, for each $k \geq 1$, the number of subgroups of index k in Γ is finite (this uses finite generation), and is equal to the number of subgroups of index k in Γ which contain N. It follows that any subgroups of finite index in Γ contains N. One uses residual finiteness to conclude that $N = \{1\}$. $\square$

Sketch of another proof (shown to me by M.R. Bridson). Consider a finitely-generated group Γ which is non-Hopfian, a homomorphism ϕ of Γ onto Γ with non-trivial kernel, and an element $\gamma \in Ker(\phi)$, $\gamma \neq 1$. Consider also a homomorphism π of Γ to a finite group A. We will indicate why $\pi(\gamma) = 1$, and thus why Γ is not residually finite.

For each integer $n \geq 1$, choose $\gamma_n \in \Gamma$ such that $\phi^n(\gamma_n) = \gamma$ and let $\pi_n : \Gamma \longrightarrow A$ be the composition of ϕ^n with π. Observe that $\phi^{n+1}(\gamma_n) = 1$, so that $\pi_m(\gamma_n) = 1$ as soon as $m > n$. If we had $\pi(\gamma) \neq 1$, this would imply $\pi_n(\gamma_n) \neq 1$, so that the infinitely many homomorphisms $\pi_m, m \geq 1$, would all be distinct; but this is absurd because Γ is finitely generated and A is finite. $\square$

As an example of application, one can prove the following theorem of Nielsen:

let F be a free group of rank n and let $S \subset F$ be a set of generators containing exactly n elements; then S is a free set of generators of F.

Sketch of one proof. Let F be free on T. Consider a bijection $\phi : T \to S$ and extend it to a homomorphism ϕ of F onto F. Use Mal'cev's Theorem above to conclude. $\square$

(This is one method of proof: see e.g. Corollary 2.13.1 in [MagKS–66]; for another method, see Corollary 3.1 in the same book.)

It is known that torsion-free hyperbolic groups are Hopfian [Sela–99]. An automatic group need not be Hopfian [Wise–96].

The theorem of Mal'cev was strengthened as follows by Hirshon [Hirsh–77]: *let Γ be a finitely-generated residually finite group and let ϕ be an endomorphism of Γ such that the index $[\Gamma : \phi(\Gamma)]$ is finite; then there exists an integer $m \geq 1$ such that the restriction of ϕ to $\phi^m(\Gamma)$ is an injection (where ϕ^m denotes the mth iterate of ϕ); in particular, if Γ is torsion free, then ϕ is an injection.* The hypothesis $[\Gamma : \phi(\Gamma)] < \infty$ cannot be removed [Wise–99].

20. Linear groups. A group G is *linear* over a field $\mathbb{K}$ if it is isomorphic to a subgroup of $GL(n, \mathbb{K})$ for some $n \geq 1$. As any finitely-generated field of characteristic zero is a subfield of $\mathbb{C}$, a finitely-generated group is linear over some field of characteristic zero if and only if it is linear over $\mathbb{C}$.

There are important consequences of linearity for finitely-generated groups, among which we mention the following.

- Any finitely-generated group which is linear over $\mathbb{C}$ is residually finite (due to Mal'cev [Mal'c–40], and already mentioned in Item 4).
- In particular, a finitely-generated group which is linear has soluble word problem (see Theorems 5.1 and 5.3 in [Mille–92]).
- The "lemma" of Selberg on the existence of torsion-free subgroups of finite index (see V.20.iii).
- The structure of finitely-generated subgroups of $GL(2, \mathbb{C})$, due to Bass (see, e.g., [Bass–84] and [Shale–99]).
- Lubotzky's result [Lubo–87]: let $\mathbb{K}$ be a field of characteristic not 2 or 3, Γ a finitely-generated infinite subgroup of $GL(n, \mathbb{K})$, and d an integer; then Γ has a finite index subgroup whose index is divisible by d.

On the other hand, it requires some work to construct a group which is *not* linear, and often even more to show that some *specific* group is not linear. Let us mention

- simple infinite finitely-generated (or finitely-presented) groups, as in Complement V.26,
- non-Hopfian Baumslag-Solitar groups of the next complement,
- the Grigorchuk group (see Item VIII.19),
- examples in [Meier–82],
- the group $Aut(F_k)$ of automorphisms of the free group of rank $k \geq 3$ [ForPr–92],
- commensurators of uniform lattices in automorphism groups of trees k-regular for $k \geq 3$ [LubMZ–94].

There are, however, many difficult problems of this kind. For example, it has long been an open problem whether the so-called "braid groups" B_n are linear (where $n \geq 4$); see [Birma–98] and [Jones]. Now B_4 is linear by [Kramm–a]; moreover, both S. Bigelow [Bigel] and D. Krammer [Kramm–b] have recently shown that B_n is linear for all $n \geq 5$.

There are group-theoretic characterizations of finitely-generated groups which are linear over $\mathbb{C}$: [Lubo–88] and Interlude B in [DiDMS–91].

Similarly, a topological group G is *linear* if there exists an injective continuous homomorphism $G \longrightarrow GL(n, \mathbb{C})$ for some $n \geq 1$. We want to point out that there exist non-linear locally compact groups which have linear lattices. More precisely, the universal covering $\tilde{G}$ of $PSL(2, \mathbb{R})$ is not linear; see for example [Bourb–72], § 8, th. 1 and § 6, ex. 2; but $\tilde{G}$ does contain linear lattices, as the two following examples show.

Consider first the inverse image in $\tilde{G}$ of the lattice $PSL(2, \mathbb{Z})$ in $PSL(2, \mathbb{R})$; this is a lattice in $\tilde{G}$. Here are two arguments showing it is a linear group. First, observe that $PSL(2, \mathbb{Z})$ has a subgroup of finite index Γ which is free of rank 2, and that it is enough to show that its inverse image $\tilde{\Gamma}$ in $\tilde{G}$ is linear; as the extension $1 \longrightarrow \mathbb{Z} \longrightarrow \tilde{\Gamma} \longrightarrow \Gamma \longrightarrow 1$ is both central and split (since Γ is free), the subgroup $\tilde{\Gamma}$ is a direct product of a free group and a cyclic group, and is consequently linear. Second, the inverse image of $PSL(2, \mathbb{Z})$ in $\tilde{G}$ is isomorphic to the braid group B_3 on three strings (see Exercise V.16), and the Bureau representation provides faithful homomorphisms from B_3 into $GL(3, \mathbb{C})$; see [Birma–98].

Consider next a cocompact lattice in $PSL(2, \mathbb{R})$ which is the fundamental group of an orientable closed surface of genus at least 2, and its inverse image $\tilde{\Gamma}$ in $\tilde{G}$; then $\tilde{\Gamma}$ is linear, as noted below at the end of Item IV.48.

21. Baumslag-Solitar groups. There are finitely-generated groups which are not Hopfian. Consider for example two integers $p, q \geq 1$, $(p, q) \neq (1, 1)$ and the *Baumslag-Solitar group*

$$\Gamma_{p,q} = \left\langle \, s \, , \, t \mid st^p s^{-1} = t^q \, \right\rangle .$$

It is convenient to distinguish several cases.

First case: either $p = 1$, or $q = 1$, or $p = q$. Then $\Gamma_{p,q}$ is residually finite, and therefore Hopfian.

Other cases. The group $\Gamma_{p,q}$ is not residually finite. Moreover, $\Gamma_{p,q}$ is Hopfian if and only if p and q are meshed.

Here, two integers $p, q \geq 1$ are *meshed* if one divides the other or if they have precisely the same prime divisors. These results are essentially those of [BauSo–62], but some corrections seem to be necessary in case one of p or q divides the other; see [ColLe–83] and [AndRV–92].

In particular $\Gamma_{2,3}$ is not Hopfian. For example, the group

$$H = \left\langle s, t \mid st^2 s^{-1} = t^3 , \left(sts^{-1} t^{-1} \right)^2 = t \right\rangle$$

is isomorphic to $\Gamma_{2,3}$ and the kernel of the obvious quotient homomorphism $\Gamma_{2,3} \longrightarrow H$ is not reduced to one element (a proof appears in Section 4.4 of [MagKS–66]).

More on Baumslag-Solitar groups in Complement IV.43; see also [FarMo–98], [FarMo–99], and [MilSh–98].

There are other families of one-relator groups which are useful to know. For p, q as above and $l \geq 2$, the group

$$\Gamma_{p,q,l} = \left\langle\ s\ ,\ t\ \middle|\ \left(st^p s^{-1} t^{-q}\right)^l\ =\ 1\ \right\rangle$$

has torsion and is Hopfian; indeed, it is a theorem that *any one-relator group with torsion is Hopfian* [Baums–99]. Also, for p, q distinct primes, the group

$$\Delta_{p,q} = \left\langle\ s\ ,\ t\ \middle|\ ts^{-1} t s^p t^{-1} st\ =\ s^q\ \right\rangle$$

is not Hopfian and all its finite quotient groups are cyclic [Baums–99].

22. Co-Hopfian groups. A group is *co-Hopfian* if any homomorphism $\Gamma \to \Gamma$ which is injective is necessarily a bijection.

There are obvious examples of co-Hopfian groups, such as finite groups, the additive group $\mathbb{Q}$ of rationals, and $\mathbb{Q}/\mathbb{Z}$ (for the latter, observe that the subset $\left\{\gamma \in \mathbb{Q}/\mathbb{Z} \mid \gamma^k = 0\right\}$ is finite for each $k \geq 1$).

Observe also that the fundamental group of a closed surface of genus at least 2 is co-Hopfian. Indeed, from the correspondence between subgroups and coverings, any proper subgroup of such a group is either of finite index, and then the fundamental group of a surface of strictly higher genus, or of infinite index, and then free (as the fundamental group of an open surface).

The same argument (together with other ingredients, including Mostow rigidity) is used to show that the fundamental group of a closed manifold which has a Riemannian structure of constant curvature -1 is co-Hopfian. For generalizations of this, see 5.4.B in [Gromo–87].

The group $SL(n, \mathbb{Z})$ is co-Hopfian for any $n \geq 3$: this follows from an elementary super-rigidity result given as Theorem 6 in [Stein–85]. More generally, an irreducible lattice in G is co-Hopfian for G a connected semi-simple real Lie group which is linear and not locally isomorphic to the product of $SL(2, \mathbb{R})$ with a compact group [PrasG–76].

It is known that there are uncountably many pairwise non-isomorphic co-Hopfian quotients of the modular group [MillS–71].

There are also obvious examples of groups which are *not* co-Hopfian, such as $\mathbb{Z}, \mathbb{Z}^k, F_k$, and $\mathbb{Q}^*$ (think of $q \longmapsto q^3$). The additive group of $\mathbb{R}$ is not co-Hopfian, since $\mathbb{R} \approx \mathbb{R} \oplus \mathbb{R}$ as infinite-dimensional vector spaces over $\mathbb{Q}$; it follows that $\mathbb{R}^* \approx \{\pm 1\} \oplus \mathbb{R}$ (this using the exponential) and $\mathbb{C} \approx \mathbb{R} \oplus \mathbb{R}$ are not co-Hopfian. The multiplicative group $\mathbb{C}^*$ is not co-Hopfian since it is isomorphic to the subgroup $\mathbb{S}^1$ of complex numbers of modulus 1; indeed, using the exponential map and the polar decomposition, we have group isomorphisms $\mathbb{S}^1 \approx \mathbb{R}/\mathbb{Z} \approx \mathbb{R}/\mathbb{Z} \oplus \mathbb{R} \approx \mathbb{S}^1 \oplus \mathbb{R}^*_+ \approx \mathbb{C}^*$.

The modular group $PSL(2, \mathbb{Z})$ is not co-Hopfian: see Example V.45 below.

23. General Burnside problem. For a long time, it was an open question whether there exist finitely-generated infinite groups which are torsion-groups (namely in which any element is of finite order); this is known as the *general Burnside problem*.

Now, we know that there exist finitely-generated infinite torsion-groups. The first constructions are due to Golod and Shafarevic (1964). References, and another construction due to Grigorchuk, will be described in the last chapter of this book (see in particular VIII.19).

24. Maximal subgroups. In a finitely-generated group,

any proper subgroup is contained in a maximal one

(Theorem 5 in [NeuBH–37]). The same property holds in certain other groups (Example 6 in the same reference).

But there are groups, such as the additive group of the rational numbers $\mathbb{Q}$, which have no maximal subgroup. Let us indicate why.

A group Γ is said to be *divisible* if, for any element $\gamma \in \Gamma$ and any integer $k \geq 2$, there exists an element $x \in \Gamma$ such that $x^k = \gamma$. Any quotient of a divisible group is clearly divisible, so that any proper normal subgroup of a divisible group has infinite index.

On the other hand, any maximal subgroup of an abelian group is necessarily of index finite and prime. This holds more generally for nilpotent groups (Theorem 5.40 in [Rotma–95]).

It follows that a group which is both divisible and nilpotent has no maximal subgroup whatsoever. Besides $\mathbb{Q}$, this applies for example to the group of upper triangular n-by-n matrices

$$
UT_n(\mathbb{Q}) \; = \;
\begin{pmatrix}
1 & \mathbb{Q} & \mathbb{Q} & \cdots & \mathbb{Q} \\
0 & 1 & \mathbb{Q} & \cdots & \mathbb{Q} \\
0 & 0 & 1 & \cdots & \mathbb{Q} \\
\cdots & \cdots & \cdots & \cdots & \cdots \\
0 & 0 & 0 & \cdots & 1
\end{pmatrix}
$$

for any $n \geq 2$.

There are known examples of countable simple groups without maximal subgroups: see Theorem 35.3 in [Ol's–89].

For a finitely-generated subgroup Γ of $GL(n, \mathbb{K})$ (where $\mathbb{K}$ is some field), the maximal subgroups of Γ are all of finite index if and only if Γ has a solvable subgroup of finite index. In particular, groups like $SL(n, \mathbb{Z})$ have maximal subgroups of infinite index, indeed uncountably many of them [MarSo–81].

A subgroup Γ_0 of a group Γ is *weakly maximal* if Γ_0 is of *infinite* index in Γ and if any subgroup of Γ which properly contains Γ_0 is of finite index in Γ. An example of a weakly maximal subgroup which is not maximal appears in Item VIII.39.

The following fact was brought to my attention by R. Grigorchuk.

*In a finitely-generated group Γ, any subgroup of infinite index
is contained in a weakly maximal subgroup.*

To prove this, we have to show that the set S of all subgroups of infinite index
of Γ, ordered by inclusion, has enough maximal elements. By Zorn's lemma, it
is enough to show that any linearly ordered family $(\Gamma_\iota)_{\iota \in I}$ in S has an upper
bound. But $\Gamma_\omega = \cup_{\iota \in I} \Gamma_\iota$ cannot have finite index in Γ, otherwise it would be
finitely generated (by Exercise 8), and one would have $\Gamma_\omega = \Gamma_\iota$ for some $\iota \in I$,
which is absurd. Thus $\Gamma_\omega \in S$ and this ends the proof.

25. Geometrically finite groups. Let Γ be a discrete subgroup of
$PSL(2, \mathbb{R})$, in other words a *Fuchsian group.* Then Γ is finitely generated if
and only if it is *geometrically finite,* which means that it has a fundamental
domain in the Poincaré half-plane which is a polygon with finitely many sides
(possibly with vertices at infinity). It is a theorem of Siegel that, if Γ is a
lattice, namely if there exists on $PSL(2, \mathbb{R})/\Gamma$ a $PSL(2, \mathbb{R})$-invariant probability measure, then Γ is geometrically finite. See N°s 3.5.4, 4.6.1 and 4.1.1 in
[Katok–92].

A Fuchsian group Γ determines a *limit set* Λ_Γ in the circle (identified with
the boundary of the hyperbolic plane); this set is important in the study of the
dynamics of Γ. There are results which show that the structure of Λ_Γ is "simple"
in an appropriate sense if and only if Γ is finitely generated (see Theorem 10.2.5
in [Beard–83], or [Stark–95]).

There is also a standard notion of geometric finiteness for discrete groups of
isometries of hyperbolic spaces of dimension ≥ 3 (in terms of polyhedrons with
finitely many faces). Such a discrete group of isometries which is geometrically
finite is then always finitely generated (a consequence of the Poincaré theorem
of Section V.B), but the converse does *not* hold.

III.B. Uncountably many groups with two generators
(B.H. Neumann's method)

This section is an exposition of a result of B.H. Neumann [NeuBH–37]. See
also Chapter X in [Kuros–56].

26. Standard facts on finite alternating groups.

For any integer $n \geq 3$, let A_n denote the alternating group of degree n and
of order $\frac{1}{2}n!$. We recall the following facts.

Fact (i). *The group A_n is generated by cycles of order 3.*

Indeed, any element of A_n is a product of an even number of transpositions.
The claim follows because

$$(x\ y)(x\ z) \ = \ (x\ z\ y) \qquad \text{and} \qquad (x\ y)(z\ t) \ = \ (x\ t\ z)(x\ y\ z)$$

for distinct elements $x, y, z, t \in \{1, 2, ..., n\}$.

Fact (ii). *If n is odd, A_n is generated by $\alpha_n = (1\ 2\ ...\ n)$ and $\beta_n = (1\ 2\ 3)$.*

Indeed, for $n \geq 5$, we have

$$\beta_n^{-1}\alpha_n(1\ 2\ x)\alpha_n^{-1}\beta_n = (1\ 2\ x + 1) \qquad \text{for } x \text{ such that } 3 \leq x < n$$

$$(1\ y\ 2)(1\ 2\ x)(1\ 2\ y) = (1\ x\ y) \qquad \text{for } x \neq y \text{ such that } x, y \geq 3$$

$$(1\ x\ t)(1\ y\ z)(1\ t\ x) = (x\ y\ z) \qquad \text{for } x, y, z, t \geq 1, \text{ all distinct}$$

and the claim follows.

Observe that one can take for β_n any cycle of the form $(j, j+1, j+2)$, for $j \in \{1, \dots, n\}$ (mod n), rather than the cycle $(1, 2, 3)$.

Fact (iii). *If a normal subgroup N of A_n contains a cycle of length 3, then N is the whole of A_n.*

Indeed, consider $(r\ s\ t) \in N \lhd A_n$ and $(x\ y\ z) \in A_n$. Upon changing from $(r\ s\ t)$ to $(r\ s\ t)^2 = (r\ t\ s)$, we may assume that there exists an even permutation $\sigma \in A_n$ such that $\sigma(r) = x$, $\sigma(s) = y$, $\sigma(t) = z$. Then $\sigma(r\ s\ t)\sigma^{-1} = (x\ y\ z) \in N$.

Fact (iv). *If $n \geq 5$ then the group A_n is simple.*

Indeed, consider a normal subgroup N of A_n and an element $\sigma \neq 1$ in N. It is enough to show that N contains an element which is a cycle of length 3.

First case: σ contains a cycle of length ≥ 4, say $\sigma = (x\ y\ z\ t\ ...\ u)...(...)$. Then

$$N \ni \left\{(x\ y\ z)\sigma(x\ z\ y)\right\}\sigma^{-1} = \left\{(y\ z\ x\ t\ ...\ u)...\right\}(t\ z\ y\ x\ u\ ...)... = (x\ y\ t)$$

(with the appropriate typographical modification if $t = u$).

Second case: σ contains cycles of lengths 2 and 3. If σ contains a unique cycle of length 3, then σ^2 is a cycle of length 3. Otherwise, $\sigma = (r\ s\ t)(x\ y\ z)...$ so that

$$N \ni (r\ y\ z)\sigma(r\ y\ z)^{-1}\sigma^{-1} = (x\ r\ y\ z\ s)$$

and the previous argument applies.

Third case: $\sigma = (x\ y)(z\ t)$. Then, for $u \notin \{x, y, z, t\}$, we have

$$N \ni (x\ y\ u)\sigma(u\ y\ x)\sigma = (x\ u\ y).$$

Fourth case: $\sigma = (x_1\ y_1)(x_2\ y_2) \dots (x_k\ y_k)$ with k even. Then

$$(x_1\ y_1\ x_2)\sigma(x_1\ x_2\ y_1)\sigma = (x_1\ x_2)(y_1\ y_2)$$

and the argument of the third case applies.

[For another proof, see Theorem 11.08 in [MacDo–88].]

Fact (v). *If $n \geq 5$ is odd,*

$$[\alpha_n^v\beta_n\alpha_n^{-v}, \beta_n] \begin{cases} = 1 & \text{if } v \in \{3, 4, ..., n-3\} \text{ or if } n|v \\ \neq 1 & \text{if } v = n - 2. \end{cases}$$

Indeed, we have for example

$$(\alpha_n)^{n-2} \beta_n (\alpha_n)^{-n+2} = (n-1\ n\ 1)$$

and this does not commute with $\beta_n = (1\ 2\ 3)$.

27. The construction of B.H. Neumann.

Let $\mathcal{U} = \{\, 5 \le u_1 < u_2 < \dots \}$ be a strictly increasing infinite sequence of odd integers. Let $X_{\mathcal{U}}$ be an infinite countable set with elements $x_{j,k}$, with $j \ge 1$ and $1 \le k \le u_j$. Define two permutations α, β of $X_{\mathcal{U}}$ by

$$\alpha = (x_{1,1}\ x_{1,2}\ \dots\ x_{1,u_1})\,(x_{2,1}\ x_{2,2}\ \dots\ x_{2,u_2})\,(x_{3,1}\ x_{3,2}\ \dots\ x_{3,u_3})\,\cdots\cdots$$
$$\beta = (x_{1,1}\ x_{1,2}\ x_{1,3})\,(x_{2,1}\ x_{2,2}\ x_{2,3})\,(x_{3,1}\ x_{3,2}\ x_{3,3})\,\cdots\cdots$$

and denote by $G_{\mathcal{U}}$ the group of permutations of $X_{\mathcal{U}}$ generated by α and β. Observe that α is of infinite order and that β is of order 3.

For each $j \ge 1$, one has a homomorphism $\phi_j : G_{\mathcal{U}} \to A_{u_j}$ which is onto by Fact (ii); one can observe that $\phi_j(\alpha) = \alpha_{u_j}$ and $\phi_j(\beta) = \beta_{u_j}$.

Fact (vi). *Any finite normal subgroup N of $G_{\mathcal{U}}$ is isomorphic to some finite product of A_{u_j} 's.*

Indeed, for each $j \ge 1$, the image $\phi_j(N)$ is an invariant subgroup of A_{u_j}. As the latter is simple, we have either $\phi_j(N) = \{1\}$ or $\phi_j(N) = A_{u_j}$. As N is finite, there exists $j_0 \ge 1$ such that $\phi_j(N) = \{1\}$ whenever $j > j_0$. It follows that $\left(\prod_{j=1}^{j_0} \phi_j\right)(N)$ is an injective image of N and a product of some of the A_{u_j} 's $(1 \le j \le j_0)$. [See Exercise 31.]

Fact (vii). *Set $c_i = \left[\alpha^{u_i - 2} \beta \alpha^{-u_i + 2}, \beta\right] \in G_{\mathcal{U}}$. Then*

$$c_i(x_{j,k}) = x_{j,k} \qquad \text{for all} \qquad j > i\,,\ 1 \le k \le u_j,$$
$$c_i(x_{i,k}) \ne x_{i,k} \qquad \text{for some} \qquad k \in \{1, \dots, u_i\}.$$

This follows from Fact (v).

Fact (vii) implies that the *normal* subgroup N_i of $G_{\mathcal{U}}$ generated by $\{c_1, \dots, c_i\}$ is isomorphic to a normal subgroup of $A_{u_1} \times A_{u_2} \times \dots \times A_{u_i}$ which is not reduced to $\{1\}$, and more precisely which is such that $\phi_i(N_i) = A_{u_i}$. By an easy induction argument on i, one then shows the first part of the following fact (the second part being true by (vi)).

Fact (viii). *For each $i \ge 1$, there exists an invariant subgroup of $G_{\mathcal{U}}$ isomorphic to A_{u_i}. Any simple finite invariant subgroup of $G_{\mathcal{U}}$ is isomorphic to one of the A_{u_i} 's.*

An immediate consequence of Fact (viii) is the following item.

28. Proposition. *If $\mathcal{U}, \mathcal{U}'$ are two distinct sequences as above, the groups $G_{\mathcal{U}}, G_{\mathcal{U}'}$ are not isomorphic.*

In particular, there exist uncountably many pairwise non-isomorphic groups with two generators.

See Items 43 and 44 for other proofs of the second statement above.

29. Two more facts. The following two facts and Proposition 30, also taken from Neumann's paper [NeuBH–37], anticipate Section V.A below on group presentations. For each $n \geq 1$, let

$$w_n = 2^{2^n} + 1$$

denote the nth Fermat number.

Fact (ix). *Let* $m, n \geq 1$;

$$\text{if } n < m, \text{ then } 3 \leq w_n - 2 \leq w_m - 3,$$
$$\text{if } n > m, \text{ then } w_m \mid (w_n - 2).$$

For the second claim, when $n > m$, we have

$$w_n - 2 = 2^{2^n} - 1 = \left(2^{2^{n-1}} + 1\right)\left(2^{2^{n-1}} - 1\right)$$
$$= \left(2^{2^{n-1}} + 1\right)\left(2^{2^{n-2}} + 1\right)\left(2^{2^{n-2}} - 1\right) = \dots$$
$$= \left(2^{2^{n-1}} + 1\right)\left(2^{2^{n-2}} + 1\right)\dots\left(2^{2^m} + 1\right)\left(2^{2^m} - 1\right)$$

and the last factor but one is precisely w_m.

Fact (x). *For* $m, n \geq 1$, *set*

$$c_{m,n} = \left[\, \alpha_{w_m}^{w_n - 2} \beta_{w_m} \alpha_{w_m}^{-w_n + 2} \, , \, \beta_{w_m} \,\right] \in A_{w_m}$$

(recall that α_{w_m} and β_{w_m} are defined in Fact (ii)). Then

$$c_{m,n} = 1 \quad \text{ if } m \neq n,$$
$$c_{m,n} \neq 1 \quad \text{ if } m = n.$$

This follows from Facts (v) and (ix).

30. Proposition. *The group*

$$G = \left\langle\, \alpha\, ,\, \beta \,\bigg|\, \left[\, \alpha^{w_n - 2}\, \beta\, \alpha^{-w_n + 2} \, , \, \beta \,\right] = 1 \quad \text{ for all } \quad n \geq 1 \,\right\rangle$$

is not finitely presented.

Proof. For each subset S of $\{1, 2, 3, \dots\}$, set

$$G_S = \left\langle\, \alpha\, ,\, \beta \,\bigg|\, \left[\, \alpha^{w_n - 2}\, \beta\, \alpha^{-w_n + 2} \, , \, \beta \,\right] = 1 \quad \text{ for all } \quad n \in S \,\right\rangle$$

and let $\psi_S : G_S \to G$ be the canonical projection. For a proper subset S of $\{1, 2, 3, \dots\}$, we claim that $Ker(\psi_S)$ is not reduced to $\{1\}$.

For any $m \in \{1, 2, 3, ...\}$, $m \notin S$, consider the diagram

$$
\begin{array}{ccc}
G_S & \xrightarrow{\ \phi_m\ } & A_{w_m} \\
\psi_S \downarrow & & \\
G & &
\end{array}
$$

where ϕ_m is defined by $\phi_m(\alpha) = \alpha_{w_m}$ and $\phi_m(\beta) = \beta_{w_m}$ (this makes sense because of Fact (x)). We have

$$
\phi_m\left(\left[\alpha^{w_m-2}\beta\alpha^{-w_m+2}, \beta\right]\right) \neq 1
$$

by Fact (x) and

$$
\psi_S\left(\left[\alpha^{w_m-2}\beta\alpha^{-w_m+2}, \beta\right]\right) = 1
$$

by definition of G, so that $Ker(\psi_S) \not\subset Ker(\phi_m)$. In particular $Ker(\psi_S) \neq \{1\}$, as claimed, so that G is a *proper* quotient of G_S.

It follows that G is not finitely presentable (see Proposition V.2 and Exercise V.11). □

Exercises for III.B

31. Exercise. Let $S_1, \ldots, S_k$ be non-abelian simple groups and let G be their direct product. Check that any normal subgroup of G is a product of some of the S_j 's.

[Hint. Let N be a normal subgroup containing some $g = (g_1, \ldots, g_k)$ with $g_j \neq 1$ for some $j \in \{1, \ldots, k\}$. Let $h = (1, \ldots, 1, h_j, 1, \ldots, 1)$; then N contains $g^{-1}h^{-1}gh$, and it follows that N contains S_j.]

32. Exercise. Let $\mathbb{F}_q$ denote the finite field of order q.

(i) Check that $A_4 \approx PSL(2, \mathbb{F}_3)$, and that $Sym(4)$ is *not* isomorphic to $SL(2, \mathbb{F}_3)$.

(ii) Check that $A_5 \approx PSL(2, \mathbb{F}_4)$.

The following isomorphisms are also well-known [Dieud–54]:

$A_5 \approx PSL(2, \mathbb{F}_5)$; indeed, two simple groups of order 60 are isomorphic;

$A_6 \approx PSL(2, \mathbb{F}_9)$; indeed, two simple groups of order 360 are isomorphic;

$A_8 \approx PSL(4, \mathbb{F}_2)$.

33. Exercise. If n is even, check that A_n is generated by (2 3 4 ... n) and (1 2 3). [Compare with Fact 26.ii.]

It is known that, for any $n \geq 3$, for any $x \in A_n, x \neq 1$, there exists $y \in A_n$ such that x and y generate A_n [Chigi–97].

34. Exercise. Anticipating Chapter V on presentations of groups, observe that Section III.B shows more precisely that

$$\left\langle s,t \mid s^3 = \left(s^{-1}t^{-1}st\right)^3 = 1 \right\rangle$$

has uncountably many pairwise non-isomorphic quotients. See also Complement 37 below.

Problems and complements for III.B

35. On the structure of $G_{\mathcal{U}}$. Let A_∞ denote the group of permutations of $\mathbb{Z}$ of finite support which are even, and let B_∞ denote the group of permutations of $\mathbb{Z}$ generated by A_∞ and the two-sided shift (= the unit translation).

The following points are reformulations of the last 10 lines of [NeuBH–37]. They show in particular that all $G_{\mathcal{U}}$'s are elementary amenable groups, in the sense of [Day–57] and [Chou–80].

(i) Check that the group A_∞ is simple [SchUl–33].

(ii) Check that there is a short exact sequence

$$1 \longrightarrow A_\infty \longrightarrow B_\infty \longrightarrow \mathbb{Z} \longrightarrow 0$$

with $p(A_\infty) = 0$ and $p(\text{shift}) = 1$.

(iii) Check that B_∞ is generated by the two-sided shift and by the 3-cycle (0 1 2). [Compare with Exercise 7.]

(iv) Show that, for any sequence $\mathcal{U}$ as in Item 27 and Proposition 28, there is a short exact sequence

$$1 \longrightarrow \bigoplus_{j=1}^{\infty} A_{u_j} \xrightarrow{\mu} G_{\mathcal{U}} \xrightarrow{\pi} B_\infty \longrightarrow 1$$

where π maps $\alpha \in G_{\mathcal{U}}$ [respectively $\beta \in G_{\mathcal{U}}$] to the two-sided shift of $\mathbb{Z}$ [respectively to the permutation $(0,1,2)$].

[Hint for (iv). Set $u_j = 2n_j + 1$ for all $j \geq 1$ and

$$X'_{\mathcal{U}} = \coprod_{j \geq 1} \{ -n_j, -n_j + 1, \ldots, -1, 0, 1, \ldots, n_j - 1, n_j \}.$$

Define two permutations α', β' of $X'_{\mathcal{U}}$, with α' acting as a long cycle and β' as the 3-cycle $(-1\ 0\ 1)$ on each factor. With minor modifications in Neumann's construction, one obtains a group $G'_{\mathcal{U}}$ of permutations of $X'_{\mathcal{U}}$, generated by $\{\alpha', \beta'\}$, which is isomorphic to $G_{\mathcal{U}}$, and a homomorphism $\pi' : G'_{\mathcal{U}} \longrightarrow B_\infty$. It is enough to show that the modified sequence

$$1 \longrightarrow \bigoplus_{j=1}^{\infty} A_{u_j} \xrightarrow{\mu'} G'_{\mathcal{U}} \xrightarrow{\pi'} B_\infty \longrightarrow 1$$

is exact.

Consider $\gamma \in G'_{\mathcal{U}}$. An integer $k \in \mathbb{Z}$ defines a point $k_j = k \in \{-n_j, \ldots, n_j\}$ as soon as $n_j \geq |k|$. Observe that the integer $\gamma(k_j) \in \{-n_j, \ldots, n_j\}$ is independent of j for j large enough. Define $\pi'(\gamma)$ by

$$\pi'(\gamma)(k) = \lim_{j \to \infty} \gamma(k_j)$$

for all $\gamma \in G'_{\mathcal{U}}$ and $k \in \mathbb{Z}$; for example, $\pi'(\alpha')$ is the two-sided shift and $\pi'(\beta')$ is the 3-cycle $(-1\ 0\ 1)$.

To show that $Im(\mu') \subset Ker(\pi')$, consider the image of the element corresponding to c_i of Fact 27.vii.

For $\gamma \in Ker(\pi')$, choose some way of writing γ as a word in α' and β', and observe that the exponents of the α' 's in this word must add up to zero. Show finally that there are constants c_v, c_h such that $\gamma(k_j) = k_j$ as soon as $j \geq c_v$ and $|k_j| \geq c_h$, so that $\gamma \in \bigoplus_{j=1}^{c_v-1} A_{u_j}$.]

(v) With the terminology of Chapter VI, show that B_∞ is of exponential growth. It follows that $G_{\mathcal{U}}$ is of exponential growth for any sequence $\mathcal{U}$ as above.

[Hint. Set $a = \pi(\alpha)$, $b = \pi(\beta)$. For any integer $j \geq 1$, the elements $a^3 b^{\epsilon_1} a^3 b^{\epsilon_2} \ldots a^3 b^{\epsilon_j}$, for $\epsilon_1, \epsilon_2, \ldots, \epsilon_j \in \{-1, 0, 1\}$, are distinct from each other. Thus the growth function $\beta(k)$ of B_∞ with respect to the generating set $\{a, b\}$ satisfies $\beta(4j) \geq 3^j$, so that B_∞ is of exponential growth. More precisely, if $\lambda = \lim_{k \to \infty} \sqrt[k]{\beta(k)}$, then $\lambda \geq \sqrt[4]{3} \approx 1.316 > 1$.

Let $\beta_{\mathcal{U}}(k)$ denote the growth function of $G_{\mathcal{U}}$ with respect to the generating set $\{\alpha, \beta\}$, and set $\lambda_{\mathcal{U}} = \lim_{k \to \infty} \sqrt[k]{\beta_{\mathcal{U}}(k)}$. We have $\lambda_{\mathcal{U}} \geq \lambda > 1$, and in particular $G_{\mathcal{U}}$ is of exponential growth.]

Problem. Compute the numbers $\lambda_{\mathcal{U}}$. (See also Problem VI.61.)

36. Further uses of Neumann's construction. Neumann's construction was profitably used for at least two other purposes: once to construct families of finite graphs with required "expanding properties" [LubWe–93], and another time to provide examples of groups with special properties regarding the growth of subgroups of finite index [LubPS–96].

37. Quotients of $SL(2, \mathbb{Z})$ and SQ-universal groups. The two-generated modular group $PSL(2, \mathbb{Z}) \approx (\mathbb{Z}/2\mathbb{Z}) * (\mathbb{Z}/3\mathbb{Z})$ has uncountably many pairwise non-isomorphic quotient groups. I do not know the simplest proof of this statement, but here are some comments and references.

(i) The group $SL(2, \mathbb{Z})$ has a subgroup of finite index which is free of rank 2. (See Example II.25. It is also known that the commutator subgroup of $SL(2, \mathbb{Z})$ is free of rank 2 and that the quotient of $SL(2, \mathbb{Z})$ by its commutator subgroup is the cyclic group of order 12; see Exercise 8 in § II.4 of [Bourb–72].)

(ii) The previous statement and Proposition 28 do not seem to imply that $SL(2, \mathbb{Z})$ has uncountably many pairwise non-isomorphic quotient groups. Indeed, there is an example of a finitely-generated group Γ and a subgroup Γ_0

of finite index in Γ (of index 2) such that Γ_0 has uncountably many pairwise non-isomorphic quotient groups and Γ *does not!* See [BehNe–81].

(iii) A group G is said to be *SQ-universal* if, for any countable group Γ, there exists an embedding of Γ in a quotient of G (thus "S" stands for "subgroup" and "Q" for "quotient"). Britton and Levin have shown that a free product $G = G_1 * G_2$ with $|G_1| \geq 2$ and $|G_2| \geq 3$ is SQ-universal (Theorem V.10.3 in [LynSc–77]). In particular, the modular group $PSL(2, \mathbb{Z}) \approx (\mathbb{Z}/2\mathbb{Z}) * (\mathbb{Z}/3\mathbb{Z})$ is SQ-universal, and so is a fortiori $SL(2, \mathbb{Z})$. More recently, Ol'shanskii has shown that any non-elementary hyperbolic group is SQ-universal [Ol's–95b].

(iv) It is known that an SQ-universal group has uncountably many pairwise non-isomorphic quotient groups. For a group Γ and a subgroup Γ_0 of finite index, it is also known that Γ is SQ-universal if and only if Γ_0 is SQ-universal; in particular, the property of being SQ-universal for the free group $\mathbb{Z} * \mathbb{Z}$ and for the modular group follow from each other. (See [NeuPM–73].)

(v) By methods like that of Section III.B (see also Exercise 34), one can show that the presentation

$$\left\langle s, t \mid s^2 = t^3 = (st)^{30} = 1 \right\rangle$$

defines a group which has uncountably many pairwise non-isomorphic quotients [BruBW–79]; this implies a fortiori the same claim for the modular group $\langle s, t \mid s^2 = t^3 = 1 \rangle$.

38. Simple and solvable quotients of the modular group. It is known that the modular group has uncountably many non-isomorphic *simple* quotients. The same group also has uncountably many non-isomorphic *solvable* quotients. (This follows from results of P. Hall, P.M. Neumann, and Schupp; references and comments in [BruBW–79]. For solvable quotients of F_2, see also Lemma 40 below.)

39. $SL(3, \mathbb{Z})$ is almost, but not quite, a quotient of $PSL(2, \mathbb{Z})$. It is known that $SL(3, \mathbb{Z})$ is almost a quotient of $PSL(2, \mathbb{Z})$. More precisely [Conde–90], the $(\mathbb{Z}/2\mathbb{Z})$-extension of $SL(3, \mathbb{Z})$ by its inverse-transpose automorphism is isomorphic to a finitely-presented group, denoted by $4^+(a^{12})$, which contains a quotient of $PSL(2, \mathbb{Z})$ as a subgroup of index 8.

This group $4^+(a^{12})$ arises in the study of finite symmetric graphs (with automorphism group transitive on oriented edges); see Chapters 17–19 in [Biggs–74] for an introduction to the subject, and [Biggs–84] for further information.

On the other hand $SL(3, \mathbb{Z})$ is not a quotient of $PSL(2, \mathbb{Z})$. This is an unpublished result from the thesis of Paula Zucca (under the supervision of Chiara Tamburini), communicated to me by Marston Conder. Another way of formulating the same result is that "$SL(3, \mathbb{Z})$ is not $(2, 3)$-generated". (In this thesis, it is proved that $GL(3, \mathbb{Z})$ is not $(2, 3)$-generated; the result for $SL(3, \mathbb{Z})$ follows easily.)

However, it is known that, for each large enough integer n, the group $SL(n, \mathbb{Z})$ is a quotient of the triangle group $\langle s, t \mid s^2 = t^3 = (st)^7 \rangle$, and a fortiori a quotient of $PSL(2,\mathbb{Z})$. "Large enough" means $n \geq 287$ in [LucTW], $n \geq 28$ in later work by the same authors, and conceivably $n \geq 5$. An earlier paper shows that $SL(n,\mathbb{Z})$ is a quotient of $PSL(2,\mathbb{Z})$ for $n \geq 28$ [TaWG–94].

III.C. On groups with two generators

There are other ways than those in Section III.B of showing that the number of groups which can be generated by two elements is uncountable.

One goes as follows. Consider the free group F_2 on two generators and the set $\mathcal{N}$ of all normal subgroups of F_2. The first step is to show that $\mathcal{N}$ is uncountable. Then define $N, N' \in \mathcal{N}$ to be equivalent, and write $N \sim N'$, if the quotient groups F_2/N and F_2/N' are isomorphic. The second step is to show that $\mathcal{N}/\sim$ is uncountable.

To make this precise, we state and prove two lemmas.

40. Lemma. *The set of normal subgroups of F_2 is uncountable.*

Proof. Let $S = (s_j)_{j \in \mathbb{Z}}$ denote a set of generators indexed by $\mathbb{Z}$ and define

$$R = \left\{ [[s_i, s_j], s_k] = 1 \right\}_{i,j,k \in \mathbb{Z}} \cup \left\{ [s_i, s_j] = [s_{i+k}, s_{j+k}] \right\}_{i,j,k \in \mathbb{Z}}$$

(with $[y,z] = y^{-1}z^{-1}yz$ for y, z in a group). Anticipating Chapter V, on presentations of groups, we define a group

$$\Gamma_0 = \langle S \mid R \rangle$$

and we denote by ϕ the automorphism of Γ_0 which maps s_i to s_{i+1} for all $i \in \mathbb{Z}$. Let

$$\Gamma = \Gamma_0 \rtimes_\phi \mathbb{Z}$$

be the corresponding semi-direct product; if t is "the" generator of $\mathbb{Z}$, we have $t\gamma_0 t^{-1} = \phi(\gamma_0)$ for all $\gamma_0 \in \Gamma_0$. Set further $u_i = [s_0, s_i]$ for all $i \in \mathbb{Z}, i \neq 0$. We have

    (a) the commutator subgroup of $[\Gamma_0, \Gamma_0]$ is the free abelian group on the set $\{u_i\}_{i \in \mathbb{Z}, i \neq 0}$,

    (b) as a consequence of the definition of R, we have $\phi(\gamma_0) = \gamma_0$ for all $\gamma_0 \in [\Gamma_0, \Gamma_0]$, so that $[\Gamma_0, \Gamma_0]$ is in the centre of Γ,

    (c) as Γ is generated by s_0 and t, it is a quotient of a free group F_2 on two two generators.

Now let X be any subset of $\mathbb{Z} \setminus \{0\}$, and let N_X denote the subgroup of $[\Gamma_0, \Gamma_0]$ generated by $\{u_x\}_{x \in X}$. By (b) above, N_X is a normal subgroup of Γ. It follows that Γ contains uncountably many different normal subgroups, and the same holds a fortiori for F_2. $\square$

41. Remark. Let $\left(D^k F_2\right)_{k \geq 0}$ denote the derived series of F_2, defined by $D^0 F_2 = F_2$ and $D^k F_2 = \left[D^{k-1} F_2, D^{k-1} F_2\right]$ for $k \geq 1$. The previous proof shows in particular that F_2 contains uncountably many normal subgroups N containing D^3F_2. In contrast, the number of normal subgroups N containing D^2F_2 is countable: this is a result of P. Hall (Corollary 2 of Theorem 3 in [HallP–54]).

The idea of our proof of Lemma 40 comes from the same paper [HallP–54].

The next lemma is also used by P. Hall in [HallP–54], and given with all details in [BruBW–79].

42. Lemma. *For a finitely-generated group* Γ, *the following two properties are equivalent:*
- *(i)* Γ *has uncountably many normal subgroups,*
- *(ii)* Γ *has uncountably many non-isomorphic quotients.*

Proof. One has to show that (i) implies (ii). Let $\aleph$ (pronounce "aleph") denote the cardinality of the set of normal subgroups of Γ and let $\beth$ (pronounce "beth") denote the cardinality of the set of isomorphism classes of quotient groups of Γ.

Let us assume that $\aleph$ is uncountable and that $\beth$ is countable; we will obtain a contradiction. By this assumption, there would exist a group Q and an uncountable subset $\mathcal{S}$ of $\mathcal{N}$ such that Γ/N is isomorphic to Q for all $N \in \mathcal{S}$. Consequently, there would exist an uncountable number of homomorphisms from Γ onto Q. But this is absurd because Γ is finitely generated and Q is countable. $\square$

43. The second part of Proposition 28 again. *There are uncountably many pairwise non-isomorphic two-generator groups.*

Proof: immediate from Lemma 40, and Lemma 42 for F_2. $\square$

Complements to III.C

44. Other proofs. Here is one more method of proof of the second part of Proposition 28. First, one observes that there are uncountably many pairwise non-isomorphic countable abelian groups. For example, given any set S of prime numbers, one can consider the direct sum T_S over $p \in S$ of the cyclic groups $\mathbb{Z}/p\mathbb{Z}$; these groups are clearly pairwise non-isomorphic (because T_S has p-torsion if and only if $p \in S$).

Second, one shows that any countable group C can be embedded in a group Γ generated by 2 elements in such a way that, for any integer $n \geq 2$, the group Γ has an element of finite order n if and only if C does. This is due to Higman, B.H. Neumann and H. Neumann in their 1949 paper [HigNN–49], and uses the construction of HNN-extension alluded to in Complement 14.

Another method is to show that, for any real number $c \geq 1$, there exists a finitely-generated group of cost c; see [Gabor–00] and IV.47.vi below. See also Remark IV.3.ix.

45. Strengthenings. Anticipating on notions defined in Section IV.B, we also mention that one knows uncountably many two-generated groups which are pairwise non-quasi-isometric (and a fortiori non-isomorphic) in the following classes:

(i)   amenable p-groups, for any given prime p, [Grigo–84], [Grigo–85].

(ii) non-amenable torsion-free groups [Bowdi–98].

A group is *icc* it all its conjugacy classes distinct from $\{1\}$ are infinite; a group is known to be icc if and only if its von Neumann algebra is a *factor* of type II_1. Does there exist an uncountable family of finitely-generated icc groups such that the corresponding factors are pairwise non-isomorphic ? (For the question without "finitely-generated", Dusa McDuff has shown that the answer is yes; see Theorem 4.3.9 in [Sakai–71].)

46. Other finite and infinite groups with two generators. It is known that any finite simple group can be generated by two elements: this results from investigations going from L.E. Dickson and G.A. Miller around 1900 to Steinberg ([Stein–62], for the "groups of Lie type") and Aschbacher and Guralnick [AscGu–84]. The proof of the general result relies on the classification of finite simple groups.

There are more recent results, of a probabilistic flavour, on the number of pairs of elements which generate a finite simple group. On this, and on other two-generation problems for finite simple groups, see [LieSh–96].

It is not easy to produce a finitely-generated infinite simple group which cannot be generated by two elements. However, it has been shown that there exist k-generator simple groups in which 2-generator subgroups are free, first by V.S. Guba for $k = 460\,000$ [Guba–86] and then by V.N. Obraztsov for $k = 3$ [Obraz–93].

The above result for finite simple groups has some analogue for compact groups, by the following result already recalled in Complement II.41: for a compact connected second countable topological group G, the set of pairs of $G \times G$ which generate a dense subgroup of G is a countable intersection of open dense subsets [SchUl–35]. The same property holds for other groups, such as the group of all Lebesgue-measure-preserving transformations of the unit interval (see [UlavN–45] and [PrasV–81]).

It is, however, an open problem to find the minimal number of elements which generate a dense subgroup in the unitary group of a factor of type II_1, with the strong topology; it should be easy to show that this number is 2 for the hyperfinite factor of type II_1. (It is *another problem* to find, in a given factor, the minimal number of unitaries which generate the factor as a von Neumann algebra; in case of the hyperfinite factor of type II_1, this number is 2 [Saitô]; see also n° 2.10 in [Harpe–95].)

III.D. On finite quotients of the modular group

A *congruence subgroup* of $SL(n, \mathbb{Z})$ is a subgroup of finite index which contains the kernel of the reduction

$$SL(n, \mathbb{Z}) \longrightarrow SL(n, \mathbb{Z}/\ell\mathbb{Z})$$

for some integer $\ell \geq 2$. One defines similarly congruence subgroups of the quotient group $PSL(n, \mathbb{Z})$.

For $n \geq 3$, it is known that any non-central normal subgroup of $SL(n, \mathbb{Z})$ is a congruence subgroup ([BasMS–67], [Menni–65]). For $n = 2$, there are two important differences:

  (i) there are non-trivial normal subgroups of infinite index
      (see Complement 37),
 (ii) there are normal subgroups of finite index
      which are not congruence subgroups.

Fact (ii) is due to F. Klein (§ 1, page 63, in [Klein–80]); see one proof in Section III.2 of [Magnu–74]; there are references to other proofs in Section VII.6.C of [Lehne–64] and in [Jones–86]. We will now sketch a variant of the proof in Magnus' book, which we learned from A. Lubotzky.

47. Definition. Let S, G be two finite groups. Say that S is *involved* in G if S is isomorphic to a quotient of a subgroup of G. (The terminology is that of Section 1.1 in [Goren–68]; other authors say that S is a *section* of G: see e.g. Section 3.2 of [DixMo–96].)

Recall that a *composition series* for a group G is a series

$$G = G_0 > G_1 > \ldots > G_m = \{1\}$$

in which either $G_i = G_{i-1}$ or G_i is a maximal normal subgroup of G_{i-1} for all $i \in \{1, \ldots, m\}$.

It is clear that any finite group has a composition series. It is a theorem of O. Schreier [Schre–28] that, if an arbitrary group G has a composition series, then the *Jordan-Hölder theorem* holds; this means in particular that the *composition factors* G_{i-1}/G_i not reduced to $\{1\}$ depend only on G, up to isomorphism and order, and not on the choice of the composition series. (We have followed the terminology of Chapter 5 in [Rotma–95], *and not* that of Bourbaki, for whom G_i is an arbitrary normal subgroup of G_{i-1}.)

48. Lemma. *Let S be a simple finite group which is involved in a finite group G. Then S is involved in a composition factor of G.*

Proof. By hypothesis, there exists a subgroup H of G of which S is a quotient. Let

$$G = G_0 > G_1 > \ldots > G_m = 1$$

be a composition series of G. The series

$$H = H \cap G_0 > H \cap G_1 > \ldots > H \cap G_m = 1$$

can be refined to a composition series

(*) $$H = H_0 > H_1 > \ldots > H_n = 1$$

of H, of which S is a composition factor by the Jordan-Hölder theorem applied to H. By definition of the series (*), there exist $i \in \{1, \ldots, m\}$ and $j \in \{1, \ldots, n\}$ such that

$$H \cap G_{i-1} > H_{j-1} > H_j > H \cap G_i$$

and $H_{j-1}/H_j \approx S$. It follows that S is a quotient of the image of H_{j-1} in G_{i-1}/G_i, and this concludes the proof. $\square$

49. Corollary. *A simple finite group S which is involved in a direct product $\prod_{i=1}^n S_i$ of simple finite groups $S_1, \ldots, S_n$ is involved in one of $S_1, \ldots, S_n$.*

We will use the next lemma and proposition for $d = 2$ only.

50. Lemma. *For an integer $d \geq 2$, a prime p and an integer $a \geq 1$, the canonical reduction $\mathbb{Z}/p^a\mathbb{Z} \longrightarrow \mathbb{Z}/p\mathbb{Z}$ provides a short exact sequence*

$$1 \longrightarrow H \longrightarrow PSL\,(d, \mathbb{Z}/p^a\mathbb{Z}) \longrightarrow PSL\,(d, \mathbb{Z}/p\mathbb{Z}) \longrightarrow 1$$

where H is a p-group.

Proof. Set first

$$G(1) = \left\{ x \in M_d\,(\mathbb{Z}/p^a\mathbb{Z}) \mid x \equiv 1 \pmod{p} \right\}$$

where $M_d(\mathcal{R})$ denotes the ring of d-by-d matrices over a ring $\mathcal{R}$.

Observe that any matrix $x \in G(1)$ is invertible. Indeed, as $x = 1 + py$ for some matrix y, the dth exterior power of x is of the form $1 + p(\ldots)$, so that $\det(x) \equiv 1 \pmod{p}$. In other words, $\det(x) = 1 + p\xi$ for some $\xi \in \mathbb{Z}/p^a\mathbb{Z}$; hence $\det(x)$ is invertible, of inverse $1 - p\xi + p^2\xi^2 \pm \ldots + (-1)^{a-1}p^{a-1}\xi^{a-1}$.

It follows that $G(1)$ is a group, indeed a normal subgroup of $GL\,(d, \mathbb{Z}/p^a\mathbb{Z})$, and that there is an exact sequence

$$1 \longrightarrow G(1) \longrightarrow GL\,(d, \mathbb{Z}/p^a\mathbb{Z}) \longrightarrow GL\,(d, \mathbb{Z}/p\mathbb{Z}).$$

As the order of $G(1)$ is clearly

$$|G(1)| = \left(p^{a-1}\right)^{d^2},$$

$G(1)$ is a p-group.

Now set $SG(1) = SL\,(d, \mathbb{Z}/p^a\mathbb{Z}) \cap G(1)$. We also have a short exact sequence

$$1 \longrightarrow SG(1) \longrightarrow SL\,(d, \mathbb{Z}/p^a\mathbb{Z}) \xrightarrow{\pi} SL\,(d, \mathbb{Z}/p\mathbb{Z}) \longrightarrow 1$$

where $SG(1)$ is a p-group (we leave it as an exercise to check the surjectivity of π). When $p = 2$, this is the short exact sequence of the lemma. When p is odd, this provides the statement of the lemma after dividing $SL\,(d, \mathbb{Z}/p^a\mathbb{Z})$ and $SL\,(d, \mathbb{Z}/p\mathbb{Z})$ by their centres. $\square$

51. Proposition. *Consider an integer $d \geq 2$, a prime p and an integer $a \geq 1$. If $d \geq 3$ or if $p \geq 5$, the composition factors of $PSL\,(d, \mathbb{Z}/p^a\mathbb{Z})$ are*

either isomorphic to $\mathbb{Z}/p\mathbb{Z}$,

or isomorphic to $PSL\,(d, \mathbb{Z}/p\mathbb{Z})$.

If $d = 2$ and $p \in \{2,3\}$, the composition factors of $PSL\,(2, \mathbb{Z}/p^a\mathbb{Z})$ are cyclic groups.

Proof. It is well known that each composition factor of a p-group is a cyclic group of order p, and that $PSL\,(d, \mathbb{Z}/p\mathbb{Z})$ is simple for $d \geq 3$ or $p \geq 5$ (Theorems 4.6, 8.13 and 8.23 in [Rotma–95]).

On the other hand, the composition factors of the groups $PSL\,(2, \mathbb{Z}/2\mathbb{Z}) \approx S_3$ and $PSL\,(2, \mathbb{Z}/3\mathbb{Z}) \approx A_4$ are cyclic.

Consequently Proposition 51 follows from Lemma 50. $\square$

52. Lemma. *For any prime $p \geq 5$, the Sylow 3-subgroups of $PSL(2, \mathbb{Z}/p\mathbb{Z})$ are cyclic.*

Proof. This is well-known, and we now recall the main steps of the argument. For a complete description of the Sylow subgroups of $PSL(2, \text{finite field})$, see § II.8 in [Huppe–67].

Set $\Gamma = PSL(2, \mathbb{Z}/p\mathbb{Z})$; this is a group of order $\frac{1}{2}(p+1)p(p-1)$. We begin by describing two cyclic subgroups of Γ (knowing that the multiplicative group of a finite field is always cyclic).

The first one is the image T_s in Γ of the diagonal matrices in $SL(2, \mathbb{Z}/p\mathbb{Z})$; this is a cyclic group of order $\frac{1}{2}(p-1)$. (The letter T holds for "torus", and the subscript s for "split".)

For the second one, consider the finite field $\mathbb{F}_{p^2}$ as a vector space of dimension 2 over $\mathbb{F}_p = \mathbb{Z}/p\mathbb{Z}$. This provides an embedding of the cyclic group $GL(1, \mathbb{F}_{p^2})$ of order $p^2 - 1$ into the group $GL(2, \mathbb{Z}/p\mathbb{Z})$. Taking matrices of determinant $+1$, one obtains an embedding of the cyclic group of order $p+1$ into the group $SL(2, \mathbb{Z}/p\mathbb{Z})$; see Theorem II.7.3.b in [Huppe–67] for details. Dividing by the centre $\left\{ \pm \begin{pmatrix} 1 & 0 \\ 0 & 1 \end{pmatrix} \right\}$ of the latter group one obtains a cyclic subgroup T_{ns} of order $\frac{1}{2}(p+1)$ of Γ.

Now, if $p \equiv 1 \pmod 3$, the largest power of 3 which divides

$$|\Gamma| = \frac{1}{2}(p+1)p(p-1)$$

divides $p - 1$, and it is clear that T_s contains a Sylow 3-subgroup of Γ. And if $p \equiv -1 \pmod 3$, similar arguments show that T_{ns} contains a Sylow 3-subgroup of Γ. Thus Sylow 3-subgroups of Γ are cyclic in all cases. $\square$

Recall from Complement II.16 that the commutator subgroup of the modular group $PSL(2, \mathbb{Z}) \approx (\mathbb{Z}/3\mathbb{Z}) * (\mathbb{Z}/2\mathbb{Z})$ is free on two generators and of finite index; recall also from Item 26.ii that alternating groups can be generated by two elements. It follows that homomorphisms such as π below do exist.

53. Theorem. *Consider the modular group* $PSL(2, \mathbb{Z})$ *and a homomorphism*

$$\pi \; : \; D\Big(PSL(2, \mathbb{Z})\Big) \longrightarrow A_n$$

from its commutator subgroup onto the alternating group A_n, *for some* $n \geq 6$.
Then $Ker(\pi)$ *is not a congruence subgroup of* $PSL(2, \mathbb{Z})$.

Proof. Let us assume that the kernel of π contains the kernel of the reduction
modulo ℓ

$$\Gamma(\ell) \doteq Ker\Big(PSL(2, \mathbb{Z}) \longrightarrow PSL(2, \mathbb{Z}/\ell\mathbb{Z})\Big)$$

for some integer $\ell \geq 2$; we will obtain a contradiction. If Δ denotes the canonical image of $D\big(PSL(2, \mathbb{Z})\big)$ in $PSL(2, \mathbb{Z}/\ell\mathbb{Z}) \approx PSL(2, \mathbb{Z})/\Gamma(\ell)$, the homomorphism π induces a homomorphism π' of Δ onto A_n, so that A_n is involved in $PSL(2, \mathbb{Z}/\ell\mathbb{Z})$.

Let $\ell = \prod_{i=1}^{k} p_i^{a_i}$ be the prime decomposition of ℓ, so that

$$PSL(2, \mathbb{Z}/\ell\mathbb{Z}) \approx \prod_{i=1}^{k} PSL\Big(2, \mathbb{Z}/\, (p_i^{a_i}\mathbb{Z})\Big).$$

A composition factor of $PSL\big(2, \mathbb{Z}/\ell\mathbb{Z}\big)$ is

(i) either a cyclic group,

(ii) or a group isomorphic to $PSL(2, \mathbb{Z}/p\mathbb{Z})$ for some prime $p \geq 5$,

by Corollary 49 and Proposition 51. It follows from Lemma 48 that A_n is involved in $PSL(2, \mathbb{Z}/p\mathbb{Z})$ for some prime $p \geq 5$.

Consider now a Sylow 3-subgroup P of A_n. As $n \geq 6$, the group P is not cyclic. As A_n is a quotient of a subgroup of $PSL(2, \mathbb{Z}/p\mathbb{Z})$, it follows that the Sylow 3-subgroups of $PSL(2, \mathbb{Z}/p\mathbb{Z})$ are not cyclic. (Indeed, if a finite group H is a quotient of a finite group G, any Sylow subgroup of H may be written as the image of a Sylow subgroup of G; for this standard fact, see e.g. [Bourb–70], chap. I, § 6, n° 6, corollaire 4.)

But this is absurd, because Sylow 3-subgroups of $PSL(2, \mathbb{Z}/p\mathbb{Z})$ are cyclic by Lemma 52, and this concludes the proof. $\square$

In Theorem 53, one may of course replace $D\big(PSL(2, \mathbb{Z})\big) \approx F_2$ by a subgroup of finite index in $D\big(PSL(2, \mathbb{Z})\big)$ (thus a free group F_k for some rank $k \geq 2$), and A_n ($n \geq 6$) by a simple k-generated finite group S "larger" in some sense than what is allowed for the kernel of $F_k \longrightarrow S$ to be a congruence subgroup.

54. Complement. Let $\mathbb{K} = \mathbb{Q}(\sqrt{-d})$ be an imaginary quadratic number field and let $\mathcal{O}$ denote its ring of integers. A *congruence subgroup* of $PSL(2, \mathcal{O})$ is a subgroup containing the kernel of the reduction $PSL(2, \mathcal{O}) \longrightarrow PSL(2, \mathcal{O}/\underline{a})$ where $\underline{a}$ is a non-zero ideal of $\mathcal{O}$. Arguments analogous to those used for Theorem 53 can be used to show that $PSL(2, \mathcal{O})$ has many subgroups of finite index which are *not* congruence subgroups (see for example Proposition 7.5.5 in [ElsGM–98]. For other groups of the type $SL(2, -)$ in which *any finite index subgroup is a congruence subgroup*, see [Serr–70b].

CHAPTER IV

FINITELY-GENERATED GROUPS
VIEWED AS METRIC SPACES

It can be non-trivial to decide whether or not a particular group is finitely generated. One set of tools for this comes from geometry. With this in mind, Section A is devoted to Cayley graphs and word lengths; it shows how a *given finitely-generated group* can be viewed as a certain *metric space*. In Section B we introduce quasi-isometries between metric spaces and show how these ideas provide *proofs* that some groups are indeed finitely generated (Theorem 23).

There are many more reasons for studying "Infinite groups as geometric objects" (we quote here the title of [Gromo–84]). In Item 0.3 of [Gromo–93], Gromov discusses several other motivations, such as

- widening the context to allow transcendental methods (analysis of infinity),
- introducing the geometric language to bring amplification and clarification,
- suggesting new concepts and constructions,
- extending the class of applicable ideas changes the class of essential examples.

And Gromov ends this discussion with the following confession. "After all, the actual reason why one approaches a problem from a geometric angle is because one's mind is bent this way. No amount of rationalization can conceal the truth."

IV.A. Word lengths and Cayley graphs

1. Definition. Let Γ be a group given as a quotient $\pi : F(S) \longrightarrow \Gamma$ of the free group on a set S. Some authors refer to these data as a *marked group*.

The *word length* $l_S(\gamma)$ of an element $\gamma \in \Gamma$ is the smallest integer n for which there exists a sequence $(s_1, s_2, \ldots, s_n)$ of elements in $S \cup S^{-1}$ such that $\gamma = \pi(s_1 s_2 \ldots s_n)$. The *word metric* d_S is defined on Γ by

$$d_S(\gamma_1, \gamma_2) = l_S(\gamma_1^{-1}\gamma_2).$$

The group Γ is thus a metric space, and Γ acts on itself from the left by isometries.

Since d_S takes integral values, this metric space is discrete, and this could impede geometric understanding. One way out (logically useless but intuitively worthwhile) is as follows.

2. Definition. Let $\pi : F(S) \longrightarrow \Gamma$ be as above. The corresponding *Cayley graph* $Cay(\Gamma, S)$ is the graph with vertex set Γ in which two vertices γ_1, γ_2 are the two ends of an edge if and only if $d_S(\gamma_1, \gamma_2) = 1$.

There is an obvious action from the left of Γ on this graph, which is transitive on the set of vertices and which has finitely many orbits on the set of edges.

Each edge of $Cay(\Gamma, S)$ can be made a metric space isometric to the segment $[0, 1]$ of the real line, in such a way that the left action of Γ produces isometries between the edges. One defines naturally the length of a path between two points (not necessarily two vertices) of the graph, and the distance between two points is defined to be the infimum of the appropriate path-lengths. In this way, $Cay(\Gamma, S)$ is made a *metric space* which is *arc-connected,* the natural inclusion $\Gamma \subset Cay(\Gamma, S)$ is an isometric embedding, and the left action of Γ on $Cay(\Gamma, S)$ is by isometries.

Cayley graphs of finite groups were introduced by Cayley (1878) and others, but the first use of infinite Cayley graphs is due to Dehn (1910), for surface groups and for the group of the trefoil knot. This is why some authors write "Dehns Gruppenbild" for "Cayley graph".

3. Remarks. (i) In the two definitions above, one assumes that

the set S is finite

unless stated otherwise. The reason for this appears below in Item 22: there is a metric structure on a finitely-generated group whose quasi-isometry class depends only on the group, and not on the choice of a *finite* set S of generators.

Let us, however, mention two situations in which it is natural to consider infinite generating sets. The first one is that of a free product $\Gamma = *_{\iota \in I} \Gamma_\iota$ as in Proposition II.1. Then Γ is generated by $S = \coprod_{\iota \in I} (\Gamma_\iota \setminus \{e_\iota\})$, and $\ell_S(\gamma)$ is the length of the reduced word representing $\gamma \in \Gamma$. For the second one, consider a group Γ; its commutator subgroup $[\Gamma, \Gamma]$ is generated by the (usually infinite) *set* of commutators S. The resulting length ℓ_S and the corresponding "stable length", $[\Gamma, \Gamma] \ni \gamma \longmapsto \lim_{n \to \infty} \frac{1}{n} \ell_S(\gamma^n)$, have been investigated in [Bavar-91]. For example, ℓ_S is unbounded on $[\Gamma, \Gamma]$ if Γ is a non-abelian free group or $SL(2, \mathbb{Z})$, but is bounded if $\Gamma = SL(n, \mathbb{Z})$ with $n \geq 3$. (On stable lengths, see also (viii) below.)

(ii) The definition of a graph used implicitly above is that of a set V of vertices and a set E of edges given with an incidence relation between them; see for example [Biggs-74] and [VaLWi-92]. Definition 2 forces the graph $Cay(\Gamma, S)$ to be *simple,* namely without loops and multiple edges.

(iii) A Cayley graph $Cay(\Gamma, S)$ is regular; its degree is the order of the set

$$\{\gamma \in \Gamma \mid \ell_S(\gamma) = 1\} = \pi(S \cup S^{-1}) \cap (\Gamma \setminus \{1\}).$$

(iv) For some investigations, it is important to define a *graph* by the following data: a vertex set V, an edge set E, two mappings $h : E \longrightarrow V$ (head) and

$t : E \longrightarrow V$ (tail), and a fixed-point free involution $e \longmapsto \bar{e}$ on E (inversion), such that $h(\bar{e}) = t(e)$ for all $e \in E$. An *orientation* of such a graph is a subset E_+ of E such that E is the disjoint union of E_+ and $\overline{E_+}$.

One can then define a directed graph $Cay'(\Gamma, S)$ with a vertex set $V = \Gamma$, the positively oriented part of the edge set $E_+ = \Gamma \times S$, and mappings h, t defined by

$$h(\gamma, s) = \gamma, \qquad t(\gamma, s) = \gamma\pi(s).$$

The underlying non-oriented graph has no loops if and only if $1 \notin \pi\left(S \cup S^{-1}\right)$; when this condition holds, the graph has no multiple edges if and only if the restriction of π to $S \cup S^{-1}$ is injective. (This is exposed in [Serr–77]; see in particular chap. I, § 2, n° 1.)

One can moreover attach naturally a "color" from S to each edge, coloring (γ, s) by s. These colored directed Cayley graphs are those of Section 1.6 in [MagKS–66], where necessary and sufficient conditions are given for a colored directed graph to be such a Cayley graph; see also Chapter 3 in [CoxMo–57].

(v) As we are mainly interested here in Cayley graphs as means of visualising metric spaces, we will keep to Definition 2.

(vi) For the word metric d_S on Γ as in Definition 1, the left action of Γ on itself is an isometry and the right action is in general not an isometry. But, for this *same* word metric, the right multiplication $\phi : \gamma \longmapsto \gamma\gamma_0$ by some element $\gamma_0 \in \Gamma$ is at *bounded distance of the identity* in the sense that

$$\sup_{\gamma \in \Gamma} d_S(\gamma, \phi(\gamma)) < \infty.$$

It is sometimes interesting to consider both the left isometric action and the right action of Γ on the metric space (Γ, d_S); see for example [CecGH–99].

(vii) For a word length $\ell_S : \Gamma \longrightarrow \mathbb{N}$ as in Definition 1 and for $\gamma_1, \gamma_2 \in \Gamma$, it is obvious that $\ell_S(\gamma_1\gamma_2) \leq \ell_S(\gamma_1) + \ell_S(\gamma_2)$. But it is not easy to understand more about lengths of products in general. In case Γ is hyperbolic (in the sense of Section V.D below) and for $\gamma \in \Gamma$ of infinite order, it is known that there are constants $c_1, c_2 > 0$ such that

$$c_1|n| \leq \ell_S(\gamma^n) \leq c_2|n|$$

for all $n \in \mathbb{Z}$ (see e.g. Proposition 21 of Chapter 8 in [GhyHp–90]). In other cases, the function $n \longmapsto \ell_S(\gamma^n)$ may behave quite differently; for example Ol'shanskii has constructed a non-trivial 2-generated torsion-free group Γ such that

$$\ell_S(\gamma^n) = O(\log|n|)$$

for all $\gamma \in \Gamma, \gamma \neq 1$ [Ol's–99b]. (See also [Ol's–97] and Complement VI.43.)

(viii) Given a group Γ and a word length (or some other type of length) $\ell : \Gamma \longrightarrow \mathbb{N}$, one defines the *stable length* or *translation number* of a group element $\gamma \in \Gamma$ by

$$\tau(\gamma) = \limsup_{n \to \infty} \frac{1}{n}\ell(\gamma^n).$$

The following properties are straightforward [GerSh–91]:

$$0 \leq \tau(\gamma) = \lim_{n \to \infty} \frac{1}{n} \ell(\gamma^n) \leq \ell(\gamma) \quad \text{for all} \quad \gamma \in \Gamma,$$
$$\tau(\gamma\gamma'\gamma^{-1}) = \tau(\gamma') \quad \text{for all} \quad \gamma, \gamma' \in \Gamma,$$
$$\tau(\gamma^k) = |k|\tau(\gamma) \quad \text{for all} \quad \gamma \in \Gamma \quad \text{and} \quad k \in \mathbb{Z},$$
$$\tau(\gamma\gamma') \leq \tau(\gamma) + \tau(\gamma').$$

If Γ is hyperbolic and if ℓ is a word length, Gromov has shown that $\tau(\gamma)$ takes rational values, with bounded denominators (see 8.5.X in [Gromo–87] and Proposition 3.1 in [Delz–96a]). Conner has exhibited examples of the form $\Gamma = \mathbb{Z}^3 \rtimes_\Phi \mathbb{Z}$, where $\Phi \in GL(3, \mathbb{Z})$ has three eigenvalues of unit modulus, with irrational values of $\tau(\gamma)$; for this and other properties of translation numbers, see [Conne–97] and references quoted there.

(ix) Let X be a connected graph with vertex set X^0 and base vertex x_0. To each automorphism γ of X there corresponds a growth function defined by

$$\xi(\gamma, k) = \max \{d(x, \gamma(x)) \mid d(x_0, x) \leq k\}$$

and a *rotation number*

$$\rho(\gamma) = \limsup_{k \to \infty} \frac{\xi(\gamma, k)}{k} \in [0, 2].$$

The *spectrum* of a group Γ of automorphisms of X is the countable subset

$$Sp(\Gamma) = \{r \in [0, 2] \mid \text{there exists } \gamma \neq 1 \text{ in } \Gamma \text{ such that } \rho(\gamma) = r\}$$

of $[0, 2]$. For example, the spectrum of a finitely-generated free abelian group [respectively non-abelian free group] acting on its standard Cayley graph is reduced to $\{0\}$ [respectively to $\{2\}$].

Trofimov has shown that, for any countable subset C of $[0, 2]$, there exists a finitely-generated group Γ for which the spectrum, with respect to an appropriate Cayley graph, contains C; see [Trofi–89], as well as [Trofi–92].

4. Convention. If the restriction to S of the projection π of Definition 1 is an injection, we will improperly identify S and its image in Γ. Accordingly, we speak of *the Cayley graph $Cay(\Gamma, S)$ of a group Γ with respect to a generating set $S \subset \Gamma$*, and it is usually assumed that $1 \notin S$.

There are situations in which one does not know a priori whether S injects or not in Γ, and that is why Definition 2 is formulated in terms of π.

5. Examples. If Γ is the infinite cyclic group $\mathbb{Z}$ and if $S = \{1\}$, then $Cay(\Gamma, S)$ is the real line with integral points as vertices. (We identify a graph and its geometric realization.)

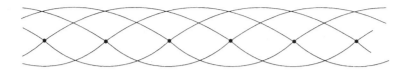

FIGURE 1.   A Cayley graph of Example 5

If $\Gamma = \mathbb{Z}^2$ and if $S = \left\{ \begin{pmatrix} 1 \\ 0 \end{pmatrix}, \begin{pmatrix} 0 \\ 1 \end{pmatrix} \right\}$, then $Cay(\Gamma, S)$ is the standard unit grid of the Euclidean plane.

Similarly for $\mathbb{Z}^n$ and its standard Cayley graph, which embeds in the Euclidean space $\mathbb{R}^n$.

If $\Gamma = \mathbb{Z} \supset S = \{2, 3\}$, then $Cay(\Gamma, S)$ is shown in Figure 1. Observe that, from a distance, this picture "looks like" a real line (see below Definition 20 of a quasi-isometry, and Example 21.iii).

6. Example. Let Γ be a free group on two generators a, b and let $S = \{a, b\}$. Then $Cay(\Gamma, S)$ is a regular tree of degree 4, of which Figure 2 represents the ball of radius 3 about the identity.

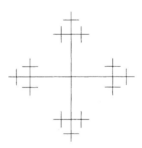

FIGURE 2.   Ball of radius 3 of Example 6

More generally, let Γ be a free product of p groups of order 2 and q infinite cyclic groups, and let S be the natural set of $p + q$ generators; set $n = p + 2q$. Then $Cay(\Gamma, S)$ is a regular tree of degree n. (These groups appear again in Examples V.42 and VI.2.vii.)

7. Example. Set $\mathbb{Z}/3\mathbb{Z} = \{1, a, a^{-1}\}$, $\mathbb{Z}/2\mathbb{Z} = \{1, b\}$, let

$$\Gamma = \mathbb{Z}/3\mathbb{Z} * \mathbb{Z}/2\mathbb{Z}$$

denote the modular group, and set $S = \{a, b\}$. If $u = a^{-1}b^{-1}ab$ and $v = ab^{-1}a^{-1}b$, recall from Problem II.16 that the set $S_0 = \{u, v\}$ generates a subgroup Γ_0 of Γ of index 6 which is free of rank 2. Figure 3 represents part of the Cayley graph $Cay(\Gamma, S)$, with special marks on the vertices in Γ_0. The reader should imagine $Cay(\Gamma_0, S_0)$ represented on the same picture.

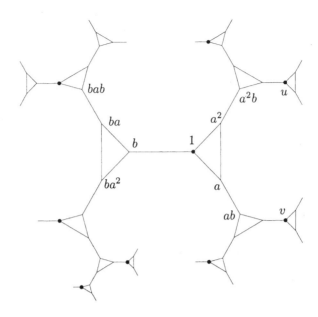

FIGURE 3.   A Cayley graph for the modular group (Example 7)

8. Example. Consider the *Heisenberg group*

$$\Gamma = \left\{ \begin{pmatrix} 1 & 0 & 0 \\ k & 1 & 0 \\ m & l & 1 \end{pmatrix} \mid k, l, m \in \mathbb{Z} \right\}$$

generated by the matrices

$$s = \begin{pmatrix} 1 & 0 & 0 \\ 1 & 1 & 0 \\ 0 & 0 & 1 \end{pmatrix} \qquad t = \begin{pmatrix} 1 & 0 & 0 \\ 0 & 1 & 0 \\ 0 & 1 & 1 \end{pmatrix} \qquad u = \begin{pmatrix} 1 & 0 & 0 \\ 0 & 1 & 0 \\ 1 & 0 & 1 \end{pmatrix}$$

satisfying

$$su = us \qquad\qquad tu = ut \qquad\qquad t^{-1}s^{-1}ts = u.$$

Each element in Γ has a unique "normal form" in the generators, given by

$$\begin{pmatrix} 1 & 0 & 0 \\ k & 1 & 0 \\ m & l & 1 \end{pmatrix} = s^k t^l u^m,$$

so that Γ is in natural bijection with $\mathbb{Z}^3$. The formulas

$$\begin{aligned}
\left(s^k t^l u^m\right) s &= s^{k+1} t^l u^{m+l} \\
\left(s^k t^l u^m\right) t &= s^k t^{l+1} u^m \\
\left(s^k t^l u^m\right) u &= s^k t^l u^{m+1}
\end{aligned}$$

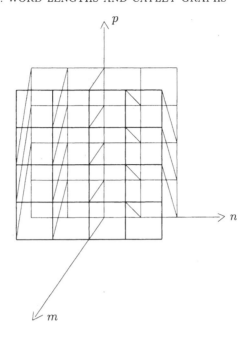

FIGURE 4.   A Cayley graph for the Heisenberg group (Example 8)

show how to draw the Cayley graph of Γ with respect to the generating set $\{s, t, u\}$.

9. Groups with isomorphic Cayley graphs. Although Cayley graphs are in some sense good pictures of groups, it is also true that different groups can have *isomorphic Cayley graphs*.

An elementary illustration of this fact is that, for any finite group F, the Cayley graph $Cay(F, F \setminus \{1\})$ is a complete graph, depending only on the order of the group F. It follows that, if F_1, F_1' [respectively F_2, F_2'] are finite groups of the same order, the free products $F_1 * F_2$ and $F_1' * F_2'$ have isomorphic Cayley graphs. See also Example 6.

For another class of examples, consider two integers $m, n \geq 2$, the regular trees T_{2m}, T_{2n} of degrees $2m, 2n$ respectively, the direct product $C_{2m,2n} = T_{2m} \times T_{2n}$, which is a square complex, and its 1-skeleton

$$X = \left(C_{2m,2n}\right)^{(1)},$$

which is a $(2m + 2n)$-regular graph. Thus X is clearly a Cayley graph of the direct product $F_m \times F_n$ of two free groups of ranks m and n. It follows from the work of Burger and Mozes ([BurMz–a], [BurMz–b]) that, for appropriate values of m and n, this X is also a Cayley graph for quite different groups, some not residually finite, and even some containing simple subgroups of finite index.

One more class of examples, from [Dioub], appears in Complement 44: there are pairs $\Gamma_F, \Gamma_{F'}$ of finitely-generated groups having isomorphic Cayley graphs, one group being solvable and the other not virtually solvable.

Exercises for IV.A

10. Exercise. Draw the Cayley graphs of the symmetric group $Sym(3)$ for the following sets of generators

$$S_1 = \{ (1\ 2)\ ,\ (2\ 3)\ \},$$
$$S_2 = \{ (1\ 2)\ ,\ (2\ 3)\ ,\ (1\ 3)\ \},$$
$$S_3 = \{ (1\ 2)\ ,\ (1\ 2\ 3)\ \},$$
$$S_4 = Sym(3) \setminus \{1\}.$$

Draw the Cayley graph of $Sym(4)$ for $S = \{ (1\ 2)\ ,\ (2\ 3)\ ,\ (3\ 4)\ \}$.

11. Exercise. Let P denote the *Petersen graph*. Its vertex set is the set of unordered pairs from $\{1, 2, 3, 4, 5\}$, and two vertices $\{i, j\}, \{k, l\}$ are connected by an edge if and only if $\{i, j\} \cap \{k, l\} = \emptyset$.

(i) Draw the Petersen graph.

(ii) Check that the automorphism group of P is the symmetric group $Sym(5)$ of order 120. (See if necessary the solution of Exercise 12.1 in [Lovás–79].)

(iii) Show that P is *not* a Cayley graph.
[Sketch of one argument: if P were the Cayley graph of some group Γ, this group would be a subgroup of $Sym(5)$ of order 10, and would in particular contain an element of order 2; but any element of order 2 in $Sym(5)$ has fixed vertices in P; this is absurd because a group acts freely on the vertex set of any of its Cayley graphs. Indication for another argument: show that the Petersen graph cannot be split into three disjoint 1-factors; recall that a 1-*factor* in a graph is a regular subgraph of degree 1, made up of disjoint edges which cover all vertices, and that several 1-factors are said to be *disjoint* if their edge-sets are disjoint; see Chapter 10 in [BigLW–76].]

12. Exercise. Show that a Cayley graph of a finitely-generated infinite group contains an isometric copy of the line, namely a sequence $(\gamma_j)_{j \in \mathbb{Z}}$ of vertices whose distances satisfy $d(\gamma_i, \gamma_j) = |i - j|$ for all $i, j \in \mathbb{Z}$.
[Hint. There exists a sequence $(x_n)_{n \geq 0}$ of vertices such that $d(1, x_n) = n$, and therefore for each $n \geq 1$ a geodesic interval I_n from 1 to x_n. Use a Cantor diagonal argument to show that there exists a geodesic ray I from 1 to infinity, say with vertices $y_0 = 1, y_1, y_2, \ldots$. Translating $J_0 = I$ by each of the y_j^{-1} 's, one obtains an infinite sequence $(J_n)_{n \geq 0}$ of geodesic rays, with J_n containing 1 at distance n from its finite end. Using again a Cantor diagonal argument, one shows that there exists a geodesic line J with vertices $(\gamma_j)_{j \in \mathbb{Z}}$ as requested, and with $\gamma_0 = 1$.]

13. Exercise. A graph X is said to have the *extension property for geodesic segments* if, for any pair x_0, x of vertices of X, there exists a vertex x' which is a neighbour of x such that $d(x_0, x') = d(x_0, x) + 1$ (where d denotes the combinatorial distance). Decide which of the following Cayley graphs have the extension property for geodesic segments:

(i) the Cayley graphs of Examples 5 through 8,

(ii) the Cayley graph defined by the direct product $\Gamma = \mathbb{Z} \times \{e, j\}$ and the generating set $S = \{(1, e), (1, j)\}$. [Thanks to A. Valette for this one.]

Complements to IV.A

14. More on Cayley graphs and the extension property for geodesic segments. (Here we anticipate group presentations, as in Chapter V, and we refer to small cancellation theory, not discussed elsewhere in these notes.) It can be shown that presentations $\Gamma = \langle S \mid R \rangle$ satisfying a small cancellation hypothesis $C'(1/6)$ provide Cayley graphs with the extension property for geodesic segments of the previous exercise (lemme 4.19 in [Champ–95]). But one-relator presentations don't necessarily do this: consider the presentation of $\mathbb{Z}$ with two generators s, t and one relator $stst = 1$.

[Check the following: first $st = ts$, second st generates the group, third there is no way to extend a geodesic segment from the origin to st. (If $\mathbb{Z} = \{\ldots, a^{-2}, a^{-1}, 1, a, a^2, \ldots\}$, think of s as a^2 and of t as a^{-3}.)]

More on this in [Bogop].

15. Schreier graphs. In the same way that finitely-generated groups are made visible to geometers as Cayley graphs, coset spaces are made visible as so-called *Schreier graphs*. In the literature, there are variations on both definitions (see Item 3). We choose here a definition of Schreier graphs which will make it possible to state a result of J.L. Gross.

Let Γ be a finitely-generated group and let Γ_0 be a subgroup of Γ. Let $S = \{s_1, \ldots, s_n\}$ be a finite set, let $F(S)$ denote the free group on S, and assume that Γ is given as a quotient $\pi : F(S) \to \Gamma$.

Define $Sch(\Gamma_0 \backslash \Gamma, S)$ to be the graph whose vertex set is the coset space $\Gamma_0 \backslash \Gamma$, and which has as many edges between two vertices $\Gamma_0 \gamma_1, \Gamma_0 \gamma_2$ as there are $s \in S \cup S^{-1}$ with $\Gamma_0 \gamma_2 = \Gamma_0 \gamma_1 \pi(s)$. Observe that $Sch(\Gamma_0 \backslash \Gamma, S)$ may have loops, in case there are pairs (γ, s) in $\Gamma \times (S \cup S^{-1})$ with $\gamma \pi(s) \gamma^{-1} \in \Gamma_0$; it may also have multiple edges, in case there are triples (γ, s, t) in $\Gamma \times (S \cup S^{-1}) \times (S \cup S^{-1})$ with $s \neq t$ and $\gamma \pi(st) \gamma^{-1} \in \Gamma_0$. There are examples of Schreier graphs in § 3.7 of [CoxMo–57].

The theorem of Gross we wish to mention is this:

let X be a connected regular graph of some degree $k \geq 2$;

if k is even, then X is isomorphic to a Schreier graph.

See Theorem 5.4 in [Lubo–95b], where the theorem is stated for finite graphs, but with a proof which carries over *cum grano salis* to infinite graphs.

Gross' theorem *does not* carry over to odd degrees. For example, the Petersen graph is not a Schreier graph because it cannot be split into three disjoint 1-factors; see [BigLW–76], and Exercise 11 above.

IV.B. Quasi-isometries

16. Definitions. Consider two metric spaces X and X', respectively with distance functions d and d'.

A mapping $\phi : X \to X'$ such that $d'(\phi(x), \phi(y)) = d(x, y)$ for all $x, y \in X$ is an *isometric embedding*. It is an *isometry* if it is also onto; the two spaces X and X' are then *isometric*. Equivalently X and X' are isometric if there exist isometric embeddings $\phi : X \to X'$ and $\psi : X' \to X$ such that $\psi\phi = id_X$ and $\phi\psi = id_{X'}$.

A *geodesic segment* between two points $x, y \in X$ is the image of an isometric embedding ϕ of the interval $[0, L]$ of the real line to X such that $\phi(0) = x$ and $\phi(L) = y$, where $L = d(x, y)$. (This is what is sometimes called a *minimizing* geodesic segment in Riemannian geometry.) There may exist several geodesic segments between two given points!

A metric space is *geodesic* if any two of its points can be joined by at least one geodesic segment.

17. Definition. A metric space is *proper* if its closed balls of finite radius are compact.

A proper space is locally compact and complete. There is a partial converse: a metric space which is locally compact, complete, *and geodesic* is necessarily proper (*théorème* 1.10 in [GroLP–81]).

18. Examples. (i) Any connected smooth Riemannian manifold is naturally a metric space. For two points x, y of the manifold, $d(x, y)$ is the infimum of the lengths of smooth curves from x to y.

A *complete* Riemannian manifold is geodesic and proper. This is the Hopf-Rinow theorem; see e.g. [Miln–63], Theorem 10.9, or [Ballm–95], Chapter I, Theorem 2.4.

(ii) Let Γ be a finitely-generated group, given as a quotient $\pi : F(S) \to \Gamma$ of the free group on a finite set S, as in Definition 1. A geodesic from γ to γ' in $Cay(\Gamma, S)$ is a simple path going through vertices $\gamma_0 = \gamma, \gamma_1, \gamma_2, \ldots, \gamma_n = \gamma'$, with $n = l_S(\gamma^{-1}\gamma')$, such that $\gamma_{j-1}^{-1}\gamma_j \in \pi(S \cup S^{-1}) \setminus \{1\}$ for $j \in \{1, \ldots, n\}$. In general, there are *several* geodesic segments connecting 1 to γ.

The space $Cay(\Gamma, S)$ is geodesic and proper. (If $\pi(S)$ were infinite, the space would still be geodesic, but would no longer be proper.)

19. Pseudo-metrics. To continue the presentation, it is sometimes convenient to consider *pseudo-metric spaces* rather than just metric spaces. A pseudo-metric space is a space X given together with a mapping $d : X \times X \to \mathbb{R}_+$ satisfying

$$(i) \quad d(x,x) \ = \ 0 \quad \text{for all } x \in X,$$
$$(ii) \quad d(y,x) \ = \ d(x,y) \quad \text{for all } x,y \in X,$$
$$(iii) \quad d(x,z) \ \leq \ d(x,y) + d(y,z) \quad \text{for all } x,y,z \in X$$

(X would be a metric space if moreover $d(x,y) = 0 \implies x = y$).

For later appearances of *pseudo*-metrics, see e.g. Item VI.42 below.

20. Quasi-isometries. Let X, X' be two pseudo-metric spaces. A mapping $\phi : X \to X'$ is a *quasi-isometric embedding* if there exist constants $\lambda \geq 1, C \geq 0$ such that

$$\frac{1}{\lambda} d(x,y) - C \ \leq \ d'(\phi(x), \phi(y)) \ \leq \ \lambda d(x,y) + C$$

for all $x, y \in X$. It is a *quasi-isometry* if there exists moreover a constant $D \geq 0$ such that any point of X' is within distance D from some point of $\phi(X)$; the two spaces X and X' are then *quasi-isometric*.

Equivalently, X and X' are quasi-isometric if there exist two mappings

$$\phi : X \to X' , \quad \psi : X' \to X$$

and constants $\lambda \geq 1, C \geq 0, D \geq 0$ such that

$$d'(\phi(x), \phi(y)) \leq \lambda d(x,y) + C \qquad d(\psi(x'), \psi(y')) \leq \lambda d'(x', y') + C$$
$$d(\psi\phi(x), x) \leq D \qquad\qquad d(\phi\psi(x'), x') \leq D$$

for all $x, y \in X$ and $x', y' \in X'$. (The proof of the equivalence relies on the axiom of choice.)

We leave it to the reader to check that quasi-isometry between pseudo-metric spaces is an equivalence relation.

21. Examples. (i) Any metric space of finite diameter is quasi-isometric to a point.

(ii) Let Γ and S be as in Definitions 1 and 2. Then the natural inclusion $\Gamma \hookrightarrow Cay(\Gamma, S)$ is a quasi-isometry.

(iii) Let Γ be a group given as $\pi : F(S) \to \Gamma$ and also as $\pi' : F(S') \to \Gamma$, where S and S' are finite sets. Let d, d' denote the corresponding metrics on Γ and let ϕ be the identity mapping of Γ, viewed as a map $(\Gamma, d) \to (\Gamma, d')$. Then ϕ is a quasi-isometry. Indeed, setting

$$\lambda_1 \ = \ \max\{ \ d'(\pi(s), 1) \mid s \in S \ \},$$
$$\lambda_2 \ = \ \max\{ \ d(\pi'(s'), 1) \mid s' \in S' \ \},$$
$$\lambda \ = \ \max\{\lambda_1, \lambda_2\},$$

one checks that

$$\frac{1}{\lambda}d(x,y) \ \leq \ d'(\phi(x),\phi(y)) \ \leq \ \lambda d(x,y)$$

for all $x,y \in \Gamma$.

In particular, the Cayley graph shown in Figure 1 is quasi-isometric to the real line.

(iv) Let $\pi : F(S) \to \Gamma$ and d be as in (iii), and let α be an automorphism of Γ. In general, $\alpha(\pi(S)) \neq \pi(S)$ and α is not an isometry with respect to d. But α is always a quasi-isometry: setting

$$\lambda \ = \ \max_{s \in S} \ell_S\big(\alpha(\pi(s))\big),$$

we have

$$\frac{1}{\lambda} \ d(x,y) \ \leq \ d(\alpha(x),\alpha(y)) \ \leq \ \lambda \ d(x,y)$$

for all $x,y \in \Gamma$.

(v) On a finite-dimensional real vector space, two distance functions corresponding to two scalar products are necessarily quasi-isometric. It follows that, on a connected real Lie group, two distance functions corresponding to two left-invariant Riemannian metrics are necessarily quasi-isometric.

(vi) On a compact manifold M, two distance functions corresponding to two Riemannian metrics are necessarily quasi-isometric. It follows that, on

either a connected covering of M
or a leaf of a foliation on M,

there is a distance function which is well-defined up to quasi-isometry.

(vii) Let M_1, M_2 be two closed manifolds of the same dimension n with the same universal covering manifold, say M. Consider a Riemannian metric on M_1 and the corresponding metric on M, so that M is in particular a metric space. Let $\underline{f} : M_1 \longrightarrow M_2$ be a homotopy equivalence. Then any lift $f : M \longrightarrow M$ of $\underline{f}$ is a quasi-isometry. (See also Example 25.v.)

If moreover M_1, M_2 are Riemannian manifolds of constant negative curvature, so that M can be identified to the hyperbolic space H^n, the quasi-isometry f induces a boundary transformation ∂f of the sphere at infinity of H^n. When $n \geq 3$, it is an important step in the proof of Mostow rigidity that there also exists an isometry of H^n which induces the same boundary transformation.

22. Proposition. *(i) On a finitely-generated group, the word metric is unique up to quasi-isometry.*

(ii) There is a well-defined notion of quasi-isometry for finitely-generated groups.

(iii) On a finitely-generated group Γ, let $d : \Gamma \times \Gamma \longrightarrow \mathbb{R}_+$ be a metric with the following properties:

the resulting metric space is proper (Definition 17)
and long-range connected (see Exercise 37 below),
Γ operates from the left on itself by isometries;

then the identity from (Γ, d) to Γ with some word metric is a quasi-isometry.

Proof. For (i), see Example 21.iii above. Statement (ii) follows, because if two groups Γ_1, Γ_2 are quasi-isometric in the sense of Definition 20 with respect to some finite generating sets S_1 of Γ_1 and S_2 of Γ_2, then they are also quasi-isometric with respect to any finite generating sets.

We leave the verification of Statement (iii) to the reader as an exercise. [Hint: if C is as in Item 37.i below, the ball S of radius C in (Γ, d) generates Γ.] $\square$

Let Γ be a group acting by homeomorphisms on a locally compact space X. Recall that the action is *proper*, or that Γ *acts properly*, if, for every compact subspace K of X, the set $\{\, \gamma \in \Gamma \mid \gamma K \cap K \neq \emptyset \,\}$ is finite. For such an action, the orbit space $\Gamma \backslash X$ is Hausdorff and locally compact. (More on proper actions in § 4 of Chapter 3 of [Bourb–60].)

The next theorem is the *fundamental observation of geometric group theory*, and is essentially Lemma 2 in [Miln–68b]. It was already known to Efremovich [Efrem–53] and Schwarzc[1] [Schwa–55] in the early 1950s.

23. Theorem. *Let X be a metric space which is geodesic and proper, let Γ be a group and let $\Gamma \times X \to X$ be an action by isometries (say from the left). Assume that the action is proper and that the quotient $\Gamma \backslash X$ is compact.*

Then the group Γ is finitely generated and quasi-isometric to X. More precisely, for any $x_0 \in X$, the mapping $\Gamma \to X$ given by $\gamma \mapsto \gamma x_0$ is a quasi-isometry.

Proof. Let $\pi : X \to \Gamma \backslash X$ denote the canonical projection. The space $\Gamma \backslash X$ has a canonical metric defined by

$$d(p, q) \;=\; \inf\{\, d(x, y) \mid x \in \pi^{-1}(p) \,,\, y \in \pi^{-1}(q) \,\}.$$

(To check that $d(p, q) = 0$ implies $p = q$, one uses the hypothesis "X is proper".) As $\Gamma \backslash X$ is compact, its diameter

$$R \;=\; \sup \big\{\, d(p, q) \mid p, q \in \Gamma \backslash X \,\big\}$$

is finite. Choose a base point $x_0 \in X$ and set

$$B \;=\; \{\, x \in X \mid d(x_0, x) \leq R \,\}.$$

Observe that $(\gamma B)_{\gamma \in \Gamma}$ is a covering of X. Set

$$S \;=\; \big\{ s \in \Gamma \mid s \neq 1 \text{ and } sB \cap B \neq \emptyset \,\big\}.$$

---

[1]Other spelling: Švarc.

Observe that $S^{-1} = S$, and that S is finite because the action is proper. Finally, set

$$r = \inf \{ d(B, \gamma B) \mid \gamma \in \Gamma , \ \gamma \notin S \cup \{1\} \}$$

(where $d(B, \gamma B)$ is defined as usual to be the minimum of the distances $d(x, y)$ for $x \in B$ and $y \in \gamma B$) and set

$$\lambda = \max_{s \in S} d(x_0, s x_0).$$

Claim 1: $r = \min \{ d(B, \gamma B) \mid \gamma \in \Gamma , \ \gamma \notin S \cup \{1\} \} > 0.$

Choose $\gamma' \in \Gamma$, $\gamma' \notin S \cup \{1\}$ and set $r' = d(B, \gamma' B)$; we have $r' > 0$ by definition of S and by compacity of B. Set

$$T = \{ \gamma \in \Gamma \mid \gamma \notin S \cup \{1\} , \ d(B, \gamma B) \leq r' \}.$$

Then T is non-empty (it contains γ') and finite (because Γ acts properly on X). As r is the infimum of the strictly positive numbers $d(B, \gamma B)$ for γ in the finite set T, the claim follows.

[Exercise: show that $r \leq 2R$.]

Claim 2: S *generates* Γ *and*

$$(1) \qquad \frac{1}{\lambda} d(x_0, \gamma x_0) \ \leq \ d_S(1, \gamma) \ \leq \ \frac{1}{r} d(x_0, \gamma x_0) + 1$$

for all $\gamma \in \Gamma$.

To prove the claim, choose $\gamma \in \Gamma$. We may assume that $\gamma \notin S \cup \{1\}$, otherwise (1) is straightforward. Let k be the integer defined by

$$(2) \qquad\qquad R + (k - 1)r \ \leq \ d(x_0, \gamma x_0) \ < \ R + kr$$

($\gamma \notin S \cup \{1\}$ implies $d(x_0, \gamma x_0) > R$ and therefore $k \geq 1$). As X is geodesic, we can choose $x_1, \ldots, x_k, x_{k+1} = \gamma x_0 \in X$ such that $d(x_0, x_1) < R$ and $d(x_i, x_{i+1}) < r$ for $i \in \{1, \ldots, k\}$. As the Γ-translates of B cover X, we can choose $\gamma_0 = 1, \gamma_1, \ldots, \gamma_{k-1}, \gamma_k = \gamma$ in Γ such that $x_i \in \gamma_{i-1} B$ for $i \in \{1, \ldots, k\}$. Set $s_i = \gamma_{i-1}^{-1} \gamma_i$ so that $\gamma = s_1 s_2 \ldots s_k$. We have $s_i \in S \cup \{1\}$ because

$$d(B, s_i B) \ \leq \ d\left(\gamma_{i-1}^{-1} x_i \, , \ s_i \gamma_i^{-1} x_{i+1} \right) \ = \ d(x_i, x_{i+1}) \ < \ r.$$

(This and the definition of r imply that $d(B, s_i B) = 0$, i.e. that $B \cap s_i B \neq \emptyset$; but we will not use this.) It follows that S generates Γ. Moreover, using the first inequality in (2), we have

$$d_S(1, \gamma) \ \leq \ k \ \leq \ \frac{1}{r} d(x_0, \gamma x_0) + 1 - \frac{R}{r}$$

and the second inequality of (1) follows. The first inequality of (1) is straightforward, by induction on $d_S(1, \gamma)$.

Observe that $\{\gamma \in \Gamma \mid \gamma x_0 = x_0\}$ is a finite subgroup of Γ which need not be reduced to $\{1\}$ (this subgroup is inside $S \cup \{1\}$). It follows that the constant "+1" in Equation (1) of Claim 2 may not be deleted. (But see the beginning of Item 46.)

Claim 3: the mapping

$$f \; : \; \begin{cases} (\Gamma, d_S) & \longrightarrow & (X, d) \\ \gamma & \longmapsto & \gamma x_0 \end{cases}$$

is a quasi-isometry.

Indeed, by left-invariance of the two metrics involved and by Claim 2, we have

$$\frac{1}{\lambda} d\left(f(\gamma_1), f(\gamma_2)\right) \; \leq \; d_S(\gamma_1, \gamma_2) \; \leq \; \frac{1}{r} d\left(f(\gamma_1), f(\gamma_2)\right) + 1$$

for all $\gamma_1, \gamma_2 \in \Gamma$. We also have $d(f(\Gamma), x) \leq R$ for all $x \in X$ because $(\gamma B)_{\gamma \in \Gamma}$ is a covering of X. $\square$

Anticipating Chapter V, one is tempted to ask under what additional conditions the situation of Theorem 23 provides a finite *presentation* of Γ. Here is one answer, from [Macbe–64] (see also [Swan–71], as well as the appendix to § I.3 in [Serr–77]). Let X be a connected topological space, let Γ be a group of homeomorphisms of X, and let V be a non-empty open subset of X such that $X = \bigcup_{\gamma \in \Gamma} (\gamma V)$; set

$$S = \{s \in \Gamma \mid V \cap sV \neq \emptyset\} \quad \text{and} \quad R = \{(s,t) \in S \times S \mid V \cap sV \cap stV \neq \emptyset\};$$

for each pair $r = (s,t) \in R$, say with $u = st \in S$, we denote by w_r the element in the free group over S represented by the 3-letter word stu^{-1}. Then S is a (possibly infinite) set of generators of Γ; moreover, if V is path-connected and if X is simply-connected, then

$$\langle S \mid (w_r)_{r \in R} \rangle$$

is a (possibly infinite) presentation of Γ.

In particular, if a group acts simplicially on a simply-connected simplicial complex and if the quotient space is compact, then the group is finitely presented. Conversely, any group which is finitely presented has such an action (Exercise V.13, together with a subdivision argument).

We refer to [Brown–84] for a detailed discussion on how to obtain finite presentations from actions with suitable hypotheses.

24. Corollary. *(i) Let Γ be a finitely-generated group and let Γ' be a subgroup of finite index in Γ. Then Γ' is finitely generated and quasi-isometric to Γ.*

(ii) Let $1 \to \Gamma' \to \Gamma \to \Gamma'' \to 1$ be a short exact sequence of groups, with Γ' finite and Γ'' finitely generated. Then Γ is finitely generated and quasi-isometric to Γ''.

(iii) Let Γ be the fundamental group of a compact Riemannian manifold Y. Then Γ is quasi-isometric to the Riemannian universal covering manifold of Y.

In particular, the fundamental group of a closed orientable surface of genus at least 2 is quasi-isometric to the hyperbolic plane.

(iv) More generally, any finitely-presented group is quasi-isometric to some four-dimensional smooth Riemannian manifold.

(v) Let G be a connected real Lie group and let Γ be a uniform lattice in G, namely a discrete subgroup of G such that the quotient space $\Gamma \backslash G$ is compact. Then Γ is finitely generated; moreover Γ is quasi-isometric to G (viewed as a metric space for some left-invariant Riemannian metric).

Proof. All claims except (iv) are straightforward particular cases of the previous proposition, respectively applied to

(i)   the natural action of Γ' on a Cayley graph of Γ,
(ii)   the natural action of Γ on a Cayley graph of Γ'',
(iii)   the natural action of Γ on the universal covering of Y,
(v)   the natural action of Γ on G.

For (i), see also Exercise III.8. For (iii), see Example 21.vi. For (v), see Example 21.v and Complement V.20.

For (iv), see Complements V.27 and V.29. $\square$

For a reformulation of Claims (i) and (ii) of this corollary, see Remark 29.i. There is an alternative for proving Claim (v), since easy arguments of general topology show that

in a locally compact group which is compactly generated,
any discrete subgroup with compact quotient is finitely generated

(see Item 0.40 in Chapter I of [Margu–91]).

25. Examples. (i) For two integers $k, l \geq 2$, the free groups of rank k and l are quasi-isometric, because F_2 has a subgroup of finite index isomorphic to F_k, and also one isomorphic to F_l (see Problem II.20); they are moreover Lipschitz equivalent: see Complement 46.

Denote by Γ_g the fundamental group of an orientable closed surface Σ_g of genus g. Then, similarly, for two integers $g, h \geq 2$, the groups Γ_g and Γ_h are quasi-isometric, because they are both subgroups of finite index, respectively $g - 1$ and $h - 1$, in Γ_2.

(ii) For a finite group F of order $k \geq 2$, the free products $\mathbb{Z} * F$ and $\mathbb{Z} * (\mathbb{Z}/2\mathbb{Z})$ are quasi-isometric. Indeed, each contains a non-abelian finitely-generated free group of finite index.

See also Item 46.ii.

(iii) The modular group $PSL(2, \mathbb{Z}) \approx (\mathbb{Z}/3\mathbb{Z}) * (\mathbb{Z}/2\mathbb{Z})$ is quasi-isometric to the free group on two generators (because F_2 is a subgroup of index 6 in the modular group, see Problem II.16). More generally, a non-trivial free product of a finite number of finite groups is quasi-isometric to F_2. Here, a free product $\Gamma_1 * \ldots * \Gamma_k$ of groups is *non-trivial* if $k \geq 2$ and $\sum_{j=1}^{k} (|\Gamma_j| - 1) > 2$.

(iv) Let G be a connected real Lie group with finite centre; let $\underline{G}$ denote the quotient of G by its centre. Any finitely-generated subgroup in G is quasi-isometric to a finitely generated subgroup in $\underline{G}$ (this applies to lattices, see Complement V.20).

(v) Let M_1, M_2 be two compact Riemannian manifolds of constant curvature -1 and of the same dimension. Then the fundamental groups of M_1 and of M_2 are quasi-isometric. This is a particular case of Corollary 24.iii, because the universal coverings of M_1 and M_2 are both isometric to the hyperbolic space H^n, with n the dimension of these manifolds; see also Example 21.vii.

(vi) A finite group is clearly *not* quasi-isometric to an infinite group. The groups $\mathbb{Z}, \mathbb{Z}^2, F_2$ are pairwise *not* quasi-isometric because the complement of a compact subset of a Cayley graph of $\mathbb{Z}$ [respectively of $\mathbb{Z}^2$, of F_2] has exactly two [respectively one, arbitrarily many] unbounded connected components. More generally, two quasi-isometric groups necessarily have the same "number of ends".

For ends, and the basic results of Freudenthal and H. Hopf, see e.g. [Epste–62] or Paragraph 5 of [ScoWa–79]. A basic result of the theory is that a group has 0, 1, 2, or infinitely many ends. Moreover, a group has 0 ends if and only if it is finite, 2 ends if and only if it has a subgroup of finite index isomorphic to $\mathbb{Z}$, and groups of infinitely many ends have been characterized by Stallings (see Item III.15); in this sense, $\mathbb{Z}^2$ behaves like "most groups" in having 1 end.

It is also true that $\mathbb{Z}^m$ and $\mathbb{Z}^n$ are not quasi-isometric when $n \neq m$, because they do not have the same "growth type" (see Definition VI.35). Growth is the most important invariant to distinguish quasi-isometry classes of groups.

There are many other invariants of quasi-isometry (Complement 50).

(vii) This as well as (viii) below were brought to my attention by Benson Farb. Recall from Example 21.v that there is a natural distance on G, well-defined up to quasi-isometry.

Let G be a real Lie group which is connected, simply-connected, and nilpotent, and let $\underline{g}$ denote its Lie algebra. Assume that there does *not* exist any *rational form* of $\underline{g}$, namely any Lie algebra $\underline{g}_0$ over $\mathbb{Q}$ such that $\underline{g} \approx \underline{g}_0 \otimes_{\mathbb{Q}} \mathbb{R}$; by a theorem of Mal'cev (see VII.26.ii), this is equivalent to assuming that G does not contain any lattice.

Does there exist a finitely-generated group Γ which is quasi-isometric to G ?

If the answer were "yes", Γ would have to be virtually nilpotent (by Gromov's theorem VII.29), and there would be no loss of generality in assuming Γ nilpotent and torsion-free (because any finitely-generated nilpotent group becomes torsion-free after quotienting it by the finite normal subgroup of its elements of finite order). By another theorem of Mal'cev (VII.26.i), Γ would be a lattice in a nilpotent connected and simply-connected real Lie group, say N, of which the Lie algebra would have a rational form. The above question may consequently be reformulated as

does there exist a nilpotent Lie group N as above
which is quasi-isometric to G ?

In fact, the two basic problems in this direction are

classify torsion-free finitely-generated nilpotent groups up to quasi-isometry,

classify simply-connected nilpotent real Lie groups up to quasi-isometry.

Theorem 3 of Section 1 in [Pansu–89] gives one indication: if two connected simply-connected nilpotent real Lie groups are quasi-isometric, then the corresponding graded Lie groups are isomorphic.

(viii) Consider now the standard semi-direct product G of the additive group $\mathbb{R}^2$ and of $SL(2,\mathbb{R})$. Clearly, the group G has lattices, for example $\mathbb{Z}^2 \rtimes SL(2,\mathbb{Z})$. Let us first recall that G does not have any uniform lattice.

Let Γ be a lattice in G. Its intersection with $\mathbb{R}^2$ is again a lattice (by Corollary 8.27 of [Raghu–72], discussed below in V.20.vi), and we may assume without loss of generality that $\Gamma \cap \mathbb{R}^2 = \mathbb{Z}^2$; also the canonical projection of Γ in $SL(2,\mathbb{R})$ is a lattice there (by the same Corollary 8.27 in [Raghu–72]). As the stabiliser $SL(2,\mathbb{Z})$ of $\mathbb{Z}^2$ in $SL(2,\mathbb{R})$ is not cocompact, Γ itself is not cocompact in G. The question is now:

does there exist a finitely-generated group Γ which is quasi-isometric to $\mathbb{R}^2 \rtimes SL(2,\mathbb{R})$?

The question can be repeated for any connected Lie group without uniform lattices. It is known that there are only countably many simply-connected Lie groups which admit lattices. (See Proposition 8.7' in [Witte–95], and [Wink–97b]; the restriction to *simply-connected* groups is necessary, by Example 8.8 of [Witte–95].)

(ix) Let G be a connected real Lie group. There are *some* well-known conditions under which G has co-compact lattices (e.g. G semi-simple [Borel–63], or G simply-connected nilpotent satisfying the Mal'cev conditions of VII.26), and some known conditions under which G has no such such lattice (e.g. G non-unimodular).

Let moreover H be a closed subgroup of G. The question whether G has a discrete subgroup Γ such that $\Gamma \backslash G / H$ is compact is more delicate. For example, for $G/H = SL(n,\mathbb{R})/SL(n-1,\mathbb{R})$ with $n \geq 3$, such a subgroup Γ exists if and only if n is even. See [Labou–96] and [Margu–97].

26. Variation on Theorem 23. Let X be a geodesic metric space and let $R \geq 0$ be a constant. A subspace Y of X is said to be *R-quasi-convex* if two points of Y can be joined by a geodesic segment in the R-neighbourhood of Y. A subspace of X is said to be *quasi-convex* if it is R-quasi-convex for some R. An action of a group on X is *quasi-convex* if its orbits are quasi-convex. (See 5.3 and 7.3 in [Gromo–87].)

For example, the orbits of an isometric action of $\mathbb{Z}$ or $\mathbb{R}$ on a Euclidean plane are always quasi-convex. But there are isometric actions of $\mathbb{Z}$ and $\mathbb{R}$ on a hyperbolic plane with non-quasi-convex orbits; more precisely, on the Poincaré half-plane $\{ z \in \mathbb{C} \mid \Im(z) > 0 \}$ with its hyperbolic metric, the horocycle of equation $\Im(z) = 1$ is not quasi-convex.

The following generalization of Theorem 23 appears in [Bourd–95].

Proposition. *Let X be a metric space which is geodesic and proper, let Γ be a group and let $\Gamma \times X \to X$ be an action by isometries.*

Assume that the action is proper and quasi-convex. Then Γ is a finitely-generated group and, for any $x_0 \in X$, the mapping $\Gamma \to X$ given by $\gamma \mapsto \gamma x_0$ is a quasi-isometric embedding.

On the proof. If the orbit of x_0 is R-quasi-convex, the set

$$S = \{ \, s \in \Gamma \mid s \neq 1 \ \text{ and } \ d(x_0, s x_0) \leq 2R + 1 \, \}.$$

generates Γ. $\square$

Proposition 26 applies for example to the so-called convex co-compact Fuchsian groups, i.e. the discrete subgroups Γ of $PSL(2, \mathbb{R})$ such that the quotient by Γ of the hyperbolic convex hull of the limit set of Γ is compact.

27. Definitions. (i) Two groups Γ_1, Γ_2 are *commensurable* if there exist two subgroups of finite index $\Gamma_1' < \Gamma_1, \Gamma_2' < \Gamma_2$ such that Γ_1' and Γ_2' are isomorphic; the indices need *not* be equal. (If two subgroups of a given group are commensurable in the sense of II.28, they are of course commensurable in the present sense; the groups of Exercise II.33.vii show that the converse does *not* hold.)

For example, non-abelian free groups and non-trivial free products of finite groups are commensurable with each other. (See Examples 25.i and 25.iii.)

(ii) Two groups Γ_1, Γ_2 are *commensurable up to finite kernels* if there exists a finite sequence

$$\Gamma_1 = \Gamma_{1,1} \longrightarrow \Gamma_{1,2} \longleftarrow \Gamma_{1,3} \longrightarrow \ldots \longleftarrow \Gamma_{1,k} = \Gamma_2$$

where each arrow indicates a homomorphism of groups with finite kernel and with image of finite index.

For example, Bieberbach groups are commensurable up to finite kernels with free abelian groups (see Item V.50).

Commensurable groups are obviously commensurable up to finite kernels, but the converse does *not* hold. For example, $SL(2, \mathbb{Q})$ and $PSL(2, \mathbb{Q})$ are clearly commensurable up to finite kernels; but they are not commensurable, because any proper normal subgroup of $SL(2, \mathbb{Q})$ is central (see, e.g., Theorem 4.9 in [Artin–57]), and in particular of infinite index, so that $SL(2, \mathbb{Q})$ does not have any proper subgroup of finite index. See Complement III.18, (x) and (xi), for finitely-generated and finitely-presented examples.

"*Virtually isomorphic*" is used by some authors for our "commensurable" (e.g. [Witte–95]) and by others for our "commensurable up to finite kernels" (e.g. [Furm–99a], [Furm–99b]).

28. Proposition. *(i) Two residually finite groups are commensurable if and only if they are commensurable up to finite kernels.*

(ii) Two groups Γ_1, Γ_2 are commensurable up to finite kernels if and only if there exists a finite sequence

$$\Gamma_1 = \Gamma_{1,1} \longrightarrow \Gamma_{1,2} \longleftarrow \Gamma_{1,3} \longrightarrow \ldots \longleftarrow \Gamma_{1,k} = \Gamma_2$$

where each arrow indicates a homomorphism of groups with finite central kernel and with image of finite index.

Proof. (i) It is sufficient to prove the following. Let $\pi : \Gamma_1 \longrightarrow \Gamma_2$ be a homomorphism of groups with finite kernel and image of finite index. If Γ_1 is residually finite, there exists a group Γ_0 which is isomorphic to a subgroup of finite index in both Γ_1 and Γ_2.

By hypothesis, there exists for each $x \in Ker(\pi), x \neq 1$ a normal subgroup Γ_x of finite index in Γ_1 such that $x \notin \Gamma_x$. Set $\Gamma_0 = \cap_{x \in Ker(\pi), x \neq 1} \Gamma_x$. On one hand Γ_0 is clearly a subgroup of finite index in Γ_1. It follows that $\pi(\Gamma_0)$ is also of finite index in Γ_2. On the other hand, the restriction $\pi|\Gamma_0$ is injective, so that Γ_0 "is" also a subgroup of finite index in Γ_2. (By Problem III.18.vi, this moreover shows that Γ_2 is residually finite.)

(ii) It is sufficient to show that, if $\pi : \Gamma_1 \longrightarrow \Gamma_2$ is a homomorphism of groups with finite kernel and image of finite index, there exists a homomorphism $\pi' : \Gamma'_1 \longrightarrow \Gamma'_2$ with finite *central* kernel and image of finite index, where Γ'_1, Γ'_2 are subgroups of finite index in Γ_1, Γ_2 respectively. For this one defines Γ'_1 [respectively π', Γ'_2] as the centralizer in Γ_1 of the kernel of π [respectively the restriction $\pi \mid \Gamma'_1$, the image $\pi(\Gamma'_1)$].

I am grateful to L. Bartholdi and V. Nekrashevych for these simple arguments. $\square$

29. Remark. (i) One may reformulate as follows part of Corollary 24:

let Γ_1, Γ_2 be two groups which are commensurable up to finite kernels; if Γ_1 is finitely generated, then Γ_2 is also finitely generated, and quasi-isometric to Γ_1.

(ii) The next proposition shows that two finitely-generated groups which are quasi-isometric need not be commensurable up to finite kernels. We have used unpublished notes of C. Pittet, and [Barbo–90]; see also [BriGe–96].

A related and older reference is the first paper on Analysis Situs where Poincaré constructs torus-bundles over the circle associated to matrices in the group $GL(2, \mathbb{Z})$, and shows that two such bundles are homeomorphic if and only if the matrices are conjugate; this appears in § 11 ("Sixième exemple") and § 14 (page 255 of the "Oeuvres Complètes") of [Poinc–95], just before Poincaré asks whether the fundamental group of a manifold determines its homeomorphism type (see also V.27 below).

(iii) Here is another type of example. For two closed hyperbolic 3-manifolds M_1, M_2, the fundamental groups $\pi_1(M_1), \pi_1(M_2)$ are always quasi-isometric, as already observed in Example 25.v. (Here, "hyperbolic" means: Riemannian manifolds with constant curvature -1. Observe that $\pi_1(M_1), \pi_1(M_2)$ are finitely generated and linear, so that Proposition 28.i applies.) There is a method due to Macbeath [Macbe–83] which shows that $\pi_1(M_1), \pi_1(M_2)$ are not necessarily commensurable.

For a cocompact discrete subgroup Γ of $SL(2, \mathbb{C})$, let $\mathbb{Q}(tr\Gamma)$ denote its *trace-field*, namely the field generated by the traces of all elements of Γ, and let $\Gamma^{(2)}$ be the subgroup of Γ generated by $\{\gamma^2 \mid \gamma \in \Gamma\}$. Macbeath (with a correction pointed out by Reid [Reid–90]) has shown that $\mathbb{Q}(tr\Gamma)$ is an algebraic number field of finite degree, and that $\mathbb{Q}\left(tr\Gamma^{(2)}\right)$ depends only on the commensurability class of Γ. (Examples in [Reid–90] show that $\mathbb{Q}(tr\Gamma)$ is *not* an invariant of the commensurability class of Γ.)

Moreover Macbeath has shown that there exists an infinity of distinct commensurability classes of cocompact discrete subgroups Γ of $SL(2, \mathbb{C})$ by producing an infinity of groups with different $\mathbb{Q}\left(tr\Gamma^{(2)}\right)$. In particular, there exists an infinity of closed hyperbolic 3-manifolds with pairwise non-commensurable fundamental groups.

It is a conjecture that there exist pairs M_1, M_2 of such manifolds with irrational ratio $vol(M_1)/vol(M_2)$ of Riemannian volumes (see Question 7.7 in [Borel–81] and Question 23 of § 6 in [Thurs–82]); for such pairs, the corresponding fundamental groups cannot be commensurable. But, as far as I know, showing the existence of pairs of this kind is still an open problem (I am grateful to Alan Reid for information on this point).

Observe that, by Mostow rigidity, two lattices in $SL(2, \mathbb{C})$ are commensurable in the sense of Definition 27 if and only if they are commensurable as subgroups (see II.28).

(iv) Consider an irreducible cocompact lattice Γ_{irr} in $G = PSL(2, \mathbb{R}) \times PSL(2, \mathbb{R})$. Any normal subgroup of Γ_{irr} is of finite index, by Margulis' normal subgroup theorem (see e.g. Chapter 8 in [Zimme–84]). Consider also a closed surface Σ_g of genus $g \geq 2$, and the direct product $\Gamma_{red} = \pi_1(\Sigma_g) \times \pi_1(\Sigma_g)$ as a reducible cocompact lattice in G. This has non-trivial normal subgroups of infinite index, for example the kernel of the abelianization $\Gamma_{red} \longrightarrow \mathbb{Z}^{2g} \oplus \mathbb{Z}^{2g}$. It follows that the groups Γ_{irr} and Γ_{red} are not commensurable; they are quasi-isometric since they are both cocompact in the same group G.

(v) Let $\underline{n}$ be a nilpotent real Lie algebra; assume that there exist two sub-$\mathbb{Q}$-Lie algebras $\underline{a}, \underline{b}$ such that $\underline{g} = \underline{a} \otimes_\mathbb{Q} \mathbb{R} = \underline{b} \otimes_\mathbb{Q} \mathbb{R}$; see Remark 2.5 in [Raghu–72] for a 6-dimensional example, with $\underline{a}$ the direct sum of two Heisenberg Lie algebras over $\mathbb{Q}$ and $\underline{b}$ obtained from the Heisenberg Lie algebra over $\mathbb{Q}(\sqrt{2})$. Let N be a connected simply-connected Lie group with Lie algebra $\underline{n}$. By the theory of Mal'cev (see Item VII.26), there exist two lattices $\Gamma_{\underline{a}}, \Gamma_{\underline{b}}$ in N such that $\underline{a}$ is the sub-$\mathbb{Q}$-Lie algebra of $\underline{n}$ generated by $\log_N\left(\Gamma_{\underline{a}}\right)$, and similarly for $\underline{b}$. Then $\Gamma_{\underline{a}}$ and $\Gamma_{\underline{b}}$ are not commensurable, but they are quasi-isometric by Corollary 24.5.

(vi) Other classes of examples appear in Complements 46 through 48.

30. Proposition. *Let $A, B \in GL(2, \mathbb{Z})$ be two matrices with traces of absolute values strictly larger than 2. The semi-direct products $\mathbb{Z}^2 \rtimes_A \mathbb{Z}$ and $\mathbb{Z}^2 \rtimes_B \mathbb{Z}$ are*

(i)   *isomorphic if and only if A is conjugate in $GL(2, \mathbb{Z})$ to B or B^{-1},*

(ii)   *commensurable if and only if there exist integers $p, q \in \mathbb{Z} \setminus \{0\}$ such that A^p and B^q are conjugate in $GL(2, \mathbb{Q})$,*

(iii) *commensurable up to finite kernels if and only if they are commensurable,*

(iv)   *quasi-isometric in all cases.*

Proof. We prove (iv) here, and give hints for (i) and (ii) in Exercises 38 and 39; the equivalence $(ii) \Longleftrightarrow (iii)$ is a particular case of Proposition 28.i. We write Γ_A for the semi-direct product $\mathbb{Z}^2 \rtimes_A \mathbb{Z}$.

Observe that the condition $|trace(A)| > 2$ means that A has two real eigenvalues distinct from 1 or -1, so that A is conjugate within $GL(2, \mathbb{R})$ to the diagonal matrix with these two eigenvalues.

Consider the solvable Lie group

$$Sol = \mathbb{R}^2 \rtimes_{diag(e^s, e^{-s})} \mathbb{R}$$

which is the group of pairs $(u, s) \in \mathbb{R}^2 \times \mathbb{R}$ with multiplication given by

$$\left(\begin{pmatrix} u_1 \\ u_2 \end{pmatrix}, s \right) \left(\begin{pmatrix} v_1 \\ v_2 \end{pmatrix}, t \right) = \left(\begin{pmatrix} u_1 + e^s v_1 \\ u_2 + e^{-s} v_2 \end{pmatrix}, s + t \right).$$

If $det(A) = +1$, the matrix A is conjugate in $GL(2, \mathbb{R})$ to $diag(e^s, e^{-s})$ for s a solution of $s + s^{-1} = trace(A)$. It follows that Γ_A may be identified with a discrete subgroup of Sol. We have a diagram

$$
\begin{array}{ccccccccc}
0 & \longrightarrow & \mathbb{Z}^2 & \longrightarrow & \Gamma_A & \longrightarrow & \mathbb{Z} & \longrightarrow & 0 \\
& & \downarrow & & \downarrow & & \downarrow & & \\
0 & \longrightarrow & \mathbb{R}^2 & \longrightarrow & Sol & \longrightarrow & \mathbb{R} & \longrightarrow & 0 \\
& & \downarrow & & \downarrow & & \downarrow & & \\
0 & \longrightarrow & \mathbb{T}^2 & \longrightarrow & \Gamma_A \backslash Sol & \longrightarrow & \mathbb{T} & \longrightarrow & 0
\end{array}
$$

showing that $\Gamma_A \backslash Sol$ is a 2-torus bundle over the circle $\mathbb{T} = \mathbb{Z} \backslash \mathbb{R}$. In particular, $\Gamma_A \backslash Sol$ is a compact manifold.

If one chooses on Sol a left-invariant Riemannian metric, the construction above and Theorem 23 show that the finitely-generated group Γ_A is quasi-isometric to the Riemannian manifold Sol.

In case $det(A) = -1$, observe that $det(A^2) = +1$ and that Γ_{A^2} is a subgroup of index 2 in Γ_A, so that Γ_A is also quasi-isometric to Sol. This shows Claim (iv). $\square$

Denote by $\mathbb{Q}(A)$ the number field generated by the eigenvalues of A. If A, B are as in the proposition with $\mathbb{Q}(A) \neq \mathbb{Q}(B)$, for example if $A = \begin{pmatrix} 2 & 1 \\ 1 & 1 \end{pmatrix}$ and $B = \begin{pmatrix} 3 & 1 \\ 1 & 1 \end{pmatrix}$, then $\mathbb{Z}^2 \rtimes_A \mathbb{Z}$ and $\mathbb{Z}^2 \rtimes_B \mathbb{Z}$ are quasi-isometric but are not commensurable.

For a comparison of $\mathbb{Z}^2 \rtimes_A \mathbb{Z}$ and $\mathbb{Z}^2 \rtimes_B \mathbb{Z}$ for *any* pair A, B of matrices in $GL(2, \mathbb{Z})$, see [Barbo-90] or Theorem 5.8 in [BriGe-96].

Exercises for IV.B

31. Exercise. Write down a quasi-isometry between two metric spaces which is not a continuous map.

32. Exercise. On $\mathbb{Z}$, let d_0 denote the usual metric, $d_0(x, y) = |x - y|$, and let d be the metric defined by $d(x, y) = |x - y| + \log(|x - y|)$ for $x, y \in \mathbb{Z}, x \neq y$. Let ϕ be the identity viewed as a map $(\mathbb{Z}, d_0) \longrightarrow (\mathbb{Z}, d)$. Check that

(i) for any $\lambda > 1$, there exists $C \geq 0$ such that ϕ is a (λ, C)-quasi-isometry,
(ii) there is no $C \geq 0$ such that ϕ is a $(1, C)$-quasi-isometry.

[This exercise is due to Gilles Robert.]

33. Exercise. For two constants $\lambda > 0, d \geq 0$, say that a function $f : \mathbb{R} \to \mathbb{R}$ is λ-*Lipschitz on d-scale* [Gromo-93, page 22] if

$$|f(s) - f(t)| \leq \lambda |s - t| + d$$

for all $s, t \in \mathbb{R}$. Given such a function, check that

$$\sigma_f : (x, y) \longmapsto (x, y + f(x))$$

defines a quasi-isometry of the Euclidean plane onto itself. For an appropriate function f, check that σ_f "is not near an isometry"; namely check that

$$\sup_{(x,y) \in \mathbb{R}^2} |\sigma_f(x, y) - g(x, y)| = \infty$$

for any isometry g of the plane.

Formulate and prove an analogous statement for

$$\sigma_f : (r, \theta) \longmapsto \left(r, \theta + \frac{1}{r} f(r)\right)$$

in polar coordinates.

34. Exercise. Let X be a locally compact space which is non-empty. Let Γ, Δ be two groups acting properly on X by homeomorphisms, with compact quotients. Assume that the actions commute. It is convenient to assume that Γ acts from the left and Δ from the right on X, so that the commutation condition reads $(\gamma x)\delta = \gamma(x\delta)$ for all $\gamma \in \Gamma$, $x \in X$ and $\delta \in \Delta$.

(i) Observe that Γ, Δ need not be finitely generated.
[Hint: consider $\Gamma = \Delta = X$.]

(ii) If Γ is finitely generated, show that Δ is also finitely generated.
[Hint. Choose a compact subset K of X such that the restrictions to K of the canonical projections $X \longrightarrow \Gamma \backslash X$ and $X \longrightarrow X/\Delta$ are both onto. Let S be a finite generating set of Γ. For each $s \in S$, let T_s be a finite subset of Δ such that $sK \subset \cup_{t \in T_s} Kt$. Set $T_I = \cup_{s \in S} T_s$ and $T_{II} = \{\delta \in \Delta \mid \delta K \cap K \neq \emptyset\}$. Check that $T = T_I \cup T_{II}$ is a finite generating set of Δ.]

(iii) Assume that Γ and Δ are finitely generated. Show that there exists a quasi-isometry $\phi : \Gamma \longrightarrow \Delta$.
[Hint. Choose a compact subset K of X as in (ii), a point $x_0 \in K$, a finite generating set S of Γ, and let T be as in (ii). For each $\gamma \in \Gamma$, show by induction on the length $\ell_S(\gamma)$ that there exists $\delta_\gamma \in \Delta$ such that $\gamma x_0 \in K\delta_\gamma$ and $\ell_T(\delta_\gamma) \leq \ell_S(\gamma)$. Show that there exists a constant $C \geq 0$ such that the map $\phi : \Gamma \longrightarrow \Delta$ defined by $\phi(\gamma) = \delta_\gamma$ satisfies

$$d_T(\phi(\gamma_1), \phi(\gamma_2)) \leq d_S(\gamma_1, \gamma_2) + C$$

for all $\gamma_1, \gamma_2 \in \Gamma$. Show similarly that there exist a finite generating set S' of Γ, a map $\psi : \Delta \longrightarrow \Gamma$, and a constant $D \geq 0$ such that

$$d_{S'}(\psi(\delta_1), \psi(\delta_2)) \leq d_T(\delta_1, \delta_2) + D$$
$$d_{S'}(\gamma, \psi\phi(\gamma)) \leq D$$
$$d_T(\delta, \phi\psi(\delta)) \leq D$$

for all $\delta_1, \delta_2, \delta \in \Delta$ and $\gamma \in \Gamma$.]

35. Exercise. Let Γ, Δ be finitely-generated groups. For constants $\lambda \geq 1$ and $C \geq 0$, let $X_{\lambda,C}$ denote the space of all mappings $\phi : \Gamma \longrightarrow \Delta$ such that

(*)
$$\frac{1}{\lambda}d(x,y) - C \leq d(\phi(x), \phi(y)) \leq \lambda d(x,y) + C$$
$$d(\phi(\Gamma), z) \leq C$$

for all $x, y \in \Gamma$ and $z \in \Delta$. Observe that, if Γ, Δ are quasi-isometric, then $X_{\lambda,C}$ is not empty for appropriate constants λ and C.

The groups Γ and Δ have natural commuting actions on $X_{\lambda,C}$, defined by $(\gamma\phi\delta)(x) = \delta^{-1}(\phi(\gamma^{-1}x))$ for all $\gamma \in \Gamma$, $\phi \in X_{\lambda,C}$, $\delta \in \Delta$ and $x \in \Gamma$. We endow $X_{\lambda,C}$ with the topology of pointwise convergence.

(i) Show that $X_{\lambda,C}$ is a locally compact space.
[Hint. For a finite subset F of Γ, a number $k > 0$ and an element ϕ_0 of $X_{\lambda,C}$, the neighbourhood

$$\{\phi \in X_{\lambda,C} \mid d(\phi_0(x), \phi(x)) \leq k \text{ for all } x \in F\}$$

is compact by Ascoli's theorem.]

(ii) Show that Γ and Δ act properly on $X_{\lambda,C}$.

(iii) Show that $\Gamma \setminus X_{\lambda,C}$ and $X_{\lambda,C}/\Delta$ are both compact.
[Hint for $X_{\lambda,C}/\Delta$. Let $(\phi_n)_{n\geq 1}$ be a sequence in $X_{\lambda,C}$. For each $n \geq 1$, let $B_S(n)$ denote the ball of radius n centred at the identity e_Γ of Γ. As Δ acts transitively on itself, there exists for each $n \geq 1$ an element $\delta_n \in \Delta$ such that $(\phi_n \delta_n)(e_\Gamma) = e_\Delta$. Observe that $(\phi_n \delta_n)(B_S(k)) \subset B_T(\lambda k + C)$ for all $k \geq 0$, and use a diagonal argument to show that $(\phi_n \delta_n)_{n\geq 1}$ has a convergent subsequence.]

Exercises 34 and 35 show the *topological criterion for quasi-isometry* which is Item 0.2.C$_2'$ of [Gromo–93]. Similarly two countable groups Γ, Δ are *measure equivalent* if they have commuting measure-preserving actions on some infinite Lebesgue measure space (Ω, μ) such that each action has a fundamental domain of finite measure; for this most interesting notion, see Item 0.5.E of [Gromo–93] and [Furm–99a], [Furm–99b].

36. Exercise on torsion. Let Γ be a group which contains a torsion-free subgroup Γ_0 of index $n < \infty$. Show that Γ does not contain any element whose order is finite and strictly larger than n.
[Solution. Let $\gamma \in \Gamma$ be of finite order. Let x_0 denote the class of 1 in Γ/Γ_0. There exist $i, j \in \{0, 1, \ldots, n\}$ with $i \neq j$ and $\gamma^i x_0 = \gamma^j x_0$, and thus also $k \in \{1, \ldots, n\}$ such that $\gamma^k \in \Gamma_0$. As γ has finite order, we have $\gamma^k = 1$.]

This implies that a group which is commensurable to a torsion-free group has no torsion of large order. Is this true for a group which is *quasi-isometric* to a torsion-free group ?

A. Dioubina [Dioub] has observed that, for F a finite group of order $n > 1$, we have a quasi-isometry between the wreath product $\mathbb{Z} \wr \mathbb{Z}$, which is torsion-free, and $(\mathbb{Z} \oplus F) \wr \mathbb{Z}$, in which any subgroup of finite index has elements of order n.

37. Exercise on the Rips complex. Let X be a metric space and let r be a positive number. The *Rips complex* of X of parameter r is the simplicial polyhedron $Rips_r(X)$ with set of vertices X itself and, for each dimension $k \geq 1$, with k-simplexes the subsets $\{x_0, x_1, \ldots, x_k\}$ of X whose diameter is bounded by r. One can view $Rips_r(X)$ as a topological space (for the weak topology). Rips complexes appear in print in [Gromo–87, 1.7 and 2.2], [Gromo–93, 1.B.c], where they are named after Ilia Rips; they appear again below in V.3 and V.4.

(i) Check that $Rips_r(X)$ is connected for r large enough if and only if X is *long-range connected*. (This means that there exists a constant C such that, given any pair (x, x') of points in X, there exists a sequence $x_0 = x, x_1, \ldots, x_n = x'$ with $d(x_{i-1}, x_i) \leq C$ for all $i \in \{1, \ldots, n\}$. See Item 0.2.A$_2$ in [Gromo–93]).

(ii) Check that $Rips_r(X)$ is finite dimensional for every $r > 0$ if and only if X is *uniformly quasi-locally bounded* (this term is defined in Item VI.26, but it is a good exercise to formulate one's own definition first).

(iii) For (Γ, S) as in Section IV.A and for the metric space X defined by the word metric on Γ, we write $Rips_r(\Gamma, S)$ instead of $Rips_r(X)$. Thus $Cay(\Gamma, S)$ is precisely the 1-skeleton of $Rips_1(\Gamma, S)$.

Draw $Rips_2(\mathbb{Z}, S)$ for $S = \{1\}$.

(iv) Let X, Y be two metric spaces and let $\phi : X \to Y$ be a quasi-isometric embedding. For any $r > 0$, there exists $R > 0$ such that ϕ induces a canonical simplicial mapping $Rips_r(X) \to Rips_R(Y)$.

(v) Let X, Y be two long-range connected metric spaces which are quasi-isometric. Check that

$$Rips_r(X) \qquad \text{is simply-connected for } r \text{ large enough}$$

if and only if

$$Rips_r(Y) \qquad \text{is simply-connected for } r \text{ large enough.}$$

38. Exercise. For $A \in GL(2, \mathbb{Z})$, recall from Proposition 30 (and its proof) that Γ_A denotes the group of pairs (u, k) with $u \in \mathbb{Z}^2, k \in \mathbb{Z}$ and with multiplication

$$(u, k)(v, l) = \left(u + A^k v, \, k + l \right).$$

We identify $\mathbb{Z}^2$ with the subgroup of pairs of the form $(u, 0)$.

The aim of this exercise is to show to what extent the matrix A is determined by the isomorphism class of the group Γ_A, at least in case $|trace(A)| > 2$.

(i) Check that $(u, k) \longmapsto (u, -k)$ defines an isomorphism $\Gamma_A \longrightarrow \Gamma_{A^{-1}}$.

(ii) Check that, for any $b \in \mathbb{Z}^2$, the map

$$(u, k) \longmapsto \begin{cases} (u + b + Ab + \ldots + A^{k-1}b, \, k) & \text{if} \quad k \geq 0 \\ (u - A^{-1}b - \ldots - A^{-|k|}b, \, k) & \text{if} \quad k < 0 \end{cases}$$

is an automorphism of Γ_A.

(iii) For $P \in GL(2, \mathbb{Z})$ and $B \doteq PAP^{-1}$, check that $(u, k) \longmapsto (Pu, k)$ defines an isomorphism $\Gamma_A \longrightarrow \Gamma_B$.

(iv) Show that the commutator subgroup $D(\Gamma_A)$ of Γ_A is the image of $\mathbb{Z}^2$ by $Id - A$. If $+1$ is not an eigenvalue of A, it follows that $(Id - A)(\mathbb{Z}^2)$ is a subgroup of finite index in $\mathbb{Z}^2$ and that

$$\{\, \gamma \in \Gamma_A \,|\, \text{there exists an integer } n \neq 0 \text{ such that} \quad \gamma^n \in D(\Gamma_A) \,\}$$

can be identified with $\mathbb{Z}^2$.

(v) Let $A, B \in GL(2, \mathbb{Z})$ be matrices whose eigenvalues are real and distinct from 1 and -1, and such that there exists an isomorphism $\Phi : \Gamma_A \longrightarrow \Gamma_B$. Using (iv), show that Φ can be written as a composition of isomorphisms as in (iii), (i), and (ii). Check that this implies that there exist $P \in GL(2, \mathbb{Z})$ and $\epsilon \in \{1, -1\}$ such that $B = PA^\epsilon P^{-1}$.

39. Exercise. Let $A \in GL(2, \mathbb{Z})$ and Γ_A be as in the previous exercise. Now, the aim is to show how much of the matrix A is determined by the commensurability class of the group Γ_A, at least in the hyperbolic case $|trace(A)| > 2$.

(i) Any subgroup of $\mathbb{Z}^2$ is clearly a subgroup of Γ_A. For each integer $d \geq 1$ and for each A^d-invariant subspace V of $\mathbb{Z}^2$, the semi-direct product $V \rtimes_{A^d} \mathbb{Z}$ is also a subgroup of Γ_A.

Check that any subgroup of Γ_A is of one of these forms.

(ii) Deduce Claim (ii) of Proposition 30 from (i) above and from Exercise 38.

Problems and complements to IV.B

40. A problem on groups quasi-isometric to free abelian groups. Let Γ be a finitely-generated group which is quasi-isometric to a free abelian group of some rank k. Then Γ contains a subgroup of finite index which is isomorphic to $\mathbb{Z}^k$. (See e.g. Chapter 1 in [GhyHp–90].) The proof known to the author uses the deep theorem of Gromov which identifies almost nilpotent groups to groups of polynomial growth (Theorem VII.29); for $k = 2$, see also [SeiTr–97].

It is a natural problem to find another proof. Yehuda Shalom has recently announced a proof which does not use Gromov's theorem; it uses the notion of measure equivalence, alluded to after Exercise 35.

41. Groups of classes of quasi-isometries. Let X be a metric space. Let $\tilde{Q}I(X)$ denote the set of quasi-isometries from X to X. Say that two quasi-isometries ϕ, ψ from X to X are equivalent if $\sup_{x \in X} d(\phi(x), \psi(x)) < \infty$, and let $QI(X)$ denote the quotient set of $\tilde{Q}I(X)$ by this equivalence relation.

Show that there is a natural multiplication in $QI(X)$ which makes it a group. (See the discussion in Sections 3.3 and 3.4 of [GroPa–91].)

It is remarkable that, for many interesting spaces, the group $QI(X)$ is canonically isomorphic to the group of *isometries* of X ! This is for example the case if X is an irreducible Riemannian symmetric space of non-compact type, not one of $SO(n, 1)^0/SO(n)$ or $SU(n, 1)/S(U(n) \times U(1))$. Among many important papers on these developments, we quote [Pansu–89], [KleLe–98] and [EskFa–97].

42. Groups quasi-isometric to lattices in semi-simple Lie groups. Let G be a semi-simple group, by which we mean here a connected real Lie group which is semi-simple, with finite centre and without compact factors. A *lattice* in G is a discrete subgroup Γ of G such that G/Γ has a G-invariant probability measure; it is known that such a lattice is necessarily finitely generated (Complement V.20 below). A lattice Γ in G is *irreducible* if, for every non-central normal subgroup N of G, the projection of Γ in G/N is dense; there are several equivalent conditions: see Corollaries 5.19 and 5.21 in [Raghu–72].

The following theorems, essentially conjectured in Item 2C of [Gromo–84] and recently exposed in [Farb–97], follow from extensive work by numerous people. At one place of his paper, Farb quotes Cannon-Cooper, Sullivan, Tukia,

Mess, Gabai, Casson-Jungreis, Pansu, Ghys – de la Harpe, Stallings, Schwartz, Farb-Schwartz, Chow, Koranyi-Reimann, Kleiner-Leeb, Eskin-Farb, and Eskin. One may of course very safely add Borel, Furstenberg, Gromov, Mostow, Selberg, Thurston, Weil, ...!

Theorem. *Let* Γ *be a finitely-generated group which is quasi-isometric to an irreducible lattice in a semi-simple group* G. *Then* Γ *is commensurable up to finite kernels with some lattice in* G.

Theorem. *Let* G, G' *be two semi-simple groups with centres reduced to one element. Let* Γ *be a lattice in* G *and let* Γ' *be a lattice in* G'.

(i)   *If* Γ *and* Γ' *are quasi-isometric, then* G *and* G' *are isomorphic, and* $G/\Gamma, G'/\Gamma'$ *are either both compact or both non-compact.*

(ii)   *If* G/Γ *is compact, any cocompact lattice in* G *is quasi-isometric to* Γ.

(iii)   *If* G/Γ *is non-compact and if* G *is not isomorphic to* $PSL(2, \mathbb{R})$, *a lattice* Λ *in* G *is quasi-isometric to* Γ *if and only if it is commensurable with* Γ *in* G, *i.e. if and only if there exists* $g \in G$ *such that* $g\Lambda g^{-1} \cap \Gamma$ *is of finite index in both* $g\Lambda g^{-1}$ *and* Γ.

(iv)   *If* G/Γ *is non-compact and if* $G \approx PSL(2, \mathbb{R})$, *any non-cocompact lattice in* G *is quasi-isometric to* Γ *(and indeed commensurable with the free group* F_2*).*

Observe that Claim (ii) is a consequence of Theorem 23. Claim (iv) has a simple proof, because a lattice Γ as in (iv) has a subgroup of finite index which is torsion-free [Selbe–60], and which is consequently the fundamental group of a non-compact surface. Claims (i) and (iii) as well as the previous theorem require considerably more work.

The previous two theorems imply for example that

for $n \geq 3$, *a finitely generated group which is quasi-isometric to* $SL(n, \mathbb{Z})$ *is commensurable up to finite kernels with* $SL(n, \mathbb{Z})$.

Thus we have some understanding of finitely-generated groups which are quasi-isometric to lattices in *semi-simple* connected real Lie groups. The situation for *solvable* Lie groups is completely unclear, even for a low-dimensional solvable group such as the group *Sol* defined in the proof of Proposition 30.

43. Quasi-isometries of Baumslag-Solitar groups. For each pair integers $p, q \geq 1$ (not both 1), recall that the *Baumslag-Solitar group* is defined by

$$BS(p, q) = \left\langle\, s, t \mid st^p s^{-1} = t^q \,\right\rangle$$

(see Complement III.21); it is solvable if and only if $p = 1$ or $q = 1$.

For two integers $m, n \geq 2$, the following are known to be equivalent [FarMo–98]:

(i) $m = a^k$ and $n = a^l$ for some integers $a \geq 2$, $k \geq 1$, $l \geq 1$,

(ii)   the groups $BS(1, m)$ and $BS(1, n)$ are commensurable,

(iii)   the groups $BS(1, m)$ and $BS(1, n)$ are quasi-isometric.

(There are other classes of solvable groups for which it is known precisely when two groups in the class are quasi-isometric; see [FarMo–99] and [FarMo].)

On the other hand, for pairs $(p_1, q_1), (p_2, q_2)$ with $2 \leq p_1 < q_1$ and $2 \leq p_2 < q_2$, it is known that $BS(p_1, q_1)$ and $BS(p_2, q_2)$ are always quasi-isometric [Whyte–a].

44. Lamplighter groups. To a finite group F, one associates the wreath product

$$\Gamma_F = \left(\bigoplus_{j \in \mathbb{Z}} F_j \right) \rtimes \mathbb{Z}$$

where F_j denotes a copy of F for each $j \in \mathbb{Z}$ and where the semi-direct product is taken with respect to the shift action. This is clearly a finitely-generated group, which is known to be finitely presented if and only if F is the group with *one* element [Baums–61].

Open problem. For two finite groups F and F' not reduced to one element, are the two groups Γ_F and $\Gamma_{F'}$ quasi-isometric ?

It is easy to check that the answer is yes if F and F' are of the same order, and more generally if there exist integers $k, k' \geq 1$ such that $|F|^k = |F'|^{k'}$. We do not know the answer for F of order 2 and F' of order 3; thus, the previous question constitutes a (possibly easy) research problem.

In particular, if $|F| = |F'| < \infty$ with F solvable and F' not, this provides an example of two quasi-isometric groups, Γ_F which is solvable and $\Gamma_{F'}$ which is not almost solvable. This very simple observation, which is due to A. Dioubina [Dioub], solves a question which had been open for more than ten years! I do not know of any pair of *finitely-presented* quasi-isometric groups, one solvable and the other not almost solvable.

Groups of the form $\Gamma_F = \left(\bigoplus_{j \in \mathbb{Z}^2} F_j \right) \rtimes \mathbb{Z}^2$ are probably more difficult to deal with, because understanding word lengths and geodesics in these groups presupposes understanding something about the "travelling salesman problem".

The group Γ_F and similar groups associated to other wreath products

$$\left(\bigoplus_{j \in A} F_j \right) \rtimes A$$

have been called *lamplighter groups* by J. Cannon [Parr–92b].

45. Rough isometries. When $\lambda = 1$, the notion of (λ, C)-quasi-isometry reduces to that of C-*isometry* (though ϵ is used more often than C in this context — see [BhaSe–97]), also called *rough isometry* in [Monod–97] and [BonSc].

Let us first mention a result answering a question which goes back to [HyeUl–47] (see [BhaSe–97] for background, references, and proof). *Let $\mathbb{R}^n$ denote the usual Euclidean space, and let $f : \mathbb{R}^n \longrightarrow \mathbb{R}^n$ be an ϵ-isometry such that $f(0) = 0$. Then there exists a unique linear isometry g of $\mathbb{R}^n$ such that*

$$\sup_{x \in \mathbb{R}^n} \|g(x) - f(x)\| \leq 2\,\epsilon.$$

Let us also mention the main theorem of [BonSc]. *Let (X,d) be a Gromov hyperbolic geodesic metric space with bounded growth at some scale, e.g. a Gromov hyperbolic connected graph with bounded valence. Then there exists an integer n and a constant $\lambda > 0$ such that $(X, \lambda d)$ is roughly isometric to a convex subset of the n-dimensional real hyperbolic space.*

46. Lipschitz equivalence. There is also a variation connected with the case $C = 0$ of Definition 20. A *Lipschitz equivalence* between two metric spaces (X,d) and (X',d') is a mapping $\phi : X \longrightarrow X'$ which is a $(\lambda, 0)$-quasi-isometry *and which is a bijection.* The mappings of Example 21, (iii) through (vi), are Lipschitz equivalences. Also, in Theorem 23, if the point x_0 is such that the isotropy group $\{\gamma \in \Gamma \mid \gamma(x_0) = x_0\}$ is reduced to 1, then the natural mapping from the group Γ (with some word metric) to the orbit Γx_0 (with the metric from the ambiant space X) is a Lipschitz equivalence.

A metric space (X, d) is *uniformly discrete* if

$$\inf \{d(x,y) \mid x,y \in X \text{ with } x \neq y\} > 0.$$

A uniformly discrete metric space (X, d) has *bounded geometry* if, for any $r > 0$, there exists a constant C_r such that subsets of X of diameter at most r have at most C_r points. For example, Cayley graphs of finitely generated groups are uniformly discrete and have bounded geometry.

The following proposition answers a question of Gromov (see Item 1.A' in [Gromo-93]). It was proved independently by several people including: Volodymyr Nekrashevych [Nekra], Panos Papasoglu [Papas-95], and Kevin Whyte [Whyte-99]; see also [Bogop]. For the notion of amenability of discrete metric spaces, see for example [CecGH-99]; let us only recall here that the Cayley graphs associated with finitely-generated groups which are non-amenable as groups (in particular groups containing non-abelian free groups) are metric spaces which are non-amenable in the sense used here.

Proposition. Two non-amenable uniformly discrete metric spaces with bounded geometry are quasi-isometric if and only if they are Lipschitz equivalent.

Proof. To show the non-trivial implication, consider two discrete non-amenable spaces X, X' and a quasi-isometry $f_1 : X \longrightarrow X'$; one has to show that X and X' are Lipschitz equivalent.

The first step is to show that there exists an *injective* quasi-isometry from X to X' at a bounded distance from f_1. From the hypothesis on f_1, the quantity $\sup\{d(x,y) \mid x,y \in X \text{ and } f_1(x) = f_2(x)\}$ is finite. As X has bounded geometry, there exists an integer b such that $|f_1^{-1}(F')| \leq b|F'|$ for any finite subset F' of X'. A fortiori, we have $|F| \leq b|f_1(F)|$ for any finite subset F of X. As X' is non-amenable, there exists a constant $r > 0$ such that $|F| \leq |\mathcal{N}_r(f_1(F))|$ for any finite subset F of X, where $\mathcal{N}_r(f_1(F))$ denotes $\{x' \in X' \mid d(x', f_1(F)) \leq r\}$.

It is now a consequence of Philip Hall's "Marriage Lemma" that there exists an injective map $g_1 : X \longrightarrow X'$ such that $g_1(x) \in \mathcal{N}_r(f(x))$ for all $x \in X$. Observe that $\sup_{x \in X} d'(f_1(x), g_1(x)) \leq r < \infty$, so that g_1 is in particular also a quasi-isometry from X to X'. For Hall's lemma, see [Mirsk–71]; for a preliminary exposition in the context of finite graphs, we recommend Chapter III of [Bollo–79].

Similarly, there exists an injective quasi-isometry $g_2 : X' \longrightarrow X$ at a bounded distance from a quasi-inverse of f_1.

Consider the bipartite graph with vertex set $V = X \sqcup X'$ and with edge set

$$E = (\{x, g_1(x)\})_{x \in X} \cup (\{g_2(x'), x'\})_{x' \in X'}.$$

The connected components of this graph are of four types:

(a) two vertices connected by one edge,
(b) cycles of even length,
(c) half-lines
(d) bi-infinite lines.

Consider a subset F of E consisting of

all edges appearing in (a) above,
every other edge in each cycle of even length appearing in (b),
every other edge including the first one in each half-line appearing in (c),
every other edge in each line appearing in (d);

there are two possibilities and an arbitrary choice involved for each cycle and for each line. Then F defines a 1-factor of the graph (V, E). (See Exercise 11 for "1-factor".) Define $h : X \longrightarrow X'$ by $h(x) = x'$ whenever $\{x, x'\} \in F$. Then h is clearly a quasi-isometry which is one-to-one and onto. (This step is a typical Schroeder-Bernstein argument.)

The proposition follows, because a quasi-isometry between uniformly discrete metric spaces which is a bijection is necessarily a Lipschitz equivalence. $\square$

(i) Using the proposition, it is straightforward to check the following assertions. Let $\Gamma_1, \Gamma_2, \Delta$ be finitely-generated groups; assume that Γ_1, Γ_2 are quasi-isometric and not amenable (hence they are Lipschitz equivalent); then the free products $\Gamma_1 * \Delta$ and $\Gamma_2 * \Delta$ are Lipschitz equivalent, and in particular quasi-isometric. Similarly, the free products $\Gamma_1 * \Gamma_1$ and $\Gamma_2 * \Gamma_2$ are quasi-isometric; we will see in the next complement that they need not be commensurable, even if Γ_1 and Γ_2 are commensurable.

(ii) **Invariance of quasi-isometry by free products.** *Let* $\Gamma_1, \Gamma_1', \Gamma_2, \Gamma_2'$ *be finitely-generated groups such that* Γ_1 *is quasi-isometric with* Γ_1' *and* Γ_2 *with* Γ_2'. *Assume that* $(|\Gamma_1| - 1)(|\Gamma_2| - 1) > 1$ *and* $(|\Gamma_1'| - 1)(|\Gamma_2'| - 1) > 1$, *so that in particular neither* $\Gamma_1 * \Gamma_2$ *nor* $\Gamma_1' * \Gamma_2'$ *is an infinite dihedral group. Then the free products* $\Gamma_1 * \Gamma_2$ *and* $\Gamma_1' * \Gamma_2'$ *are Lipschitz equivalent.*

This answers a question which appears in [Papas–95]. If the groups $\Gamma_1, \Gamma_1', \Gamma_2, \Gamma_2'$ are non-amenable, it is a straightforward consequence of the previous proposition. The arguments for the general case are more involved, but of a similar nature [PapWh].

(iii) Here is another consequence of the proposition, pointed out by K. Whyte:

the sign of the Euler-Poincaré characteristic of a group is not an invariant of quasi-isometry.

This answers the question at the beginning of Section 3 in [Gromo–84], where Gromov observes that this characteristic can be expressed in terms of the L^2-cohomology of the group, and that the vanishing of this cohomology *is* a quasi-isometry invariant; the question is repeated in Item 8.A_4 of [Gromo–93]. Indeed, consider for example the three groups

$$\Gamma_+ = (F_3 \times F_3) * F_3$$
$$\Gamma_0 = (F_3 \times F_3) * F_4$$
$$\Gamma_- = (F_3 \times F_3) * F_5 .$$

As F_3, F_4, F_5 are subgroups of finite index in the same non-amenable group F_2, they are Lipschitz equivalent with each other; thus Γ_+, Γ_0, and Γ_- are also Lipschitz equivalent, and a fortiori quasi-isometric, with each other. Recall that the Euler-Poincaré characteristic $\chi(\Gamma)$ of a group Γ, when it is defined, is a rational integer (or, in a larger setting, a rational number [Serr–71]) which satisfies, among other properties:

$\chi(\Gamma_1 \times \Gamma_1) = \chi(\Gamma_1)\chi(\Gamma_2),$
$\chi(\Gamma_1 * \Gamma_2) = \chi(\Gamma_1) + \chi(\Gamma_2) - 1,$
$\chi(\Gamma_0) = [\Gamma : \Gamma_0]\chi(\Gamma)$ if Γ_0 is of finite index in Γ.

In particular,

$$\chi(\Gamma_+) = 1 , \quad \chi(\Gamma_0) = 0 , \quad \text{and} \quad \chi(\Gamma_-) = -1$$

and the three groups $\Gamma_+, \Gamma_0, \Gamma_-$ are not commensurable.

(iv) Similarly, for any $r \geq 2$, the group $(F_3 \times F_3) * F_r$ has a presentation with $6 + r$ generators and 9 relations, and this group is efficient. In particular one computes the deficiencies

$$def((F_3 \times F_3) * F_4) = 1,$$
$$def((F_3 \times F_3) * F_3) = 0,$$
$$def((F_3 \times F_3) * F_2) = -1.$$

so that the sign of the deficiency is not a quasi-isometric invariant. (See Item V.31 for the notions of deficiency and efficient group.)

(v) Using the same proposition, Papasoglu [Papas–95] has observed that the quotient of two L_2-Betti numbers is not invariant under quasi-isometry; this answers another question from Item 8.A_4 of [Gromo–93].

(vi) Consider again two uniformly discrete metric spaces with bounded geometry which are quasi-isometric, but assume now that the spaces are amenable; then they need not be Lipschitz equivalent. Indeed, there exists in the Euclidean plane $\mathbb{E}^2$ a subset L which is a *separated net* (this means that there exist two positive constants ϵ, C such that $d(x,y) \geq \epsilon$ for all $x, y \in L, x \neq y$ and $d(z, L) \leq C$ for all $z \in \mathbb{E}^2$) and which is *not* Lipschitz-equivalent to the lattice $\mathbb{Z}^2$; this carries over to $\mathbb{E}^n$ for any $n \geq 2$. See [BurKl–98] and [McMul–98].

(vii) **Open problem.** Let Γ be a finitely-generated infinite group which is amenable and let F be a finite group; when are Γ and $\Gamma \times F$ Lipschitz equivalent ? (Easy exercise: show they are if Γ is free abelian.)

More generally, does there exist a pair Γ_1, Γ_2 of two infinite finitely-generated groups which are quasi-isometric and not Lipschitz equivalent ?

47. Other examples of pairs of quasi-isometric groups which are not commensurable. (Much of (i) through (v) below is taken from [Whyte–99].) For each integer $g \geq 2$, let Γ_g denote the fundamental group of a closed orientable surface of genus g.

(i) Let $g, h \geq 2$. Recall from Example 25.i that Γ_g and Γ_h are commensurable, and in particular quasi-isometric. For $g \neq h$, check that Γ_g and Γ_h are not isomorphic. If Γ_g is isomorphic to a subgroup of Γ_h, show that $g - 1$ is an integral multiple of $h - 1$.

[Hint. Use coverings of surfaces and recall that, for finite coverings, the Euler characteristic is multiplicative.]

(ii) Check that $\Gamma_3 * \Gamma_2 * \Gamma_2$ is a subgroup of index 2 in $\Gamma_2 * \Gamma_2$. More generally, for $g \geq 3$ and $h \geq 2$, check that $\Gamma_g * \Gamma_h * \ldots * \Gamma_h$ (with $g - 1$ factors Γ_h) is a subgroup of index $g - 1$ in $\Gamma_2 * \Gamma_h$.

[Hint. Consider a space obtained by attaching $g - 1$ surfaces of genus h, each by some point, to a surface of genus g. Make it a finite covering of a space obtained by attaching a surface of genus h to a surface of genus 2.]

(iii) It follows from the previous complement and from (i) and (ii) above that the groups $\Gamma_{g_1} * \Gamma_{g_2} * \ldots * \Gamma_{g_r}$ and $\Gamma_{h_1} * \Gamma_{h_2} * \ldots * \Gamma_{h_s}$ are quasi-isometric (indeed Lipschitz equivalent) whenever $r, s \geq 2$ and $g_1, \ldots, g_r, h_1, \ldots, h_s \geq 2$.

(iv) Let Γ be a subgroup of $\Gamma_g * \Gamma_g$ of finite index, say m. Check that there are integers $g_1, \ldots, g_p \geq 2$ such that

$$\Gamma \approx \Gamma_{g_1} * \ldots * \Gamma_{g_p} * F_{m+1-p} \quad \text{and} \quad \sum_{i=1}^{p} (g_i - 1) = 2(g-1)m.$$

[Sketch. Let X be a space obtained from the disjoint union of two closed orientable surfaces Σ', Σ'' of genus g and an interval $I = [0, 1]$ by identifying $0 \in I$ with some point of Σ' and $1 \in I$ with some point of Σ''. Then Γ is the fundamental group of the total space Y of an m-sheeted covering $\pi : Y \to X$. The connected components of $\pi^{-1}(\Sigma')$ are orientable closed surfaces, say of genus $g_1, \ldots, g_{p'}$, and the connected components of $\pi^{-1}(\Sigma'')$ are orientable closed surfaces, say of genus $g_{p'+1}, \ldots, g_p$. Using multiplicativity of the Euler characteristic, we have

$$\sum_{i=1}^{p'} (g_i - 1) = \sum_{j=p'+1}^{p} (g_j - 1) = (g-1)m.$$

Let G_Y denote the graph obtained from Y by collapsing to a point each of these connected surfaces; then G_Y is a connected graph with p vertices and m edges (these being in bijection with the connected components of $\pi^{-1}(I)$), so that the fundamental group of G_Y is free on $m + 1 - p$ generators. The statement of (iv) is now a straightforward consequence of the Seifert – van Kampen theorem.

Check: we have $\chi(X) = 2(2 - 2g) - 1$; using $\sum_{i=1}^{p}(g_i - 1) = 2(g - 1)m$, we have also $\chi(Y) = \sum_{i=1}^{p}(2 - 2g_i) - m = -4gm + 3m$; thus $\chi(Y) = m\chi(X)$.]

(v) For two integers g, h with $h > g \geq 2$, show that $\Gamma_g * \Gamma_g$ and $\Gamma_h * \Gamma_h$ are not commensurable up to finite kernels.

[Sketch. By Proposition 28, it is enough to show that these groups are not commensurable. Now, if

$$\Gamma \approx \Gamma_{g_1} * \ldots * \Gamma_{g_p} * F_{m+1-p} \quad \text{and} \quad \Delta \approx \Gamma_{h_1} * \ldots * \Gamma_{h_q} * F_{n+1-q}$$

are respectively subgroups of finite index m and n of $\Gamma_g * \Gamma_g$ and $\Gamma_h * \Gamma_h$, then

(*) $$\sum_{i=1}^{p} \frac{g_i - 1}{g - 1} = 2m \quad \text{and} \quad \sum_{j=1}^{q} \frac{h_j - 1}{h - 1} = 2n.$$

If Γ and Δ were isomorphic, it would follow from the uniqueness part of Grushko's theorem (Complement III.13) that $p = q$, $m = n$ and that, possibly after renumbering, $(g_1, \ldots, g_p) = (h_1, \ldots, h_q)$. But this is incompatible with the equalities (*) if $g \neq h$.]

It is known that the groups $\Gamma = \Gamma_{g_1} * \Gamma_{g_2} * \ldots * \Gamma_{g_r}$ and $\Delta = \Gamma_{h_1} * \Gamma_{h_2} * \ldots * \Gamma_{h_s}$ are commensurable if and only if

$$\frac{\chi(\Gamma)}{r - 1} = \frac{2\sum_{i=1}^{r}(1 - g_i) - (r - 1)}{r - 1} = \frac{2\sum_{j=1}^{s}(1 - h_j) - (s - 1)}{s - 1} = \frac{\chi(\Delta)}{s - 1}$$

(independently due to K. Whyte, Theorem 1.8 of [Whyte–99], and D. Gaboriau, unpublished).

(vi) G. Levitt [Levit–95] and D. Gaboriau [Gabor–00] have introduced a new class of invariants for groups, for their actions on probability spaces, and more generally for appropriate equivalence relations. In particular, the *cost* of a countable group Γ is an extended real number $\mathcal{C}(\Gamma) \in [0, \infty]$. Here are some of its properties:

- $\mathcal{C}(\Gamma) < 1$ if and only if Γ is finite, and then $\mathcal{C}(\Gamma) = 1 - \frac{1}{|\Gamma|}$,
- $\mathcal{C}(\Gamma) = 1$ if Γ is infinite and amenable,
- $\mathcal{C}(\Gamma_0) - 1 = [\Gamma : \Gamma_0](\mathcal{C}(\Gamma) - 1)$ if Γ_0 is of finite index in Γ,
- $\mathcal{C}(\Gamma_1 \times \Gamma_2) = 1$ for infinite groups Γ_1, Γ_2 if at least one of them contains an element of infinite order,
- $\mathcal{C}(\Gamma_1 * \Gamma_2) = \mathcal{C}(\Gamma_1) + \mathcal{C}(\Gamma_2)$ under appropriate conditions (called "fixed price") on Γ_1, Γ_2,
- in particular $\mathcal{C}(F_n) = n$ for a free group F_n of rank n,
- $\mathcal{C}(\Gamma_g) = 2g - 1$ for $\Gamma_g = \pi_1$(orientable closed surface of genus g),
- for any $c \in [1, \infty]$, there exists a finitely-generated group of cost c.

Moreover, for a group Γ having an Euler-Poincaré characteristic $\chi(\Gamma)$ as in [Serr–71], for example for groups containing subgroups of finite index which are fundamental groups of finite complexes with contractible universal coverings, the quotient $\chi(\Gamma)/(1-\mathcal{C}(\Gamma))$ is an invariant of commensurability (Corollary VI-32 in [Gabor–00]). This can be illustrated in several ways, as we now indicate (we are grateful to Damien Gaboriau for correspondence on this material).

(vii) Quasi-isometric finitely-generated groups with equal cost and different Euler-Poincaré characteristics are not commensurable. This applies to the family of groups defined by

$$\Gamma_{p,q,r,s} = (F_{p+1} \times F_{q+1}) * (F_{r+1} \times F_{s+1})$$

where p, q, r, s are integers ≥ 1. These groups all have cost 2, and we have

$$\frac{\chi(\Gamma_{p,q,r,s})}{1 - \mathcal{C}(\Gamma_{p,q,r,s})} = \frac{(-p)(-q) + (-r)(-s) - 1}{-1} = 1 - pq - rs.$$

(viii) Similarly, quasi-isometric finitely-generated groups which have equal Euler-Poincaré characteristics and different costs are not commensurable. Here is an example, shown to me by Gaboriau. One first chooses integers $p, q \geq 1$ and $n \geq 2$, and then sets

$$\begin{aligned}
\Lambda_1 &= F_{p+1} \times F_{q+1} \\
\Lambda_2 &= F_{pq} \\
\Lambda &= \Lambda_1 * \Lambda_2 \\
\Lambda_0 &= \text{some subgroup of index } n \text{ in } \Lambda.
\end{aligned}$$

By straightforward applications of results of [Gabor–00], one obtains

$$\begin{aligned}
\mathcal{C}(\Lambda_1) &= 1 & \chi(\Lambda_1) &= pq \\
\mathcal{C}(\Lambda_2) &= pq & \chi(\Lambda_2) &= 1 - pq \\
\mathcal{C}(\Lambda) &= 1 + pq & \chi(\Lambda) &= 0 \\
\mathcal{C}(\Lambda_0) &= 1 + npq & \chi(\Lambda_0) &= 0.
\end{aligned}$$

One now chooses a finitely-generated group Δ with "fixed price" and which has an Euler-Poincaré characteristic $\chi(\Delta) \neq 1$, for example a non-abelian free group or the fundamental group of an orientable closed surface of genus at least two; then

$$\begin{aligned}
\mathcal{C}(\Lambda * \Delta) &= 1 + pq + \mathcal{C}(\Delta) \neq \mathcal{C}(\Lambda_0 * \Delta) = 1 + npq + \mathcal{C}(\Delta) \\
\chi(\Lambda * \Delta) &= \chi(\Delta) - 1 = \chi(\Lambda_0 * \Delta)
\end{aligned}$$

so that the quasi-isometric groups $\Lambda * \Delta$ and $\Lambda_0 * \Delta$ are not commensurable.

48. Example à la Milnor of a pair of quasi-isometric groups which are not commensurable. Choose an integer $g \geq 2$; denote by Σ_g a closed surface of genus g and by $T_1\Sigma_g$ the space of its tangent vectors of length 1. Thus $T_1\Sigma_g$ is the total space of a circle bundle $\mathbb{S}^1 \hookrightarrow T_1\Sigma_g \longrightarrow \Sigma_g$, and one has a corresponding short exact sequence

$$(*) \qquad 0 \longrightarrow \mathbb{Z} \longrightarrow \pi_1(T_1\Sigma_g) \longrightarrow \pi_1(\Sigma_g) \longrightarrow 1$$

of fundamental groups. As Σ_g is orientable, i.e. as there exist coherent orientations on the fibres of the circle bundle, this short exact sequence is a *central extension*.

The groups $\pi_1(T_1\Sigma_g)$ and $\mathbb{Z} \times \pi_1(\Sigma_g)$ provide our last example.

Proposition (Milnor et al.). *(i) The universal covering of $PSL(2,\mathbb{R})$ is quasi-isometric to the direct product of the real line and the hyperbolic plane.*

(ii) The groups $\pi_1(T_1\Sigma_g)$ and $\mathbb{Z} \times \pi_1(\Sigma_g)$ are quasi-isometric.

(iii) The groups $\pi_1(T_1\Sigma_g)$ and $\mathbb{Z} \times \pi_1(\Sigma_g)$ are not commensurable.

(iv) The groups $\pi_1(T_1\Sigma_g)$ and $\mathbb{Z} \times \pi_1(\Sigma_g)$ are not commensurable up to finite kernels.

Comments on proofs. (i) Let $G = PSL(2,\mathbb{R})$; denote by $\tilde{G}$ its universal covering and by

$$p : \tilde{G} \longrightarrow G$$

the canonical projection. Recall from Example 21.v that $\tilde{G}$ and G are metric spaces, in a way which is well-defined up to quasi-isometry.

The group G itself acts naturally on the total space T_1H^2 of the unit tangent bundle of the hyperbolic plane H^2; as this action is simply transitive, one may identify G with T_1H^2. As the fibers $\mathbb{S}^1$ of the canonical projection $T_1H^2 \longrightarrow H^2$ are compact of constant diameter, this projection is clearly a quasi-isometry. Thus G and H^2 are quasi-isometric for obvious reasons.

G is homeomorphic to a direct product of the circle $PSO(2)$ and the contractible space

$$\left\{ \begin{pmatrix} a & b \\ 0 & a^{-1} \end{pmatrix} \mid a \in \mathbb{R}_1^* , \ b \in \mathbb{R} \right\}$$

(the proof uses Gram-Schmidt orthogonalization). The kernel of p, which is the fundamental group of G, is infinite cyclic; and $p^{-1}(PSO(2))$ is a one-parameter subgroup of $\tilde{G}$ isomorphic to $\mathbb{R}$. But one does need an argument to show that $\tilde{G}$ and $G \times \mathbb{R}$ are quasi-isometric. One strategy is to look for a mapping

$$\Phi : \tilde{G} \longrightarrow \mathbb{R}$$

such that

$$(p,\Phi) : \tilde{G} \longrightarrow G \times \mathbb{R}$$

is a quasi-isometry. Such a mapping appears in Lemma 3 of [Miln–58], and this is why we attribute Claim (i) to Milnor; the claim appears later in several places: as an observation of G. Mess in [Ghys–90], as an example of S. Gersten in Item 8.A_4 of [Gromo–93] or as Corollary 3.8 in [Gers–92a], and as Theorem 8.3 in [AloBr–95], among other places.

Here is one way of defining Φ. There is a natural action of $\tilde{G}$ on the affine real line $\mathbb{R}$ which is compatible with the action of G on the circle $\mathbb{S}^1$ (= projective line) and with the covering map $\mathbb{R} \longrightarrow \mathbb{S}^1$. An appropriate generator T of $Ker(p)$ acts on $\mathbb{R}$ as $x \longmapsto x+1$. Now the *Poincaré translation number* provides a mapping $\Phi : \tilde{G} \longrightarrow \mathbb{R}$ such that

$$\Phi(gT) = \Phi(g) + 1 \quad \text{for all} \quad g \in \tilde{G},$$

$$\Phi(gh) - \Phi(g)\Phi(h) \quad \text{is bounded on} \quad \tilde{G} \times \tilde{G},$$

$$\Phi(g^n) = n\Phi(g) \quad \text{for all} \quad g \in \tilde{G}, n \in \mathbb{Z}$$

(indeed the translation number is defined for mappings $\mathbb{R} \longrightarrow \mathbb{R}$ much more general than those in $\tilde{G}$); mappings which satisfy the second condition are called *quasi-morphisms*, and play an important role in bounded cohomology. For the Poincaré translation (and rotation) number, see for example n° II.2 in [Herma–79]; the original paper is [Poinc–85]. The first and fundamental remark of [BarGh–92] is that Φ is *uniquely defined by the three conditions above*; see also [Ghys–99].

(ii) As $\pi_1(T_1\Sigma_g)$ is a discrete subgroup of $\tilde{G}$ with compact quotient, it is quasi-isometric to $\tilde{G}$. As $\pi_1(\Sigma_g)$ is quasi-isometric to H^2, the direct product $\mathbb{Z} \times \pi_1(\Sigma_g)$ is quasi-isometric to $\mathbb{R} \times G$. Claim (ii) of the proposition follows.

(iii) Given two groups A, B and a finite index subgroup Δ of the direct product $\Gamma = A \times B$, there is a finite index subgroup Γ_0 of Δ of the form $A_0 \times B_0$ for A_0 [respectively B_0] a finite index subgroup of A [respectively B]. (One may set $A_0 = A \cap \Gamma$ and $B_0 = B \cap \Gamma$.)

In particular any finite index subgroup of $\mathbb{Z} \times \pi_1(\Sigma_g)$ contains a finite index subgroup of the form $\mathbb{Z} \times \pi_1(\Sigma_h)$ with Σ_h a finite covering of Σ_g. [It can be shown that any finite index subgroup of $\mathbb{Z} \times \pi_1(\Sigma_g)$ is itself a direct product, but this is of no use here.]

Observe that the vector space $H^1(\mathbb{Z} \times \pi_1(\Sigma_h), \mathbb{R})$ of all homomorphisms from $\mathbb{Z} \times \pi_1(\Sigma_h)$ to $\mathbb{R}$ splits as a direct sum

$$H^1(\mathbb{Z}, \mathbb{R}) \oplus H^1(\pi_1(\Sigma_h), \mathbb{R}) \approx \mathbb{R} \oplus \mathbb{R}^{2h}$$

and is consequently of odd dimension.

Claim (iii) follows from this observation and from the fact that $H^1(\Delta, \mathbb{R})$ is of even dimension for any subgroup Δ of finite index in $\pi_1(T_1\Sigma_g)$, as we proceed to explain.

Consider first a presentation of $\pi_1(\Sigma_g)$ with $2g$ generators $a_1, b_1, \ldots, a_g, b_g$ and one relation $a_1b_1a_1^{-1}b_1^{-1} \ldots a_gb_ga_g^{-1}b_g^{-1}$ (see Example V.46). Choose elements $\tilde{a}_1, \tilde{b}_1, \ldots, \tilde{a}_g, \tilde{b}_g \in \pi_1(T_1\Sigma_g)$ which project to $a_1, b_1, \ldots, a_g, b_g \in \pi_1(\Sigma_g)$ respectively, and set

$$e = \tilde{a}_1\tilde{b}_1\tilde{a}_1^{-1}\tilde{b}_1^{-1} \ldots \tilde{a}_g\tilde{b}_g\tilde{a}_g^{-1}\tilde{b}_g^{-1},$$

which is an element of the kernel $\mathbb{Z}$ of (*). It is well known that $|e| = 2g - 2$, and that e is the *Euler class* of the extension (*), or equivalently of the (oriented!) tangent bundle of Σ_g; in particular $e \neq 0$. It follows that any homomorphism from $\pi_1(T_1\Sigma_g)$ to $\mathbb{R}$ vanishes on the centre $\mathbb{Z}$, so that

$$Hom\,(\pi_1(T_1\Sigma_g), \mathbb{R}) \approx Hom\,(\pi_1(\Sigma_g), \mathbb{R}) \approx \mathbb{R}^{2g}$$

is of even dimension.

Consider more generally a subgroup Δ of finite index in $\pi_1(T_1\Sigma_g)$. Its image in $\pi_1(\Sigma_g)$ is a subgroup of finite index, thus a subgroup of the form $\pi_1(\Sigma_h)$ for some $h \geq g$, and fits into a central extension

(**) $$0 \longrightarrow \mathbb{Z} \longrightarrow \Delta \longrightarrow \pi_1(\Sigma_h) \longrightarrow 1$$

which is the pull-back of (*). As the corresponding covering $\Sigma_h \longrightarrow \Sigma_g$ is of finite degree, the induced mapping in cohomology

$$\mathbb{Z} \approx H^2\,(\pi_1(\Sigma_g), \mathbb{Z}) \longrightarrow H^2\,(\pi_1(\Sigma_h), \mathbb{Z}) \approx \mathbb{Z}$$

is just the multiplication by this finite degree. It follows that the Euler class of the extension (**), which is a non-zero multiple of $2g - 2$, is again non-zero. The same argument as above shows that $H^1(\Delta, \mathbb{R})$ is a vector space of even dimension.

[See the appendix by D. Toledo in [Gers–92a] for a related discussion. See also page 466 of [Scott–83], where it is shown that the group $\tilde{G}$ does not contain any subgroup isomorphic to $\pi_1(\Sigma_h)$.]

(iv) This claim is a consequence of Proposition 28.i and of the following proposition. $\square$

Proposition (Toledo, Millson, Gersten). *With notation of the previous proposition, the groups $\pi_1(T_1\Sigma_g)$ and $\mathbb{Z} \times \pi_1(\Sigma_g)$ are linear.*

On the proof. The claim for $\mathbb{Z} \times \pi_1(\Sigma_g)$ is straightforward. The proof of the linearity of $\pi_1(T_1\Sigma_g)$ has the following story. Around 1991, Domingo Toledo showed this using Lubotzky's criterion for linearity [Lubo–88]. A few weeks later, John Millson provided a faithful homomorphism $\pi_1(T_1\Sigma_g) \longrightarrow SL(2, \mathbb{R}) \times Heis(2g+1)$ in terms of theta functions, where $Heis(2g+1)$ denotes the Heisenberg group which is a connected nilpotent Lie group with abelianized group $\mathbb{R}^{2g}$ and with one-dimensional centre. Then Steve Gersten came up with the following argument. As far as I know, none of these proofs is published.

One shows first that $\pi_1(T_1\Sigma_g)$ has a subgroup $\pi_1(E(1))$ of finite index (in fact index $2g - 2$) which has a presentation with $2g$ generators $\tilde{a}_1, \tilde{b}_1, \ldots, \tilde{a}_g, \tilde{b}_g$ and with relations to the effect that the product of commutators

$$\tilde{a}_1 \tilde{b}_1 \tilde{a}_1^{-1} \tilde{b}_1^{-1} \ldots \tilde{a}_g \tilde{b}_g \tilde{a}_g^{-1} \tilde{b}_g^{-1}$$

is central. Let $Heis(3, \mathbb{Z})$ denote the "discrete Heisenberg group" of Example 8, with presentation $\langle s, t \mid t^{-1}s^{-1}ts$ is central$\rangle$. Consider the homomorphism

$$(p, \phi) : \pi_1(E(1)) \longrightarrow SL(2, \mathbb{R}) \times Heis(3, \mathbb{Z})$$

where p is now the composition of the canonical projection $\pi_1(E(1)) \longrightarrow \pi_1(\Sigma_g)$ with an embedding in $SL(2, \mathbb{R})$, and where ϕ is defined by $\phi(\tilde{a}_j) = s$, $\phi(\tilde{b}_j) = t$ for all $j \in \{1, \ldots, n\}$. It is easy to check that the homomorphism (p, ϕ) is injective.

Thus $\pi_1(E(1))$ is linear, so that $\pi_1(T_1\Sigma_g)$ is also linear. $\square$

[More generally, denote for each integer $k \geq 0$ by $E(k)$ the circle bundle on Σ_g with Euler class k, so that $T_1\Sigma_g = E(2g - 2)$. Then $\pi_1(E(k))$ has a presentation with $2g + 1$ generators $\tilde{a}_1, \tilde{b}_1, \ldots, \tilde{a}_g, \tilde{b}_g, c$ and with relations

c is central,
$$\tilde{a}_1\tilde{b}_1\tilde{a}_1^{-1}\tilde{b}_1^{-1} \ldots \tilde{a}_g\tilde{b}_g\tilde{a}_g^{-1}\tilde{b}_g^{-1} = c^k.$$
Also $\pi_1(E(1))$ is a subgroup of index k in $\pi_1(E(k))$ for any $k \geq 1$.]

Here is a research problem from D. Toledo: *Let G be a non-linear connected real Lie group and let Γ be a lattice in G. Can one show that Γ is linear if and only if it is residually finite ?*

49. On quasi-convex subgroups. Let Γ be a finitely-generated group given together with a finite set of generators and with the corresponding word metric $d : \Gamma \times \Gamma \longrightarrow \mathbb{N}$. According to Item 26, a subgroup Γ_0 of Γ is *quasi-convex* if there exists a constant $R \geq 0$ such that any geodesic in Γ between two elements of Γ_0 is in the R-neighbourhood of Γ_0.

A subgroup Γ_0 of a finitely-generated group Γ can be quasi-convex with respect to one system of generators and not with respect to another. For example, the cyclic subgroup generated by $(1, 1)$ in $\mathbb{Z}^2$ is quasi-convex with respect to $\{(1, 0), (0, 1), (1, 1)\}$ but not with respect to $\{(1, 0), (0, 1)\}$. However, inside a hyperbolic group, quasi-convexity of subgroups is a notion independent of any choice of generating set (see Lemma 7.3.A in [Gromo–87]). It is an easy exercise to verify the following examples:

• finite subgroups are quasi-convex,
• subgroups of finite index are quasi-convex,
• if Γ_0, Γ_1 are finitely generated (with finite generating sets S_0, S_1 resp.),
  then Γ_0 is quasi-convex in $\Gamma_0 * \Gamma_1$ (with $S_0 \cup S_1$ as finite generating set).

We have also

• quasi-convex subgroups are finitely generated,
• a finitely-generated free group is *locally quasi-convex*
  (i.e. any finitely-generated subgroup of such a free group is quasi-convex),
• the fundamental group of a closed surface of genus at least 2
  is locally quasi-convex,
• if Γ_0, Γ_1 are two quasi-convex subgroups of a finitely generated group Γ,
  the intersection $\Gamma_0 \cap \Gamma_1$ is also quasi-convex in Γ.

This can be used, for example, to show the following (generalization of a) theorem of Howson's:

in a free group or in the fundamental group of a closed surface of genus $g \geq 2$, the intersection of two finitely-generated groups is again finitely generated.

See for example [Short–91], [Pitte–93], [Arzha–98], and [Kapov–99].

Infinite cyclic subgroups are always quasi-convex inside hyperbolic groups, but not necessarily inside arbitrary groups (see Remark 3.vii). More on quasi-convex subgroups of hyperbolic groups in [Arzha].

Let us indicate two examples of subgroups of hyperbolic groups which are finitely generated and not quasi-convex.

(i) Consider a 3-manifold M which is a bundle over the circle with fibers closed surfaces of genus at least 2, and which has a Riemannian metric of constant curvature -1 (for the existence of such manifolds, due to Jorgensen, see [Sulli–81]). The fundamental group Γ of M is a semi-direct product of a surface group Γ_0 with $\mathbb{Z}$, and the subgroup Γ_0 is *not* quasi-convex in the hyperbolic group Γ; see [Pitte–93].

(ii) Given a finitely-presented group Δ, Rips [Rips–82] has constructed a short exact sequence

$$1 \longrightarrow N \longrightarrow \Gamma \overset{\pi}{\longrightarrow} \Delta \longrightarrow 1$$

with Γ hyperbolic and with N generated by 2 elements ("generated" as a group, not just as a normal subgroup of Γ). If Δ has a subgroup Δ_0 which is finitely generated and not finitely presentable, $\Gamma_0 = \pi^{-1}(\Delta_0)$ is also finitely generated and not finitely presentable; it follows that Γ_0 is a finitely-generated subgroup of the hyperbolic group Γ which is not quasi-convex. (This example was shown to me by G. Arzhantseva.)

On non-quasi-convex subgroups of hyperbolic groups, see also Proposition 3.9, by Lustig and Mihalik, in [Shor8–91], as well as [MihTo–94].

Quasi-convexity as described here is a notion concerning a *pair* $\Gamma_0 < \Gamma$ of groups. This should not be confused with Cannon's notion of *almost convexity*, which refers to the geometry of balls inside *one* finitely-generated group given together with a finite generating set (see [Canno–87], [Cann4–89], and [MilSh–98]).

50. Geometric properties. A property $(\mathcal{P})$ of finitely generated groups is *geometric* if, for a pair (Γ_1, Γ_2) of finitely-generated groups which are quasi-isometric, Γ_1 has Property $(\mathcal{P})$ if and only if Γ_2 has Property $(\mathcal{P})$. Examples of geometric properties include the following.

- Γ is finite (obvious).
- Γ has an infinite cyclic subgroup of finite index, i.e. Γ has two ends. (See for example Exercise 16 in Chapter 1 of [GhyHp–90].)
- The number of ends. In particular, for torsion-free finitely-generated groups, being a free product is invariant by quasi-isometries, by a result of Stallings [Stall–68] (see also [Dunwo–69]).
- Γ has a free subgroup of finite index. (One observes first that having infinitely many ends is a geometric property; see Theorem 19 in Chapter 7 of [GhyHp–90]; see also [Woess–89].)
- Γ has a free abelian subgroup of finite index (Complement 40).
- Γ has a nilpotent subgroup of finite index ([Gromo–81] and Section VII.C).
- Γ is finitely presented (Proposition V.4).

- There exists an Eilenberg-MacLane space $K(\Gamma, 1)$ of which the n-skeleton is a finite simplicial complex — this gives one geometric property for each $n \geq 2$ (see Item 1.C_2' in [Gromo–93] or the previous property for $n = 2$). Moreover, for Γ satisfying this property, for an integer n and a commutative ring R with unit, the cohomology group $H^n(\Gamma, R\Gamma)$ is also invariant by quasi-isometries [Gers–93].
- Γ is hyperbolic (Item V.55).
- Γ has a growth function in some equivalence class (Proposition VI.27); in particular Γ has polynomial growth, or intermediate growth, or exponential growth (Section VI.C).
- Γ is amenable [Følne–55].
- Γ is accessible (an immediate corollary of [ThoWo–93]).

The list could be extended considerably (see [Gromo–93], as well as Complement 42 above).

Among properties of a group Γ which are *not* geometric, but are nevertheless invariant by commensurability, let us list the following.

- Being commensurable with a solvable group (Item 44).
- The sign of the Euler-Poincaré characteristic (Complement 46).
- The finite dimensionality of the second space $H_b^2(\Gamma; E)$ of bounded cohomology for every finite-dimensional module E (this is an unpublished result due to the authors of [BurMn–99]).
- Being commensurable with a just infinite group (see Remark 29.iv).
- Being commensurable with a simple group.
- Being commensurable with a group which has a non-abelian free quotient.

For examples showing that these two last properties are not invariant by quasi-isometry, we refer to recent work of Burger and Mozes already mentioned in Item 9.

There are also standard open problems. For example, the following properties are not known to be geometric:

- Γ has a polycyclic subgroup of finite index,
- Γ has Kazhdan Property (T),
- Γ has uniformly exponential growth (see Item VII.20).

In connection with the first of these problems, it is known that two polycyclic groups which are quasi-isometric have the same Hirsch number (Theorem 5.3 in [BriGe–96]). The *Hirsch number* of a polycyclic group is the number of infinite cyclic quotients in a subnormal series with cyclic quotients for the group.

CHAPTER V

FINITELY-PRESENTED GROUPS

The first section of this chapter introduces *finitely-presented* groups, and the second explains an important theorem of Poincaré which provides presentations for discrete groups of isometries of spaces of constant curvature (in particular of hyperbolic spaces). The third section, about fundamental groups of Riemannian manifolds, is a preparation for Section D, which contains some comments on Gromov's hyperbolic groups.

V.A. Finitely-presented groups

1. Definition. A *presentation* of a group Γ is a pair consisting of

a free group F on a basis $(s_j)_{j \in J}$ and a homomorphism π of F *onto* Γ,

a family $(r_\iota)_{\iota \in I}$ in F which generates $Ker(\pi)$ as a normal subgroup

(thus $Ker(\pi)$ is the group generated by the $wr_\iota w^{-1}$'s for $\iota \in I$ and $w \in F$). One denotes such a presentation by

$$\left\langle\ (s_j)_{j \in J}\ \middle|\ (r_\iota)_{\iota \in I}\ \right\rangle$$

or often by

$$\Gamma\ =\ \left\langle\ (s_j)_{j \in J}\ \middle|\ (\ r_\iota\ =\ 1\ )_{\iota \in I}\ \right\rangle.$$

A *finite presentation* is one for which the sets I and J are finite; a group is *finitely presented* if it has a finite presentation.

2. Proposition. *Let* $\left\langle\ (s_j)_{j \in J}\ \middle|\ (r_\iota)_{\iota \in I}\ \right\rangle$ *be a presentation (not necessarily finite) of a finitely-presented group* Γ. *Then there exist a finite subset* J_0 *of* J *and a finite family* $(\tilde{r}_i)_{1 \leq i \leq m}$ *of elements of the free group over* J_0 *such that* $\left\langle\ (s_j)_{j \in J_0}\ \middle|\ (\tilde{r}_i)_{1 \leq i \leq m}\ \right\rangle$ *is a finite presentation of* Γ.

Proof. Let B be a finite subset of Γ which generates Γ. Since $(s_j)_{j \in J}$ generates Γ, each $b \in B$ is a finite word, say w_b, in the s_j 's. Let J_0 be the subset of J of those indices j such that s_j appears in at least one of the w_b 's. Then J_0 is finite and $(s_j)_{j \in J_0}$ generates Γ. From now on, we write $\{s_1, \ldots, s_n\}$ for $(s_j)_{j \in J_0}$.

For each $j \in J$, we can write s_j as a word in $s_1, \ldots, s_n$ and their inverses. By substitution, for each $\iota \in I$, we obtain from r_ι a new word $\tilde{r}_\iota$ written with $s_1, \ldots, s_n$ and their inverses. Thus

$$\left\langle\ s_1\ ,\ \ldots\ ,\ s_n\ \middle|\ (\tilde{r}_\iota)_{\iota \in I}\ \right\rangle$$

is clearly a presentation of Γ.

By hypothesis, we also have a *finite* presentation

$$\Big\langle\ b_1\ ,\ \dots\ ,\ b_q\ \Big|\ a_1\ ,\ \dots\ ,\ a_p\ \Big\rangle$$

of the same group.

For each $l \in \{1,\dots,q\}$, we can write b_l as a word in $s_1,\dots,s_n$ and their inverses. For each $k \in \{1,\dots,p\}$, we can write a_k as a word in $b_1,\dots,b_q$ and their inverses, hence after substitution as another word in $s_1,\dots,s_n$ and their inverses; since a_k represents 1, it is also a word v_k in the conjugates of the $\tilde{r}_\iota$'s and their inverses. Let I_0 be the subset of I of those indices ι such that $\tilde{r}_\iota$ appears in one of the v_k 's. Then I_0 is clearly finite, and we write $I_0 = \{1,\dots,m\}$.

Now, for each $\iota \in I \setminus I_0$, we can write $\tilde{r}_\iota$ both as a word in the conjugates of the a_k 's and their inverses, and also (using the v_k 's) as a word in the conjugates of $\tilde{r}_1,\dots,\tilde{r}_m$ and their inverses. This shows that the relation $\tilde{r}_\iota$ is redundant for each $\iota \in I \setminus I_0$. $\square$

This is why there is a notion of *finite presentability* of a finitely-generated group Γ which makes sense without any reference to a particular surjection of some free group onto Γ.

The number of isomorphism classes of finitely-presented groups is clearly countable. In particular, "most" finitely-generated groups are *not* finitely-presented (see Section III.B).

There are several interesting ways of sharpening the last observation. For example, it is known that the direct product $F_2 \times F_2$ of two free groups of rank two contains continuously many non-isomorphic subgroups, but that any finitely-presented subgroup of $F_2 \times F_2$ is a finite extension of a direct product of two free groups of finite rank (see [BauRo–84] and [BriWi–99]).

There has been some work on finding presentations of groups which are "large" in many senses, in particular of uncountable groups: these include groups like $SL(2,\mathbb{Q}_p)$, $GL(2,k[T])$, groups with Tits systems (see § II.1 in [Serr–77]), Kac-Moody groups [KacPe–85], or Cremona groups [Wrigh–92], to quote only these. But presentations in the present book will be for *countable groups* only.

3. Geometric reformulation of finite presentability. Let Γ be a group given together with a free group F on a finite set S and a surjective homomorphism $\pi : F \to \Gamma$; let $Cay(\Gamma, S)$ be the corresponding Cayley graph. Observe that the fundamental group of $Cay(\Gamma, S)$ is free (because fundamental groups of graphs are free), most often of infinite rank. For a real number $r > 0$, denote by $\mathcal{L}(r)$ the subset of $L \doteq \Pi_1\big(Cay(\Gamma, S), 1\big)$ represented by loops of the form $\phi\psi\phi^{-1}$, where ϕ is a path in $Cay(\Gamma, S)$ from 1 to some vertex γ, and where ψ is a closed loop, starting and ending at γ, of length at most r.

Then Γ is finitely presented if and only if $\mathcal{L}(r)$ generates L for r large enough, namely if and only if the Rips complex $Rips_r(\Gamma, S)$ of Exercise IV.37 is simply-connected for r large enough.

4. Proposition. *A finitely-generated group which is quasi-isometric to a finitely-presented group is itself finitely presented.*

Proof. This is a consequence of the reformulation above and of Exercise IV.37. □

5. Corollary. *(i) Let Γ be a group and let Γ' be a subgroup of finite index in Γ. Then Γ' is finitely presented if and only if Γ is finitely presented.*

(ii) Let $1 \to \Gamma' \to \Gamma \to \Gamma'' \to 1$ be a short exact sequence of groups, with Γ' finite. Then Γ is finitely presented if and only if Γ'' is finitely presented.

The so-called Reidemeister-Schreier method provides an actual presentation of a given subgroup of a finitely-presented group. The classical exposition is probably that of Section 2.3 in [MagKS–66]; for an exposition which is shorter, but does not conceal all the relevant geometry, see e.g. n° 4.3.8 in [Still–80].

The two claims of the corollary are parallel to the first two claims of Corollary IV.24. For finite presentability of groups appearing in IV.24.iii and IV.24.v, see Example 7 and Complement 20 below, respectively.

6. Example: finite groups. Any finite group is finitely presented, for example because the multiplication table of a finite group Γ provides a presentation of this group with one generator for each $\gamma \in \Gamma$ and one relation for each $(\gamma_1, \gamma_2) \in \Gamma \times \Gamma$.

This illustrates a limiting case of Proposition 4, because a finite group is quasi-isometric to the group with one element. It also illustrates the following fact: in a free group of finite rank, any subgroup of finite index is again a free group of finite rank (see Problems II.17 and II.20).

In particular, there are numerous presentations of the group with one element! The following example appears in Section 1.2 of [MagKS–66]:

$$\Gamma = \left\langle a, b, c \mid a^3 = b^3 = c^4 = 1 \,, ac = ca^{-1} \,, aba^{-1} = bcb^{-1} \right\rangle.$$

Indeed, we have successively

$$b^3 = 1 \quad \text{and} \quad ab^3a^{-1} = bc^3b^{-1} \quad \Longrightarrow \quad c^3 = 1$$

$$c^4 = 1 \quad \text{and} \quad c^3 = 1 \quad \Longrightarrow \quad c = 1$$

$$aba^{-1} = bcb^{-1} \quad \text{and} \quad c = 1 \quad \Longrightarrow \quad b = 1$$

$$a^3 = 1 \quad \text{and} \quad a = a^{-1} \quad \Longrightarrow \quad a = 1$$

so that $\Gamma = \{1\}$. For other presentations of the group with one element, see for example [BurMa–93] and [MilSc–99].

Here is another pair of examples, the "homophonic groups". Consider the group Γ_{English} given with a set of 26 generators $\{a, b, c, \ldots, x, y, z\}$ and with all relations $A = B$ where A and B are words with the same pronunciation in English; according to [MeSWZ–93], the group Γ_{English} and the analogous group Γ_{French} are both reduced to one element.

This raises problems of pronunciation, as observed by J. Wiegold in Math. Reviews 952:00027: the question whether *maid* = *made* is a relation or not depends on who pronounces these two words.

7. Example: fundamental groups of compact complexes and manifolds. The fundamental group of a connected finite CW-complex is finitely presented: this is a straightforward application of the Seifert – Van Kampen theorem. It follows that the fundamental group of a compact smooth (or PL) manifold is finitely presented. (See Chapter VII in [Masse–67] or § VI.5 in [Godbi–71].)

It is also true, but considerably more difficult to show, that the fundamental group of a compact topological manifold is finitely presented. (This is because such a manifold, in general *not* a finite CW-complex, can be shown to have the *homotopy type* of a finite CW-complex; see § III.4 in [KirSi–77]. I believe it is even true that such a manifold is *homeomorphic* to a finite CW-complex — but the attaching maps of the cells are of course in general not PL !)

Any finitely-presented group is the fundamental group of a compact smooth manifold of dimension 4 : see Complements 27 and 29 below.

Observe that higher homotopy groups of finite CW-complexes need not be finitely generated. For example, if X is the wedge of a circle and a 2-sphere, then $\pi_2(X)$ and $\pi_3(X)$ are both isomorphic to the torsion-free abelian group of infinite rank $\mathbb{Z}^{(\infty)}$.

Besides homotopy groups, there are other groups defined for manifolds for which proving finite presentation (or indeed finite generation) is a non-trivial problem. For example, the mapping class group of a closed orientable surface is finitely generated (a non-trivial result of Dehn [Dehn–38], see also [Licko–64]), and finitely presented (a more recent result of Hatcher and Thurston [HatTh–80]). Another example is that of the Torelli group, mentioned in Complement III.16.

Some analogues of mapping class groups for manifolds of higher dimensions are *not* finitely generated. (See [McCul–80], [McCul–84]; this is related to Complement 22.)

8. Early examples. Historically, the earliest known examples of presentations of infinite groups were finite presentations for groups related to the theory of functions of one complex variable, to number theory and to geometry (modular group and its subgroups, Fuchsian groups, knot groups, braid groups, ...); see [ChaMa–82]. It seems that it was B.H. Neumann who, in 1937, first pointed out the existence of finitely-generated groups which cannot be finitely presented [NeuBH–37]; see Section III.B, and in particular Proposition 30 there.

9. Example: $SL(n, \mathbb{Z})$. Consider the group $SL(n, \mathbb{Z})$ for some $n \geq 3$. For each pair (i, j) with $1 \leq i, j \leq n$ and $i \neq j$, let $x_{i,j}$ denote the "elementary matrix" in $SL(n, \mathbb{Z})$ which differs from the unit matrix by the single entry 1 in the ith row and j^{th} column. Then $SL(n, \mathbb{Z})$ has a presentation with generators

$(x_{i,j})_{1 \leq i,j \leq n, \, i \neq j}$ and with relations

$$x_{i,j} x_{k,l} x_{i,j}^{-1} x_{k,l}^{-1} = \begin{cases} x_{i,l} & \text{if } i \neq l \text{ and } j = k, \\ 1 & \text{if } i \neq l \text{ and } j \neq k \end{cases}$$

and

$$\left(x_{1,2} x_{2,1}^{-1} x_{1,2} \right)^4 = 1.$$

This is due to Nielsen and Magnus; for the proof, we refer to [Miln–71]. For a very nice exposition of "some consequences of the elementary relations in SL_n", see [Stein–85].

Let X denote the space of n-by-n real symmetric matrices of determinant 1 with all eigenvalues > 0. The transitive action

$$\begin{cases} PSL(n, \mathbb{R}) \times X & \longrightarrow & X \\ (g, x) & \longmapsto & gx(^t g) \end{cases}$$

shows that X is a Riemannian symmetric space of non-compact type, associated with the Riemannian pair $\left(PSL(n, \mathbb{R}), PSO(n) \right)$; in particular X is a contractible non-positively curved Riemannian manifold of dimension $\frac{n(n+1)}{2} - 1$. The quotient $PSL(n, \mathbb{Z}) \setminus X$ is *not* compact, but X has a deformation retract Y of codimension $n - 1$ (equal to the real rank of X) on which $PSL(n, \mathbb{Z})$ acts properly with compact quotient; see among others [Ash–77], [Soulé–78], and [Bavar]. By the end of Item IV.23, this provides another proof (which could apply to other groups) that $PSL(n, \mathbb{Z})$ is finitely presented; this shows more, for example that the homological dimension of a torsion-free subgroup of finite index in $PSL(n, \mathbb{Z})$ is exactly $\frac{n(n-1)}{2}$.

10. Cayley topology. Given an integer $k \geq 1$, there are several spaces which describe in some sense the set of isomorphism classes of k-generated groups.

As any k-generated group Γ is the quotient of the free group F_k of rank k by a normal subgroup, one of these is the space $\mathcal{N}(k)$ of all normal subgroups of F_k. Observe that the group $Aut(F_k)$ of automorphisms of F_k has a natural action on $\mathcal{N}(k)$, that the group of inner automorphisms acts trivially, and consequently that $\mathcal{N}(k)$ has a natural action of $Out(F_k) = Aut(F_k)/F_k$.

For $N_1, N_2 \in \mathcal{N}(k)$, set

$$v(N_1, N_2) = \max \left\{ n \in \mathbb{N} \mid N_1 \cap B_n = N_2 \cap B_n \right\},$$

where B_n denotes the ball of radius n about the identity in F_k for the usual word length on F_k. It is straightforward to check that

$$d(N_1, N_2) = exp\left(- v(N_1, N_2) \right)$$

defines a metric (indeed an ultrametric) on the space $\mathcal{N}(k)$. The corresponding topology is called the *Cayley topology*. This topology, and similar topologies on closely related spaces, have been used in [Grigo–83], [Grigo–84], [Champ], and [Stepi–96]. Here are a few elementary properties of this topology.

The space $\mathcal{N}(k)$ is separable. Finitely-*presented* groups on k generators define a countable dense subset of $\mathcal{N}(k)$. Moreover, *if $\Gamma = F_k/N$ is finitely presented, there exists a neighbourhood of N in $\mathcal{N}(k)$ in which all points N' define groups $\Gamma' = F_k/N'$ which are quotients of Γ.* Indeed, if Γ has a presentation with the generators of F_k and a finite number of relations, all of length at most n (say), then any quotient F_k/N' with $N' \cap B_n = N \cap B_n$ is clearly a quotient of Γ.

In particular, N *is an isolated point in $\mathcal{N}(k)$ if the group $\Gamma = F_k/N$ is finite.* There are other isolated points, for example normal subgroups of F_k corresponding to finitely-*presented* groups on k generators which are simple.

The space $\mathcal{N}(k)$ is compact. (This is essentially Proposition 6.1 of [Grigo–84].) Indeed, let $(N_j)_{j\geq 1}$ be a sequence of normal subgroups of F_k. As the balls B_n are all finite, one can use a diagonal argument to extract a subsequence $(N_{j_i})_{i\geq 1}$ such that $N_{j_i} \cap B_n$ is eventually independent of i, for all $n \geq 1$. We set $N[n] = \lim_{i\to\infty} (N_{j_i} \cap B_n)$ for $n \geq 1$; observe that $N[n] \subset N[n+1]$. Finally we set $N = \cup_{n\geq 1} N[n]$. It is obvious that N is a normal subgroup of F_k which is adherent to the sequence $(N_j)_{j\geq 1}$. It follows that $\mathcal{N}(k)$ is compact.

Let $\mathcal{FN}(k)$ be the subspace of elements N for which the group Γ is finite, and let $\overline{\mathcal{FN}(k)}$ denote its closure. Consider also an element N in $\mathcal{N}(k)$, and set $\Gamma = F_k/N$. Then:

if Γ is residually finite, then $F_k/N \in \overline{\mathcal{FN}(k)}$,

if Γ is finitely presented and if $F_k/N \in \overline{\mathcal{FN}(k)}$, then Γ is residually finite.

See Complement III.18 for the notion of residual finiteness and, if necessary, Proposition 1 of [Stepi–96] for a proof of these two facts.

Exercises for V.A

11. Exercise on finite presentations. Let Γ be a group which has a presentation with n generators and m relations, and which is also generated by some set T of l generators. Show that Γ has a presentation with set of generators T and k relations, with $k \leq m + l$.

[This is Lemma 8 in [NeuBH–37].]

12. Exercise on presentations with positive relations. Show that any finitely-presented group

$$\Gamma = \langle s_1, \ldots, s_n \mid r_1, \ldots, r_m \rangle$$

has a finite presentation

$$\langle s_0, s_1, \ldots, s_n \mid r_0', r_1', \ldots, r_m' \rangle$$

such that each r_j' is written with $s_0, \ldots, s_n$ (and without $s_0^{-1}, \ldots, s_n^{-1}$).

[Hint: set $r_0' = s_0 s_1 \ldots s_n$. This trick is due to Dyck (1882); see [ChaMa–82], page 182.]

13. Exercise on Cayley 2-complexes. Let $\Gamma = \langle s_1, \ldots, s_n | r_1, \ldots r_m \rangle$ be a finitely-presented group. Define an oriented finite CW-complex $\mathcal{C}$ of dimension 2 as follows:

the 0-skeleton of $\mathcal{C}$ is a point,
the 1-skeleton $\mathcal{C}_1$ of $\mathcal{C}$ is a wedge of n oriented circles,
the complex $\mathcal{C}$ is obtained from $\mathcal{C}_1$ by attaching one 2-cell for each relation.

More precisely, if $r_i = s_{j_1}^{\epsilon_1} \ldots s_{j_p}^{\epsilon_p}$, the attaching map of the 2-cell corresponding to r_i goes first around the j_1^{th} circle of the wedge (positively if $\epsilon_1 = +1$ and negatively if $\epsilon_1 = -1$), then around the j_2^{th} circle, and so on.

It is a consequence of the Seifert – Van Kampen theorem that Γ is the fundamental group of $\mathcal{C}$. The universal covering $Cay_2(s_\bullet, r_\bullet)$ of $\mathcal{C}$ is the *Cayley 2-complex* of the given presentation.

(i) Check that the 1-skeleton of $Cay_2(s_\bullet, r_\bullet)$ is the Cayley graph of Γ with respect to the generating set $\{s_1, \ldots, s_n\}$.

(ii) Determine the Cayley 2-complex for the presentation

$$\mathbb{Z}^2 = \langle\, s_1, s_2 \mid s_1^{-1} s_2^{-1} s_1 s_2 = 1 \,\rangle.$$

(iii) Determine the Cayley 2-complex for the presentation

$$\pi_1 \begin{pmatrix} \text{closed surface} \\ \text{of genus } g \geq 2 \end{pmatrix} = \langle\, a_1, b_1, \ldots, a_g, b_g \mid a_1^{-1} b_1^{-1} a_1 b_1 \ldots a_g^{-1} b_g^{-1} a_g b_g = 1 \,\rangle.$$

(iv) Observe that the Cayley complexes in (ii) and (iii) are contractible. It is a result of Lyndon [Lyndo–50] that this holds, more generally, if $\Gamma = \langle s_1, \ldots, s_n \mid r \rangle$ is a one-relator group which is torsion-free, namely a one-relator group such that the relator r is not a proper power in the free group on $\{s_1, \ldots, s_n\}$.

14. Exercise on presentations with short relations. Show that any finitely-presented group has a finite presentation where the relations are all of length ≤ 3. About presentations of this kind, see Complement III.13.

[Hint: subdivide the Cayley 2-complex to obtain a simplicial complex.]

15. Exercise on finite presentability of extensions. Show that finite presentability is stable by extensions.

More precisely, consider a finitely-presented group B given together with a homomorphism ρ from a free group $F(\{b_1, \ldots, b_m\})$ of rank m onto B and a finite set $v_1, \ldots, v_q$ generating $Ker(\rho)$ as a normal subgroup, as well as a finitely-presented group C given together with a homomorphism σ from a free group $F(\{c_1, \ldots, c_n\})$ of rank n onto C and a finite set $w_1, \ldots, w_r$ generating $Ker(\sigma)$ as a normal subgroup. Consider also a short exact sequence

$$1 \longrightarrow B \overset{\phi}{\longrightarrow} A \overset{\psi}{\longrightarrow} C \longrightarrow 1$$

of groups.

Let S denote the disjoint union of the bases $\{b_1, \ldots, b_m\}$ and $\{c_1, \ldots, c_n\}$. For each $j \in \{1, \ldots, n\}$, choose an element $c_j' \in A$ such that $\psi(c_j') = \sigma(c_j)$. Define a homomorphism π from the free group on S to A by mapping b_i to $\phi(\rho(b_i))$ for any $i \in \{1, \ldots, m\}$ and c_j to c_j' for any $j \in \{1, \ldots, n\}$.

Check first that π is onto, so that in particular the group A is finitely generated.

Then show that A is finitely presented.

[Hint: write $q + r + mn$ relations, one for each v_i, one for each w_j, and one for each conjugate $c_j'^{-1} b_i c_j'$. See for example Chapter 10 of [Johns–90] for the details.]

16. Exercise on the "universal covering" of the modular group. Show that the three presentations

$$\Gamma_1 = \langle\, x, y, z \mid x = yzy^{-1} \,,\, y = zxz^{-1} \,,\, z = xyx^{-1} \,\rangle$$
$$\Gamma_2 = \langle\, x, y \mid xyx = yxy \,\rangle$$
$$\Gamma_3 = \langle\, a, b \mid a^3 = b^2 \,\rangle$$

define isomorphic groups.

[Hint: if $a = xy$ and $b = xyx$, then $b^2 = (ax)\left(x^{-1}a^2\right) = a^3$.]

As the Lie group $SL(2, \mathbb{R})$ has the homotopy type of $SO(2)$, its universal covering group fits into a central short exact sequence

$$1 \longrightarrow \mathbb{Z} \longrightarrow \widetilde{SL}(2, \mathbb{R}) \xrightarrow{\pi} SL(2, \mathbb{R}) \longrightarrow 1 \,.$$

Check that $\pi^{-1}\big(SL(2, \mathbb{Z})\big)$ is isomorphic to the group Γ_3, and that one obtains a central short exact sequence

$$1 \longrightarrow \mathbb{Z} \longrightarrow \Gamma_3 \longrightarrow SL(2, \mathbb{Z}) \longrightarrow 1$$

by restriction of the exact sequence above. (See also complement III.20.)

Standard names for "the" group appearing in this exercise are

the braid group on three strings,

the Steinberg group $St(2, \mathbb{Z})$,

the fundamental group of the complement in a 3-sphere of a trefoil knot.

The isomorphism $\Gamma_1 \approx \Gamma_2$ is related to a well-known presentation of the group of a knot given by a plane diagram: associate one generator to each segment of the diagram and one appropriate relation to each crossing; it is then standard that each relation is a consequence of the others (see Example 3.7 in [BurZi–85]). There is a neat geometric explanation of the isomorphism of $St(2, \mathbb{Z})$ with the group of the trefoil knot at the end of § 10 in [Miln–71].

17. Exercise on long words representing the identity. Say that a finite presentation $\Gamma = \langle S \mid R \rangle$ has "Property (ALW)" if there are arbitrarily long words w in $S \cup S^{-1}$ representing $1 \in \Gamma$ such that no non-empty subword of w represents $1 \in \Gamma$. Show that Property (ALW)

holds for $\mathbb{Z}^2 = \langle s, t \mid sts^{-1}t^{-1} = 1 \rangle$,

holds for the presentation of Γ_g given in Example 46 below,

does not hold for a free presentation of a free group,

does not hold if Γ is a finite group.

[Incidentally, the proof of Lemma 1 in [MulSc–83] requires a correction.]

18. Exercise for anyone who has ever dropped a stitch. Consider the group defined by the presentation

$$\Gamma = \langle\, b_0, b_1, b_2, \ldots \mid b_1 b_0 b_1^{-1} = b_2 b_1 b_2^{-1} = b_3 b_2 b_3^{-1} = \ldots \,\rangle$$

with infinitely many generators and infinitely many relations.

(i) For any finite subset F of the integers, check that $(b_i)_{i \in \mathbb{N}, i \notin F}$ generates Γ.

(ii) Check that there exists a homomorphism of Γ onto $\mathbb{Z}$, so that in particular Γ is infinite.

(iii) Check that Γ is not abelian.

[Hint. Define a homomorphism π from Γ onto the permutation group of $\{1, 2, 3\}$ by $\pi(b_i) = (1, 2)$ if i is even and $\pi(b_i) = (2, 3)$ if i is odd.]

I suspect that Γ is not finitely generated, but I haven't seen or found a proof[1].

The group Γ is the fundamental group of the complement in $\mathbb{R}^3$ of a wild knot which should be easy to figure out for "anyone who has ever dropped a stitch" [Fox–49].

Problems and complements for V.A

19. On nilpotent, polycyclic, and solvable groups. Show that a *nilpotent group* which is finitely generated is necessarily finitely presented. [Hint: this follows from the analogous property for abelian groups.] More generally, a finitely-generated *polycyclic group* is finitely presented (this is stated in [HallP– 54] as a "no doubt well-known result").

There exist finitely-generated *solvable groups* which are not finitely presented, such as the *wreath product* $\mathbb{Z} \wr \mathbb{Z}$ of Example III.5. (More on this in, e.g., Chapter 2 of [Streb–84].) Recall that the wreath product $G \wr H$ of two groups G and H is the semi-direct product $B \rtimes H$ where B is a direct sum of copies of G indexed by H and where H acts on B by permutation of coordinates.

20. A theorem on lattices in Lie groups. *Any lattice Γ in a connected real Lie group G is finitely presented.*

---

[1] Added in proof (February 2000): Anna Dioubina has shown me a proof.

Recall that a *lattice* in G is a discrete subgroup Γ such that the quotient space $\Gamma \setminus G$ has a G-invariant Borel probability measure.

This result, simple to state, has no simple proof. I am grateful to M. Burger for explanations on the following points.

(i) If $\Gamma \setminus G$ is compact (and thus a compact *manifold*), the lattice is finitely presented. (See Example 7 above, as well as Theorem 6.15 in [Raghu–72]; see also (vii) below.) One may thus assume from now on that $\Gamma \setminus G$ is not compact, or in other words that the lattice Γ is *not uniform*.

(ii) Consider first a discrete subgroup Γ in a product $G = G_1 \times \ldots \times G_k$ of non-compact simple real Lie groups with trivial centres, and make the crucial assumption that Γ is *torsion-free*.

If K denotes a maximal compact subgroup of G, it follows from the crucial assumption that $M \doteq \Gamma \setminus G/K$ is an analytic manifold which has a Riemannian structure of curvature ≤ 0. It is a theorem that such a manifold is diffeomorphic to the interior of a compact manifold C with boundary, so that in particular $\Gamma \approx \Pi_1(M) \approx \Pi_1(C)$ is finitely presented; the complement in M of the interior of C consists of points with small injectivity radius and, with an argument of Morse theory, the theorem shows that deleting these points does not change the diffeomorphism type of M. The central ingredient of the proof is the "Margulis Lemma" recalled below. The analyticity hypothesis cannot be removed in general (it can if the curvature κ satisfies $-1 \leq \kappa \leq -\alpha^2 < 0$ for some constant α).

For this important step, see [BalGS–85], in particular Lemma 7.11 (for the use of the analyticity hypothesis), Theorem 8.3 (the Margulis Lemma), Remark 8.4, and Theorem 13.1 (the very result quoted above).

The Margulis lemma. *Given an integer $n \geq 2$, there exist constants $\mu(n) > 0$ and $I(n) \geq 1$ such that the following holds.*

Let M be a Hadamard manifold (i.e. a connected simply-connected complete Riemannian manifold of non-positive sectional curvature) which satisfies the curvature condition $-1 \leq \kappa \leq 0$. Let Γ be a discrete group of isometries of M ("discrete" means that, for any compact subset C of M, the number of group elements γ such that $\gamma C \cap C \neq \emptyset$ is finite).

Let x be a point of M and let $\Gamma_{\mu(n)}(x)$ denote the subgroup of Γ generated by those $\gamma \in \Gamma$ for which $d(\gamma, \gamma x) \leq \mu(n)$, where d denotes the Riemannian distance on M.

Then $\Gamma_{\mu(n)}(x)$ is virtually nilpotent. More precisely, $\Gamma_{\mu(n)}(x)$ contains a nilpotent subgroup with index bounded by $I(n)$.

(iii) Let Γ be a discrete subgroup of a product G of simple groups, as in (ii). As G has a faithful linear representation (for example the adjoint representation), one may view Γ as a countable subgroup of $GL(N, \mathbb{C})$ for some integer $N \geq 1$.

Now make the *important assumption* that Γ is *finitely generated*. A result of Selberg shows that a finitely-generated subgroup of $GL(N, \mathbb{C})$ has a subgroup of finite index which is torsion-free. (See[2] [Selbe–60], or Theorem 4.1 of Chapter 5 in [Casse–86]; for a stronger result, see § 17 in [Borel–69].) Thus Γ is finitely presented for the same reason as in (ii).

(iv) The next theorem to quote is that, for a product $G = G_1 \times \ldots \times G_k$ of simple groups as in (ii), a discrete subgroup Γ of G *which is a lattice* is finitely generated.

In case the real rank of G_j is at least 2 for each $j \in \{1, \ldots, k\}$, the group G and thus also the lattice Γ have the Property (T) of Kazhdan (see [Kazhd–67], [HarVa–89]). In particular, Γ is finitely generated. (Quoting Kazhdan's Property (T) is not only smart, but also *unavoidable*. It is true that, on the one hand, there are proofs of finite generation for arithmetic groups, and that, on the other hand, all groups which occur are arithmetic; but known proofs of the arithmeticity theorem of Margulis do require the a priori knowledge that the groups are finitely generated.)

In case at least one G_j has real rank 1, a delicate analysis due to Garland and Raghunathan again shows that Γ is finitely presented (see [Raghu–72], in particular Remark 13.21).

It would be interesting to see whether the argument alluded to in (ii) could be extended to the present case of the *orbifold* $M = \Gamma \setminus G/K$ (this is a research problem).

(v) It is now easy to show that any lattice in a connected real Lie group G which is semi-simple is finitely presented, because the centre Z of G and the compact factors of the decomposition of G/Z into simple groups (if any) are harmless.

(vi) Let Γ be a lattice in an arbitrary connected real Lie group G. Let R denote the solvable radical of G; let $G = LR$ be a Levi decomposition of G and let C be the maximal connected normal compact subgroup of the semi-simple part L; we have a short exact sequence

$$1 \longrightarrow CR \longrightarrow G \overset{\pi}{\longrightarrow} S \longrightarrow 1$$

and $S = L/C$ is semi-simple without compact factor.

Then $\Lambda \doteq \pi(\Gamma)$ is a lattice in S and $(CR) \cap \Gamma$ is a cocompact lattice in CR. This is due to L. Auslander and H.C. Wang. Part of the proof consists in showing that Λ has the so-called *Property (S)* of Selberg, for which we refer to the beginning of Chapter V in [Raghu–72], so that one may apply Theorem 3.4 in [Wang–63] to show that $\pi(\Lambda)$ is a lattice. See also Theorem 1.13 and Corollary 8.27[3] in [Raghu–72], as well as the clear discussion in Section 4 of

---

[2]This lemma, from a paper published in 1960, has a curious history. On one hand, it follows easily from the work of Mal'cev on finitely-generated linear groups [Mal'c–40]; this is well-known in the former USSR, and appears for example explicitly on pages 394–395 of [Merzl–87]. On the other hand, in the (comparatively) very particular case of Fuchsian groups, it seems to have been first conjectured by W. Fenchel, then proved by J. Nielsen and others ([BunNi–51], [Fox–52]), and then exposed again in [Menni–67].

[3]There is some discussion about a technical point in Chapter 8 of [Raghu–72]; see [Wu–88].

[Wang–72]. A later proof of the Auslander-Wang theorem is alluded to in the introduction of [Zimme–87].

We now have a short exact sequence

$$1 \longrightarrow (CR) \cap \Gamma \longrightarrow \Gamma \longrightarrow \Lambda \longrightarrow 1$$

and we know from (i) and (v) that $(CR) \cap \Gamma$ and Λ are finitely presented. It follows that Γ is finitely presented (Exercise V.15), and the "proof" of the theorem stated at the beginning of this complement is complete.

(In case Γ is a lattice in a *solvable* connected real Lie group, it is an older theorem of Mostow that Γ is co-compact, and in particular finitely presented; see Theorem 3.1 in [Raghu–72]. In the general case discussed above, it is not possible to replace the subgroup CR by the solvable radical R, as the following example shows: let G be the direct product of $\mathbb{R}$ and $SO(3)$, and let Γ be the infinite cyclic subgroup of G generated by $(1, \rho)$, where $\rho \in SO(3)$ is a rotation by some angle which is an irrational multiple of 2π; then Γ is a lattice in G, but its intersection with the solvable radical $R = \mathbb{R}$ of G is not a lattice in R because it is reduced to $\{1\}$.)

(vii) For a lattice Γ in a Lie group G over a field $\mathbb{Q}_p$ of p-adic numbers (or over any local field of characteristic zero which is neither $\mathbb{R}$ nor $\mathbb{C}$), the quotient $\Gamma \backslash G$ is necessarily compact by a 1965 result of Tamagawa (see Proposition IX.3.7 in [Margu–91]). Thus Γ is finitely generated since, as already stated in Item IV.24, *in a locally compact group which is compactly generated, any discrete subgroup with compact quotient is finitely generated* (see Item 0.40 in Chapter I of [Margu–91]).

As Γ is finitely generated, the result of Selberg quoted in (iii) above implies that Γ has a subgroup of finite index Γ_0 which is torsion-free. This Γ_0 acts on the building attached to G, which is a contractible space; the corresponding quotient is a finite simplicial complex, and an Eilenberg-McLane space with fundamental group Γ_0. It follows that Γ_0 is finitely presented, and consequently Γ is also finitely presented.

In the function field case $\mathbb{F}_p[[T]]$, there are Lie groups with lattices which are not finitely generated, as the next complement indicates.

More on this in Section IX.3 of [Margu–91] and in [Lubo–95d].

(viii) Going back to lattices in semi-simple groups which are "arithmetic" (a term already used in Item II.28 and in (iv) above, but with a definition which goes beyond these notes — see e.g. [A'CaB–94] or [Zimme–84]), the fact that they are finitely presented also follows from the so-called "reduction theory". This carries over to the larger class of "S-arithmetic groups"; see for example [Abels–86], and [PlaRa–91], Theorems 4.2 & 5.11.

21. Examples of lattices which are not finitely presentable. If q denotes a prime power and $\mathbb{F}_q$ the field of order q, the group $SL(3, \mathbb{F}_q[T])$ is not finitely presentable [Behr–79], though $SL(n, \mathbb{F}_q[T])$ is finitely presented for $n \geq 4$ [RehSo–76].

Compare with the fact already mentioned in III.4: $SL(2, \mathbb{F}_q[T])$ is not finitely generated though $SL(n, \mathbb{F}_q[T])$ is finitely generated for $n \geq 3$.

For each integer $n \geq 2$, define a group Γ to be *of type (F_n)* if Γ is finitely presented and if the trivial $\mathbb{Z}[\Gamma]$-module $\mathbb{Z}$ admits a projective resolution

$$\ldots \longrightarrow P_{n+1} \longrightarrow P_n \longrightarrow \ldots \longrightarrow P_1 \longrightarrow P_0 \longrightarrow \mathbb{Z} \longrightarrow 0$$

with P_k finitely generated for $k \leq n$. Then it is known that, for $q \geq 2^{n-2}$, the group $SL(n, \mathbb{F}_q[T])$ if of type (F_{n-2}) but not of type (F_{n-1}) [Abels–91].

22. Finite presentability of automorphism groups. It is a theorem of Nielsen that the group of automorphisms of a free group of finite rank is finitely presented [Niels–24]. The proof consists in showing that two bases of Γ are always related by a finite number of "Nielsen transformations"; see Section 3.5 in [MagKS–66]. (For a survey of automorphism groups of free groups and other groups, see [Roman–92].)

Nielsen's theorem extends to a much larger setting, as a consequence of work of Sela which is beyond the scope of this book. *Let Γ be a torsion-free hyperbolic group which is not a non-trivial free product* (see Complement III.13); *then the automorphism group of Γ is finitely presented* (see Theorem 1.9 in [Sela–97b]).

Let us also mention that a polycyclic group (and in particular a finitely-generated nilpotent group) has a group of automorphisms which is finitely presented (see [Ausla–69], as well as page 123 of [Segal–83]).

There are however examples of finitely-presented groups Γ with $Aut(\Gamma)$ not finitely generated ([Lewin–67], [McCul–80]).

23. Finite presentability and homology of groups. For a group Γ, there are *integral homology groups* $H_n(\Gamma, \mathbb{Z})$, $n = 0, 1, 2, \ldots$, which are abelian groups. One of several equivalent definitions is to set $H_n(\Gamma, \mathbb{Z}) = H_n(X, \mathbb{Z})$ for X a path-connected CW-complex with $\Pi_1(X) \approx \Gamma$ and $\Pi_i(X) = \{0\}$ for $i \geq 2$; such a space is called an *Eilenberg-MacLane space* $K(\Gamma, 1)$. (This definition of course presupposes a proof that such spaces exist, and a proof that their homology does not depend on any choice.) We shall retain the following properties of the integral homology groups of a group Γ.

(i) We have $H_0(\Gamma, \mathbb{Z}) = \mathbb{Z}$.

(ii) There exists a natural isomorphism of the first group $H_1(\Gamma, \mathbb{Z})$ with the abelianized group $\Gamma / [\Gamma, \Gamma]$. In particular, if Γ is finitely generated, so is $H_1(\Gamma, \mathbb{Z})$.

(iii) Given a short exact sequence

(*) $$1 \longrightarrow R \longrightarrow F \longrightarrow \Gamma \longrightarrow 1$$

with F free, there exists a natural isomorphism

$$H_2(\Gamma, \mathbb{Z}) \approx (R \cap [F, F]) / [R, F]$$

(a formula of H. Hopf). In particular, if Γ has a finite presentation (*) with F free on n generators and R generated as a normal subgroup of F by m

relators $r_1, \ldots, r_m$, then $H_2(\Gamma, \mathbb{Z})$ is finitely generated, and can be generated by m elements.

[For the last claim, it suffices to show that $R/[R, F]$ is generated by $r_1[R, F]$, $\ldots, r_m[R, F]$. But any element of R is the product of elements of the form xrx^{-1} with $x \in F$ and $r \in \{r_1^{\pm 1}, \ldots, r_m^{\pm 1}\}$; as $xrx^{-1} \equiv r \pmod{[R, F]}$, the claim is straightforward.]

More precisely, one can show that the minimal number of generators of $H_2(\Gamma, \mathbb{Z})$ is bounded by $m - (n - \dim_{\mathbb{Q}}(H_1(G, \mathbb{Z}) \otimes_{\mathbb{Z}} \mathbb{Q})$. See Complement V.31 below, and Lemma 1.2 in [Epste–61].

For an interpretation of $H_2(\Gamma, \mathbb{Z})$ in terms of "commutator relations", see [Mille–52].

(iv) For a free product with amalgamation $\Gamma = A *_C B$ (see Complement III.14), we have a long exact sequence (we write $H_n(\Gamma)$ for $H_n(\Gamma, \mathbb{Z})$)

$$\ldots \to H_2(A) \oplus H_2(B) \to H_2(\Gamma) \to H_1(C) \to H_1(A) \oplus H_1(B) \to H_1(\Gamma) \to 0$$

called the *Mayer-Vietoris sequence*. In particular, if A, B are finitely presented and C is not finitely generated, then Γ is finitely generated and is not finitely presentable.

For example, if C denotes the kernel of the homomorphisms from the free group F on two letters a, b to $\mathbb{Z}$ which sends a to 0 and b to 1, the group

$$F *_C F = \big\langle a, b, c, d \mid b^n ab^{-n} = d^n cd^{-n} \quad \text{for all} \quad n \geq 0 \big\rangle$$

is finitely generated and not finitely presentable. (The given presentation is of course not finite; but it requires some argument to show that this group has no finite presentation whatsoever, and the appropriate Mayer-Vietoris sequence provides one argument.)

For all this, see for example [Brown–82].

(v) Several strengthenings of finite presentability of a group Γ have been investigated. One is to require that higher homology groups are also finitely generated. For H_3, there is an example of Stallings showing that $H_3(\Gamma, \mathbb{Z})$ is *not necessarily* finitely generated for a finitely-generated group Γ [Stall–63]. For other finiteness properties, see Complement 21, Chapter VIII of [Brown–82], [BesBr–97], and [BuxGo].

(vi) Ordinary (co)-homology groups of a group are *isomorphism* invariants. There are other kinds of (co)-homology groups which are *quasi-isometry* invariants; one is the "coarse homology", for which we refer to [Roe–93] and [BloWe–97]. Here is a sample application.

For integers $a, b, c, d \geq 2$ with $\{a, b\} \neq \{c, d\}$, the groups $\mathbb{Z}^a * \mathbb{Z}^b$ and $\mathbb{Z}^c * \mathbb{Z}^d$ are not quasi-isometric. (This was suggested to me by Panos Papasoglu and later confirmed by Gabor Elek.)

24. Recursively-presented groups. Finitely-generated subgroups of finitely-presented groups have been characterized by G. Higman as the so-called *recursively-presented groups*. Arguments rely on an interesting mixture of group theory and mathematical logic; see Chapter 12 in [Rotma–95].

25. Universal finitely-presented group. A corollary of the previous result of Higman is that there exists a *universal finitely-presented group* $\mathcal{U}$, that is, a finitely-presented group which contains an isomorphic copy of every finitely-presented group as a subgroup (see Theorem 12.29 in [Rotma–95]).

This has the following interesting consequence (brought to my attention by S. Gersten). Set $\Gamma_1 = \mathcal{U} \times \mathbb{Z}$ and $\Gamma_2 = \mathcal{U} * \mathbb{Z}$. Then, clearly, Γ_1 is isomorphic to a subgroup of Γ_2 and Γ_2 is isomorphic to a subgroup of Γ_1. However Γ_1 and Γ_2 are *not* quasi-isometric, e.g. because Γ_1 has one end and Γ_2 has infinitely many.

26. On simple groups. In 1951, Graham Higman showed the existence of *finitely-generated* infinite simple groups; he asked whether there exist finitely-*presented* infinite simple groups (see [Higma–51], or n° I.1.4 in [Serr–77]). In 1965, R.J. Thompson constructed such a group, which may be viewed as a homeomorphism group of the triadic Cantor set [McKTh–73]. Since 1965, many more finitely-presented infinite simple groups have been discovered; see e.g. [Higma–74] and papers by E.A. Scott and K.S. Brown in [BauMi–92], as well as [CanFP–96], [Brin–96], [Brin–99], and [Buril–99] on Thompson's group(s!).

Burger and Mozes have recently constructed finitely-presented simple groups which are torsion-free (see [Mozes–98], [BurMz–a], and [BurMz–b]). These groups are free amalgams of the form $F *_G F$ with F and G free groups of finite ranks and $[F : G] < \infty$.

There are several results and conjectures involving simple groups and decision problems (such as the three Dehn's problems: the word problem, the conjugacy problem, and the isomorphism problem). We will only mention the following statements.

- *Any countable group can be embedded in a 2-generator simple group.* See Corollary 3.10 in [Mille–92], as well as our Item III.16.i.
- *A recursively-presented (in particular a finitely-presented) simple group has soluble word problem.* See Theorem 6.2 in [Mille–92].
- *A finitely-generated group Γ has a soluble word problem if and only if there exist a simple group Γ_1 and a finitely-presented group Γ_2 such that $\Gamma < \Gamma_1 < \Gamma_2$* ([BooHi–74], Theorem 6.4 in [Mille–92], and Theorem 12.30 in [Rotma–95]).
- The previous statement holds with Γ_1 simple *and finitely generated* [Thomp–80].
- *Conjecture:* A finitely-generated group Γ can be embedded in a finitely-presented simple group if and only if Γ has soluble word problem (see [Scott–92]).

27. Finitely-presented groups as fundamental groups of closed 4-manifolds, first way. (i) Consider the curve C in $\mathbb{R}^5$ parametrised by (t, t^2, t^3, t^4, t^5), with $t \in \mathbb{R}$. Show that 6 distinct points on C are never contained in one affine hyperplane of $\mathbb{R}^5$.

[Hint: use the fact that a polynomial in $\mathbb{R}[t]$ of degree 5 has at most 5 distinct real roots.]

(ii) Deduce from (i) that any 2-dimensional finite simplicial complex can be embedded in $\mathbb{R}^5$. More generally, similar arguments show that a finite simplicial d-complex embeds in $\mathbb{R}^{2d+1}$. (See Section 0.2.3 in [Still–80]; in n° 9.4.1, Stillwell quotes a 1912 paper by Dehn.)

(iii) With the notation of Exercise 13, show that a Cayley 2-complex $K = Cay_2(s_\bullet, r_\bullet)$ of a finitely-presented group Γ embeds in $\mathbb{R}^5$.
[Hint: subdivide, to obtain a simplicial complex, and apply (ii).]

(iv) We keep the same notation. Let $j : K \longrightarrow \mathbb{R}^5$ be an embedding and let M denote the boundary of a small neighbourhood N of the image of j; thus M is a smooth 4-manifold without boundary. Show that the fundamental group of M is isomorphic to Γ.
[Hint. Observe that N retracts by deformation onto $j(K)$, so that K and N have the same fundamental group. As $j(K)$ is of codimension 3 in N, the complement $N \setminus j(K)$ has the same fundamental group as N, by transversality. Observe finally that $N \setminus j(K)$ retracts by deformation onto M.]

This is one way of showing that any finitely-presented group is the fundamental group of some closed 4-manifold, and in particular of answering the first two questions of the following list, copied from the end of § 14 in the *Analysis situs* of Poincaré [Poinc–95]. The third question was answered by J.W. Alexander, who showed that two 3-dimensional "lens spaces" with the same fundamental group can be non-homeomorphic [Alexa–19].

"Il pourrait être intéressant de traiter les questions suivantes.

1° Étant donné un groupe G défini par un certain nombre d'équivalences fondamentales, peut-il donner naissance à une variété fermée à n dimensions ?

2° Comment doit-on s'y prendre pour former cette variété ?

3° Deux variétés d'un même nombre de dimensions, qui ont même groupe G, sont-elles toujours homéomorphes ?[4]

Ces questions exigeraient de difficiles études et de longs développements. Je n'en parlerai pas ici.

Je veux toutefois attirer l'attention sur un point. [...] un groupe G peut être beaucoup plus complexe qu'un autre groupe G' et cependant correspondre à une valeur plus petite du nombre P_1."

[The number P_1 is the first Betti number of an appropriate 4-manifold with fundamental group G, i.e. the dimension of the vector space $(G/[G,G]) \otimes_{\mathbb{Z}} \mathbb{Q}$.]

28. Free groups as fundamental groups of closed 4-manifolds. For each integer $n \geq 1$, the free group on n generators appears as the fundamental group of the compact 4-manifold which is the connected sum of n copies of $\mathbb{S}^1 \times \mathbb{S}^3$.

29. Finitely-presented groups as fundamental groups of closed 4-manifolds, second way. The construction sketched below was known in 1934.

---

[4]Was it not clear for Poincaré in 1895 that the 4-sphere and the complex projective plane are two simply-connected manifolds which are not homeomorphic ?

See the final problems in Section 52 of [SeiTr–34] (page 187 of the English translation [SeiTr–80]); alternatively, see the last Note in Chapter 4 of [Masse–67].

If Γ has a presentation with generators $s_1, \ldots, s_n$ and relations $r_1, \ldots, r_m$, consider first a 4-manifold with fundamental group free on n generators. For each $i \in \{1, \ldots, m\}$, represent the relator r_i by an embedded loop c_i, and consider a tubular neighbourhood C_i around this loop, with $C_1, \ldots, C_m$ pairwise disjoint. Observe that each C_i is homeomorphic to $\mathbb{S}^1 \times \mathbb{D}^3$ and that the boundary ∂C_i is homeomorphic to $\mathbb{S}^1 \times \mathbb{S}^2$. Replace each C_i by a copy of the simply-connected $\mathbb{D}^2 \times \mathbb{S}^2$.

30. Groups which are fundamental groups of closed 3-manifolds.

The group $\mathbb{Z}^4$ is *not* the fundamental group of any closed smooth manifold of dimension 3.

[Sketch. Let Γ be a finitely-presented group and let X be a connected finite CW-complex with fundamental group Γ. To obtain an Eilenberg-McLane space $K(\Gamma, 1)$ from X, one has to attach cells of dimension at least 3. It follows that the inclusion $X \hookrightarrow K(\Gamma, 1)$ induces homomorphisms $H_j(X, \mathbb{Z}) \to H_j(\Gamma, \mathbb{Z})$ on homology groups which are *onto* for $j \leq 2$.

Assume now that X is a smooth closed 3-manifold and that $\Gamma = \mathbb{Z}^4$. Then $H_1(X, \mathbb{Z})$ is $\mathbb{Z}^4$ by definition, and thus $H_2(X, \mathbb{Z})$ is isomorphic to $\mathbb{Z}^4$ by Poincaré's duality. We have also $H_2(\Gamma, \mathbb{Z}) = H_2(K(\Gamma, 1), \mathbb{Z}) = H_2(4\text{-torus}, \mathbb{Z}) = \mathbb{Z}^6$. This is absurd because $\mathbb{Z}^4$ does not surject onto $\mathbb{Z}^6$.]

There is a theorem of Reidemeister (1936) which shows that an abelian group Γ is the fundamental group of some closed 3-manifold if and only if it is in the following list: $\mathbb{Z}$, $\mathbb{Z}^3$, $\mathbb{Z} \oplus \mathbb{Z}/2\mathbb{Z}$, $\mathbb{Z}/r\mathbb{Z}$ for some $r \geq 2$ [Epste–61]. [Exercise: show that these groups do indeed occur.]

For the *solvable* groups which are fundamental groups of some closed 3-manifold, see [EvaMo–72]. In this paper, the authors also prove 'that a non-finitely-generated abelian group which appears as the fundamental group of some (non-compact!) 3-manifold is necessarily isomorphic to a subgroup of the additive group of the rationals.

There is a list of *finite* groups which are fundamental groups of closed 3-manifolds, due to H. Hopf (1925–26), for which we refer to [Miln–57]. I. Hambleton and R. Lee have recently shown that this list is complete.

31. Fundamental groups of 3-manifolds, coherence, and deficiency.

A finitely-generated group which is the fundamental group of a manifold of dimension 3 is necessarily finitely *presented*. Moreover it is a *coherent* group, i.e. all its finitely-generated subgroups are finitely presented [Scott–73]. (More on coherent groups in [Serr–73] and [Wise–98].)

The *deficiency* of a finite presentation $\langle s_1, \ldots, s_n \mid r_1, \ldots, r_m \rangle$ is the difference $n - m$; the *deficiency of a finitely-presented group* Γ is the maximum $def(\Gamma)$ of the deficiencies of all its finite presentations. Here are a few facts about deficiencies.

- The deficiency of the free group on n generators is n [Epste–61].
- The deficiency of a finite group is ≤ 0 (Theorem 8.16 in [Macdo–88]).
- The deficiency of a finitely-presented amenable group is ≤ 1; more precisely, a group of deficiency ≥ 2 contains a non-abelian free group ([BauSh–90], or [BauPr–78], or pages 82–83 in [Gromo–82]).
- Most lattices in connected Lie groups have deficiency ≤ 0 (see [Lott–99] for a precise statement).
- There exist finitely-presented groups of deficiency $-n$ for any $n \geq 0$ (see e.g. Theorem II in [Levin–78]).
- Deficiencies are *not* additive under free products (there is an example in Section 3.1 of [HogMe–93]).
- We have $def(\Gamma) \leq rank(H_1(\Gamma, \mathbb{Z})) - s(H_2(\Gamma, \mathbb{Z}))$, where H_* denotes the homology of Γ (Complement 23); for an abelian group A, the *rank* is the integer $dim_{\mathbb{Q}}(A \otimes_{\mathbb{Z}} \mathbb{Q})$, and $s(A)$ denotes the minimal number of generators for A.
  The group Γ is *efficient* if this inequality is an equality [Epste–61].
- There are infinite groups which are not efficient.
  (See the last remark in [HauWe–85]; see also Proposition 2.9 in [Lusti–93], [Lusti–95], and [BaiPr–97].)

The fundamental group Γ of a compact 3-manifold without boundary is efficient and satisfies $def(\Gamma) \geq 0$. (See Lemma 2.3 in [Epste–61]; see also Chapter IV of [BeyTa–82].) For applications of ℓ^2-homology to deficiencies of groups, see [Eckma].

3-manifold groups have other interesting property from the point of view of geometric group theory: see e.g. [Gers–94].

32. Open problem. Does there exist a smooth projective complex variety whose fundamental group is infinite and simple ? (See Section 1.2.1 in [ABCKT–96].)

33. Open problems. Anticipating Section V.C below, let us quote the following questions.

Fundamental open question. Let Γ be a finitely-generated group "of finite dimension" (i.e. here: which is the fundamental group of a finite-dimensional polyhedron with contractible universal cover). Does there exist a complete Riemannian manifold without boundary *with curvature $\kappa \leq 0$* and with fundamental group isomorphic to Γ ? (Item 0.5.C$_2$ in [Gromo–93]).

Another basic open question. How does one characterize the finitely-generated groups which admit a discrete cocompact action on some metric geodesic space of *strictly negative curvature* ? (Item 7.B in [Gromo–93]).

V.B. The Poincaré theorem on fundamental polygons

This theorem is a very powerful tool for exhibiting explicit presentations of important groups appearing in geometry. We discuss first the particular case of the theorem for groups generated by reflections only.

34. Hyperbolic polygons. In the hyperbolic plane H^2, consider a compact convex polygon P with $n \geq 3$ vertices $x_1, \ldots, x_n$ enumerated in cyclic order (indices are taken modulo n). Assume that, for $j = 1, \ldots, n$, the interior angle at x_j is of the form π/p_j for some integer $p_j \geq 2$. By the Gauss-Bonnet theorem and easy constructions, such a polygon exists if and only if

$$\sum_{j=1}^{n} \frac{1}{p_j} < n - 2.$$

Let σ_j denote the reflection of H^2 with respect to the side of P limited by x_j and x_{j+1}, and let Γ denote the group of isometries of H^2 generated by $\{\sigma_1, \ldots, \sigma_n\}$. We have the obvious relations

$$\sigma_j^2 = 1 \qquad\qquad j \in \{1, \ldots, n\}$$
$$(\sigma_{j-1}\sigma_j)^{p_j} = 1 \qquad\qquad j \in \{1, \ldots, n\}$$

(recall that 0 is identified with n).

35. Proposition. *The generators $\{\sigma_1, \ldots, \sigma_n\}$ and the $2n$ relations above constitute a presentation of Γ.*

Moreover, Γ acts properly on H^2 and P is a fundamental domain for this action.

Proof. (We follow Milnor [Miln–75]; see also [Ivers–92], and references after Theorem 40 below.) Let $\hat{\Gamma}$ be the abstract group defined by a presentation with generating set $\{\hat{\sigma}_1, \ldots, \hat{\sigma}_n\}$ and with relations

$$\hat{\sigma}_j^2 = 1 \qquad\qquad j \in \{1, \ldots, n\}$$
$$(\hat{\sigma}_{j-1}\hat{\sigma}_j)^{p_j} = 1 \qquad\qquad j \in \{1, \ldots, n\}.$$

Let $\pi : \hat{\Gamma} \to \Gamma$ denote the homomorphism such that $\pi(\hat{\sigma}_j) = \sigma_j$ for $j \in \{1, \ldots, n\}$. As π is onto by definition of Γ, we have to show that π is injective.

Form a cellular complex as follows. Start with the product $\hat{\Gamma} \times P$ consisting of a union of disjoint polygonal cells $\hat{\gamma} \times P$. For each $\hat{\gamma} \in \hat{\Gamma}$ and for each $j \in \{1, \ldots, n\}$, paste the jth edge of $\hat{\gamma} \times P$ onto the jth edge of $\hat{\gamma}\hat{\sigma}_j \times P$; let K denote the cellular complex defined this way. Using the relations $\hat{\sigma}_j^2 = 1$, one sees that precisely two polygonal cells are pasted together along each edge of K. It is easy to see that K is connected and that there is a natural action of $\hat{\Gamma}$ on K.

Consider the canonical mapping $\tilde{D} : \hat{\Gamma} \times P \rightarrow H^2$ which sends $(\hat{\gamma}, x)$ to $\pi(\hat{\gamma})(x)$. There is an induced map $D : K \rightarrow H^2$, which is clearly equivariant for $\pi : \hat{\Gamma} \rightarrow \Gamma$.

Claim: D is a homeomorphism. For each vertex $(\hat{\gamma}, x_j)$ of K, we have identifications

$$(\hat{\gamma}, x_j) = (\hat{\gamma}\hat{\sigma}_{j-1}, x_j) = (\hat{\gamma}\hat{\sigma}_{j-1}\hat{\sigma}_j, x_j) = (\hat{\gamma}\hat{\sigma}_{j-1}\hat{\sigma}_j\hat{\sigma}_{j-1}, x_j) = \ldots$$
$$= (\hat{\gamma}\hat{\sigma}_{j-1}\hat{\sigma}_j \ldots \hat{\sigma}_{j-1}, x_j)$$

where the last term has p_j generators $\hat{\sigma}_{j-1}$ and $p_j - 1$ generators $\hat{\sigma}_j$. As $(\hat{\sigma}_{j-1}\hat{\sigma}_j)^{p_j} = 1$, there are precisely $2p_j$ polygonal cells of K which fit cyclically around $(\hat{\gamma}, x_j)$. (These are distinct; indeed $\hat{\sigma}_{j-1}, \hat{\sigma}_{j-1}\hat{\sigma}_j, \hat{\sigma}_{j-1}\hat{\sigma}_j\hat{\sigma}_{j-1}, \ldots$ are distinct in $\hat{\Gamma}$ because their images by π are distinct elements in Γ.) It follows first that the star neighbourhood of $(\hat{\gamma}, x_j)$, consisting of $2p_j$ polygonal cells, maps homeomorphically onto a neighbourhood of $D(\hat{\gamma}, x_j)$ in H^2, and second that the map D is a local homeomorphism about $(\hat{\gamma}, x_j)$.

It is also true that D is a local homeomorphism about points of the form $(\hat{\gamma}, x)$ with x a point which is either inside K or on an edge of K, but not a vertex.

Using the fact that K is tiled by copies of P, it is easy to check that K has a natural metric for which it is complete, and then that every path in H^2 can be lifted to a path in K. It follows that the map D is a covering. As H^2 is simply-connected, the map D is a homeomorphism, as claimed.

The proposition is a straightforward consequence of the claim. □

In the proof above, D is the so-called *developing map*. The proposition shows in particular that Γ is a *Coxeter group,* in the sense of [Bourb–68].

36. Example: hyperbolic triangular groups. Let a, b, c be integers such that

$$2 \leq a \leq b \leq c \qquad \text{and} \qquad \frac{1}{a} + \frac{1}{b} + \frac{1}{c} < 1.$$

The corresponding group

$$T^*_{a,b,c} = \langle\, r, s, t \mid r^2 = s^2 = t^2 = (st)^a = (tr)^b = (rs)^c = 1 \,\rangle$$

acts by isometries on the hyperbolic plane H^2 with fundamental domain a triangle with interior angles $\frac{\pi}{a}, \frac{\pi}{b}, \frac{\pi}{c}$, and therefore with area $\pi - \frac{\pi}{a} - \frac{\pi}{b} - \frac{\pi}{c}$. It is easy to show that the subgroup $T_{a,b,c}$ of $T^*_{a,b,c}$ of orientation-preserving isometries, which is of index 2 in $T^*_{a,b,c}$, has a presentation

$$T_{a,b,c} = \langle\, u, v \mid u^c = v^a = (uv)^b = 1 \,\rangle$$

(set $u = rs$ and $v = st$). It acts on H^2 with fundamental domain a pair of triangles as above, forming a quadrilateral of area $2\pi \left(1 - \frac{1}{a} - \frac{1}{b} - \frac{1}{c}\right)$. The $T_{a,b,c}$ are sometimes called the *von Dyck groups* (Section 9.4 in [Johns–90]).

The smallest area occurs for

$$(a, b, c) = (2, 3, 7) \quad \Longrightarrow \quad 2\pi \left(1 - \frac{1}{a} - \frac{1}{b} - \frac{1}{c} \right) = \frac{2\pi}{21}.$$

This is the minimum value for the area of H^2/Γ where Γ runs over all discrete subgroups of the group of orientation preserving isometries of H^2 (a group isomorphic to $PSL(2, \mathbb{R})$); see Theorem 2 in Section 3.4 of [Siege–71]. The group $T(2, 3, 7)$ is important in the theory of Riemann surfaces; for example, it contains a normal subgroup of index 168 which is the fundamental group of a surface of genus 3, and the quotient is isomorphic to *the* simple group $PSL(2, \mathbb{F}_7) \approx PSL(3, \mathbb{F}_2)$ of order 168 (see [Dieud–54]). It follows that there is a Riemann surface of genus $g = 3$, called the *Klein's quartic*, whose group of conformal automorphisms is precisely this group of order $168 = 84(g - 1)$; this is remarkable, because the group of conformal automorphisms of a Riemann surface of genus $g \geq 2$ is a finite group of order *at most* $84(g - 1)$ by a theorem of Hurwitz (Theorem 3 in Section 3.9 of [Siege–71]), and the Klein's quartic is the unique surface of genus 3 with a group of 168 automorphisms. (See Chapter II of [Magnu–74], [Bost–92], [Bavar–93], and references there.)

Triangular groups may be defined more generally for $a, b, c \in \{2, \ldots, \infty\}$ (see Item 39 below). We have for example $T_{2,3,\infty} \approx PSL(2, \mathbb{Z})$ and $T_{\infty,\infty,\infty} \approx \mathbb{Z} * \mathbb{Z}$.

37. Variation: spherical triangular groups. If $p_1, \ldots, p_n$ are integers, each at least 2, such that $\sum_{j=1}^{n} \frac{1}{p_j} > n - 2$, then $n \leq 3$. Possibly after reordering, the triple (p_1, p_2, p_3) is one of

$$(2, 3, 3), \qquad (2, 3, 4), \qquad (2, 3, 5), \qquad (2, 2, k)$$

for $k \geq 2$. Considerations similar to those above hold for groups of isometries of the 2-sphere generated by reflections in the sides of spherical triangles with angles $(\pi/p_1, \pi/p_2, \pi/p_3)$. One finds presentations of these finite groups, which are the symmetry groups of the regular polyhedrons (respectively tetrahedron, octahedron, icosahedron, dihedrons). For example, the group of isometries leaving invariant a regular icosahedron has a presentation with 3 generators $\sigma_1, \sigma_2, \sigma_3$ and 6 relations

$$\sigma_1^2 = \sigma_2^2 = \sigma_3^2 = (\sigma_2 \sigma_3)^2 = (\sigma_3 \sigma_1)^3 = (\sigma_1 \sigma_2)^5 = 1.$$

It is well known that this group is isomorphic to $\mathcal{A}_5 \times (\mathbb{Z}/2\mathbb{Z})$, where $\mathcal{A}_5$ denotes the alternating group on 5 letters. (See [Coxet–61], Section 15.8; much more on this in the monograph [Coxet–48].)

38. Variation: Euclidean triangular groups. Similarly, if $p_1, \ldots, p_n$ are integers, each at least 2, such that $\sum_{j=1}^{n} \frac{1}{p_j} = n - 2$, then either $n = 4$ and $p_1 = \ldots = p_4 = 2$, or $n = 3$ and (p_1, p_2, p_3) is one of

$$(2, 3, 6), \qquad (2, 4, 4), \qquad (3, 3, 3)$$

possibly after re-ordering. One obtains presentations for three familar groups of isometries of the Euclidean plane.

39. Cycle and cusp conditions. For a more general formulation of the Poincaré theorem, consider a convex region P of H^2 limited by a finite number of sides, these being now finite *or infinite* intervals in geodesics. There may be some vertices at infinity; we distinguish between *proper* vertices at infinity, where two sides end (such as x_1 in Figure 1) and *improper* vertices at infinity, where only one side ends (such as x_4 in Figure 1).

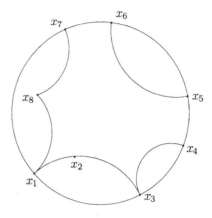

FIGURE 1. Proper and improper vertices at infinity

Consider also a map $e \mapsto e'$ from the set of sides of P onto itself which is involutory, namely which is such that $(e')' = e$, and such that e' and e are of the same hyperbolic length, for any side e. "Sides" are defined in such a way that there are 6 sides in the example of Figure 1. For each side e of P, let s_e be an isometry of H^2 which maps e onto e', and maps points *inside* P near e to points *outside* P near e'; we assume moreover that $s_{e'} = (s_e)^{-1}$.

Chains and cycles. Let x_1 be a vertex of P and let e_1 be a side of P ending at x_1. Denote by s_1 the isometry attached to e_1.

Set $x_2 = s_1(x_1)$. If there is only one side of P ending at x_2, stop there; otherwise, let e_2 be the side ending at x_2 which is *not* $s_1(e_1)$, and let s_2 denote the isometry attached to e_2. Set $x_3 = s_2(x_2)$, and go on iterating to obtain a *chain* which is either finite (with the last vertex an improper vertex at infinity) or infinite and periodic (because P has a finite number of vertices). In the periodic situation, the *period* corresponding to (x_1, e_1) is the least integer $p \geq 1$ such that $(x_{p+1}, e_{p+1}) = (x_1, e_1)$; the set $(x_1, x_2, \ldots, x_p)$ is called a *cycle* of vertices. The *cycle transformation* at x_1 is then defined to be the composition $s_p s_{p-1} \ldots s_1$; it fixes x_1 and can be shown to preserve the orientation of H^2 (Lemma VII.1.4 of [Ivers–92]). Observe that the cycle transformation at x_j is conjugate to that at x_1 for each $j \in \{1, \ldots, p\}$.

Cycle conditions. For each cycle $(x_1, x_2, \ldots, x_p)$ of vertices at finite distance, there is an integer $m \geq 1$ such that $m \sum_{j=1}^{p} \alpha_j = 2\pi$, where α_j is the interior angle of P at x_j.

Cusp conditions. For each cycle $(x_1, x_2, \ldots, x_p)$ of proper vertices at infinity, the cyclic transformation $s_p s_{p-1} \ldots s_1$ is parabolic.

Relations. Given data consisting of a polygon P, an involution $e \mapsto e'$ on the set of its sides and an assignment $e \mapsto s_e$ as above, *in such a way that the cycle and the cusp conditions hold,* define

the reflection relations: $\qquad (s_e)^2 = 1 \qquad$ whenever $\qquad e' = e$,

the cycle relations: $\qquad (s_p \ldots s_1)^m = 1 \qquad$ for any cycle $\qquad (x_1, \ldots, x_p)$

$\qquad\qquad\qquad\qquad\qquad\qquad\qquad\qquad$ with $\quad m \quad$ as above.

40. Theorem (Poincaré). *With the previous notation, let Γ be the group generated by the s_e 's. Then the reflection relations and the cycle relations constitute a presentation of Γ.*

Moreover, the action of Γ on H^2 is properly discontinuous, and P is a fundamental domain for this action.

Proposition 35 corresponds to the case of a compact polygon P with $e' = e$ for all sides e of P.

For the (delicate!) proof of this theorem and for various generalizations (such as hyperbolic spaces H^n of higher dimensions), we refer to the literature, in particular to [Poinc–82] for the original paper, to [Rham–71] and [Maski–71] for proofs of the theorem in dimension 2, to § 20 of Chapter 5 in [DouNF–82] and to [Beard–83] for general discussions about Fuchsian groups (one quite short, the other much more complete), to [EpsPe–94] for a proof in higher dimensions (this reference includes a discussion of the existing literature), to [Harpe–91] for some examples of reflection groups in higher dimensions, and to [FalZo–99] for *complex* hyperbolic spaces.

We would like to single out Iversen's book, whose central theme is the Poincaré theorem, for its clarity [Ivers–92].

41. Example. Let P be limited by two half-geodesics starting from a vertex x, with angle $2\pi/p$, and let the involution $e \mapsto e'$ be the identity. There is a unique cycle relation, and one obtains the presentation

$$\langle\ s_1, s_2 \mid s_1^2 = s_2^2 = (s_1 s_2)^p = 1\ \rangle$$

of the dihedral group of order $2p$.

42. Example. Let the polygon P have only improper vertices, as in Figure 2. Let n be the number of sides of P and let p be the number of sides e such that $e' = e$; we have $n = p + 2q$ for some integer $q \geq 0$. There are no cycle

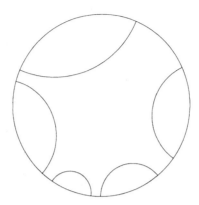

FIGURE 2.   A Poincaré polygon with improper vertices only

relations, and the group Γ of the Poincaré theorem is here a free product of p copies of $\mathbb{Z}/2\mathbb{Z}$ and of q copies of $\mathbb{Z}$.

43. Example. Let P be the hexagon of Figure 3. For each $j \in \{1, 2, 3\}$, the angle at x_j is $2\pi/p_j$ for some integer $p_j \geq 2$, and the angle at y_j is β_j. We assume that e'_j and e_j have the same length; we denote by s_j the rotation of angle $2\pi/p_j$ with fixed point x_j, mapping e'_j onto e_j. We assume moreover that $\beta_1 + \beta_2 + \beta_3 = 2\pi/m$ for some integer $m \geq 1$.

The group generated by the s_j 's has the presentation

$$\Gamma = \langle\, s_1\,,\ s_2\,,\ s_3 \mid s_1^{p_1} = s_2^{p_2} = s_1^{p_3} = (s_1 s_2 s_3)^m = 1\,\rangle\,.$$

Each vertex x_j gives rise to a cycle of period 1, and there is one cycle of period 3 which is (y_1, e'_2), (y_2, e'_3), (y_3, e'_1).

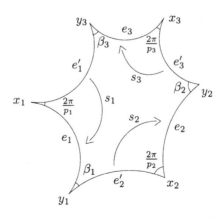

FIGURE 3.   Hexagon for Example 43

44. Example: Hecke groups. Consider an integer $q \geq 3$, the number $\lambda = 2\cos(2\pi/q)$, and the subgroup Γ_λ of $PSL(2,\mathbb{R})$ generated by

$$a_\lambda = \begin{bmatrix} 1 & \lambda \\ 0 & 1 \end{bmatrix} \quad \text{and} \quad j = \begin{bmatrix} 0 & 1 \\ -1 & 0 \end{bmatrix}$$

as in Section II.B. If P is as in the left-hand part of Figure 4, observe that a_λ maps the side (∞, x) onto the side (∞, y), and that j exchanges the sides (i, x) and (i, y). The vertices i and ∞ give rise to cycles of period 1, whereas x and y give rise to cycles of period 2. Poincaré's theorem shows that the Hecke group generated by a_λ and j has presentation

$$\Gamma_\lambda = \Big\langle\, a_\lambda, j \,\Big|\, (a_\lambda j)^q = j^2 = 1 \,\Big\rangle \approx \mathbb{Z}/q\mathbb{Z} * \mathbb{Z}/2\mathbb{Z}$$

as in Example II.28.

Consider now the polygon P' as in the right-hand part of Figure 4 (the dotted half-circle indicates the half-circle of the left-hand part); part of the boundary of P' lies on two circles with a vertical tangent at 0, containing $e^{i\pi/q}$ and $e^{-i\pi/q}$ respectively. The two sides going to ∞ are paired by a_λ, and the two sides going to 0 are paired by $a'_\lambda = j a_\lambda j$. Cycles are again of periods 1 and 2.

If q is odd, the cycle conditions are *not fulfilled* (but then a_λ and a'_λ generate Γ_λ: see Exercise II.34).

If $q = 2q'$ is even, the Poincaré theorem shows that a_λ and a'_λ generate a group with presentation

$$\Gamma'_\lambda = \Big\langle\, a_\lambda, a'_\lambda \,\Big|\, (a_\lambda a'_\lambda)^{q'} = 1 \,\Big\rangle \approx (\mathbb{Z}/q'\mathbb{Z}) * \mathbb{Z}.$$

Computing areas of P and P' shows immediately that the index $[\Gamma_\lambda : \Gamma'_\lambda]$ is 2.

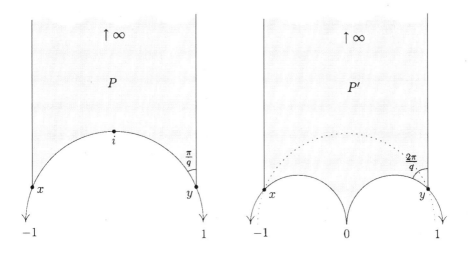

FIGURE 4.   Polygons P and P' for two Hecke groups of Example 44

45. Example: the modular group is a proper subgroup of itself.
In the previous example, when $q = 3$, so that $\lambda = 1$, one obtains the modular group $PSL(2, \mathbb{Z}) \approx \mathbb{Z}/3\mathbb{Z} * \mathbb{Z}/2\mathbb{Z}$.

Consider now an integer $n \geq 2$, the translation

$$\tau_n = \begin{bmatrix} 1 & n \\ 0 & 1 \end{bmatrix} : z \longmapsto z + n$$

of H^2, and the half-turn

$$h_n = \tau_n j \tau_n^{-1} = \begin{bmatrix} -n & n^2 + 1 \\ -1 & n \end{bmatrix}$$

around the point $i+n$. We consider the polygon P with the four following sides:

the vertical half-line from ∞ to $exp\left(i\frac{2\pi}{3}\right)$,
the half-geodesic from $exp\left(i\frac{2\pi}{3}\right)$ to 0,
the half-geodesic from $n - 1$ to $n + i$,
the half-geodesic from $n + i$ to $n + 2$

(the second side is on the unit circle translated by -1 while the third and fourth sides together make up half the unit circle translated horizontally by n). This polygon has the two vertices $exp\left(i\frac{2\pi}{3}\right)$ and $n + i$ in H^2, as well as four vertices at infinity (on ∂H^2). The hyperbolic isometry

$$g = j a_3 = \begin{bmatrix} 0 & -1 \\ 1 & 1 \end{bmatrix}$$

of order 3 maps the first side above onto the second and the half-turn h_n exchanges the third and the fourth sides.

It follows from the Poincaré theorem that g and h_n generate a group with fundamental domain P of infinite area and with presentation

$$\langle g, h_n \mid g^2 = h_n^3 = 1 \rangle \approx \mathbb{Z}/3\mathbb{Z} * \mathbb{Z}/2\mathbb{Z},$$

which is a proper subgroup of the modular group isomorphic to the modular group itself. In particular

the modular group is not co-Hopfian

(see Complement III.22).

46. Example: orientable surface of genus $g \geq 2$. To simplify notation, we assume that $g = 2$.

Let P_2 be a convex compact octagon in the hyperbolic plane with vertices $x_1, \ldots, x_8$ enumerated in cyclic order (indices are to be taken modulo 8). We denote by α_j the interior angle of P_2 at x_j, by e_j the oriented side with head x_{j+1} and tail x_j, and by $\overline{e_j}$ the side obtained from e_j by reversing the orientation. Assume that there exist

a hyperbolic isometry	a_1	mapping	e_1	to	$\overline{e_3}$,
a hyperbolic isometry	b_1	mapping	e_2	to	$\overline{e_4}$,
a hyperbolic isometry	a_2	mapping	e_5	to	$\overline{e_7}$,
a hyperbolic isometry	b_2	mapping	e_6	to	$\overline{e_8}$.

Assume moreover that

$$\sum_{j=1}^{8} \alpha_j = 2\pi.$$

Then Poincaré's theorem implies that a_1, b_1, a_2, b_2 generate a group with presentation

$$\Gamma_2 = \left\langle\, a_1, b_1, a_2, b_2 \mid a_1 b_1 a_1^{-1} b_1^{-1} a_2 b_2 a_2^{-1} b_2^{-1} = 1 \,\right\rangle$$

for which P_2 is a fundamental domain. The polygon P_2, after identification of its 8 sides according to the isometries a_1, b_1, a_2, b_2, is a closed orientable surface of genus 2; it follows that Γ_2 is the fundamental group of such a surface. (A nice picture for P_2 giving a surface appears as Figure 109 of N° 1.4.2 in [Still–80].)

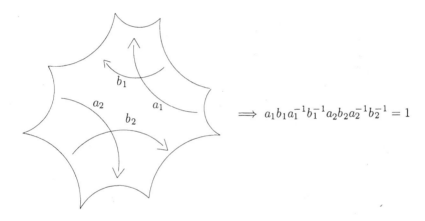

$$\Longrightarrow a_1 b_1 a_1^{-1} b_1^{-1} a_2 b_2 a_2^{-1} b_2^{-1} = 1$$

FIGURE 5.   A polygon P_2 for Example 46

More generally, the fundamental group of a closed orientable surface of genus $g \geq 2$ has a presentation

$$\Gamma_g = \left\langle\, a_1, b_1, \ldots, a_g, b_g \mid a_1 b_1 a_1^{-1} b_1^{-1} \ldots a_g b_g a_g^{-1} b_g^{-1} = 1 \,\right\rangle$$

and such a surface can be obtained by appropriate identifications from a convex polygon P_g with $4g$ vertices.

[Exercise: show that the same group also has a presentation

$$\Gamma_g = \left\langle\, a_1, a_2, \ldots, a_{2g-1}, a_{2g} \mid a_1 a_2 \ldots a_{2g} a_1^{-1} a_2^{-1} \ldots a_{2g}^{-1} = 1 \,\right\rangle$$

by considering other identifications of the sides of P_{2g}.

In fact, the number of distinct ways of identifying the sides of an oriented polygon with $4g$ sides in order to obtain an oriented surface of genus g grows quickly with g [BacVd]: it is for example 1, 4, 131, and

$$380\ 751\ 174\ 738\ 424\ 280\ 720$$

for, respectively, $g = 1, 2, 3$, and 10.]

Observe that there is a large variety of polygons like P_g above, even up to hyperbolic isometries. They form a space which can be shown to be "of dimension $6g - 6$", as already mentioned in Item II.33.viii. Distinct polygons provide distinct embeddings of Γ_g into the group $PSL(2, \mathbb{R})$ of orientation-preserving isometries of the hyperbolic plane, embeddings which are in general not conjugate with each other. Some polygons deserve special mention, such as those providing embeddings of Γ_g into $PSL(2, \mathbb{Q})$, or indeed embeddings

$$\Gamma_g < PSL\left(2, \mathbb{Z}\left[\frac{1}{2}\right]\right),$$

as worked out by Magnus in [Magnu–73] (see [Maski–94] for further work).

47. Example: non-orientable surface of genus $g \geq 3$. One shows as in the previous item that the fundamental group of a closed non-orientable surface of genus $g \geq 3$ is a subgroup of the isometry group of the hyperbolic plane with presentation

$$\Gamma_g^{n.o.} = \langle \ s_1, \ldots, s_g \mid s_1^2 s_2^2 \ldots s_g^2 = 1 \ \rangle.$$

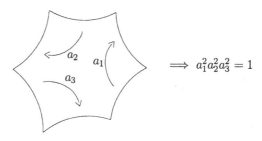

FIGURE 6.   A polygon for a non-orientable surface of genus 3

48. The case of an equilateral triangle (see Example 3.4 in [EpsPe–94]). Let P be an equilateral triangle in an Euclidean plane E^2 (or in a hyperbolic plane H^2), and let E denote the set of the three sides of P. There are 20 ways of defining data

$$e \in E \ \longmapsto \ e' \in E \qquad \text{such that} \quad (e')' = e$$
$$e \in E \ \longmapsto \ s_e \in Isom(E^2) \qquad \text{such that} \quad s_{e'} = (s_e)^{-1}$$

as at the beginning of Item 39.

Indeed the involution $e \longmapsto e'$ has either one fixed point (3 cases) or three fixed points (1 case). Whether $e' = e$ or $e' \neq e$, there are two choices for s_e, one preserving the orientation of E^2 and one reversing it. Counting gives $2 \times 2 \times 2 = 8$ sets of data for the case $e' = e$ for all $e \in E$, and $3 \times 2 \times 2 = 12$ sets of data for the cases in which the involution $e \longmapsto e'$ has exactly one fixed side. Altogether we have $8 + 12 = 20$ possible sets of data. We leave it to the reader to determine which sets of data are compatible with the cycle conditions.

49. On some 3-dimensional cases. In the Euclidean space $\mathbb{R}^3$, consider a regular tesselation $\mathcal{C}$ by cubes and a tetrahedron P which occurs in the barycentric subdivision of one cube of $\mathcal{C}$. The reflections with respect to the faces of P generate a group with presentation

$$\left\langle s_0, s_1, s_2, s_3 \;\middle|\; \begin{array}{c} s_0^2 = s_1^2 = s_2^2 = s_3^2 = (s_0 s_2)^2 = (s_0 s_3)^2 = (s_1 s_3)^2 = 1 \\ (s_0 s_1)^4 = (s_1 s_2)^3 = (s_2 s_3)^4 = 1 \end{array} \right\rangle$$

acting on $\mathbb{R}^3$ with P as fundamental domain. It is also the group of all isometries of $\mathbb{R}^3$ leaving $\mathcal{C}$ invariant. Related examples occur in the theory of Coxeter groups, e.g. in Item 76 of [Harpe–91].

There are many other examples of presentations obtained from Poincaré's theorem, for groups acting on 3-dimensional hyperbolic space, in [ElsGM–98], for example in Section 7.3 for $PSL(2, \mathbb{Z}[i])$, $PSL(2, \mathbb{Z}[\zeta])$, and $PSL(2, \mathbb{Z}[\sqrt{5}])$, where $\zeta = -\frac{1}{2} + \frac{\sqrt{-3}}{2}$, or in Chapter 10 for unit groups of quadratic forms.

V.C. On fundamental groups and curvature in Riemannian geometry

This section is a digression to motivate the definition of hyperbolic groups in Section V.D.

It is difficult to reach a good understanding of what is *precisely* the curvature of a Riemannian manifold. It is much easier to describe some qualitative consequences of *rough conditions* on the curvature. Here are four classical results which support this claim; for proofs, see e.g. Chapters 8 and 9 in [CheEb–75] and Chapter 10 in [Eberl–96]. All manifolds here are assumed to be smooth and without boundary.

50. Bieberbach groups. *Let V be a complete n-Riemannian manifold with sectional curvature satisfying*

$$K \geq 0.$$

Then there exist a finite normal subgroup F_1 of the fundamental group $\pi_1(V)$ and a short exact sequence of the form

$$1 \longrightarrow \mathbb{Z}^k \longrightarrow \pi_1(V)/F_1 \longrightarrow F_2 \longrightarrow 1$$

where $k \leq n$ and where F_2 is a finite group.

In particular $\pi_1(V)$ is commensurable up to finite kernels with a free abelian group.

51. The Bonnet theorem. *If V is a complete Riemannian manifold with sectional curvature satisfying*

$$K \geq \kappa^2$$

for some constant $\kappa > 0$, then the diameter of V is bounded by π/κ and V is compact.

In particular, the fundamental group $\pi_1(V)$ is finite.

The condition $K \geq \kappa^2$ means that $K(\sigma) \geq \kappa^2$ for any 2-plane σ in the tangent space to any point of V. The second part of the theorem follows from the first, because the universal covering of V is also compact.

If V is a complete Riemannian manifold with sectional curvature satisfying $K > 0$ and if V is non-compact, then V is contractible [GroMe–69].

For a discussion of the conditions $K > 0$ and $K \geq 0$, see § 3 in [Gromo–91].

There is also a spectacular theorem due to Synge: under the hypotheses of the Bonnet theorem, if moreover the dimension of V is even, then V is simply-connected in the orientable case and $\pi_1(V)$ has two elements in the non-orientable case [Synge–36]. (The analyticity hypothesis in Synge's paper is not needed: see Theorem 5.9 in [CheEb–75].)

More recently, one has also understood much more about the groups $\pi_1(V)$ appearing in the odd-dimensional case [Rong–94].

52. The Cartan-Hadamard theorem. *If V is a complete Riemannian n-manifold with sectional curvature satisfying*

$$K \leq 0$$

then the universal covering of V is diffeomorphic to $\mathbb{R}^n$ and $\pi_1(V)$ contains no element of finite order other than the identity.

In particular, if V is also compact, then the fundamental group $\pi_1(V)$ is infinite.

This was proved by Hadamard for surfaces (1898) and by E. Cartan in his book "Leçons sur la géométrie des espaces de Riemann" (1926). For manifolds V_1, V_2 as in Theorem 52 which are also compact and with isomorphic fundamental groups, Farell and Jones have shown two striking results:

(i) V_1 and V_2 are necessarily homeomorphic [FarJ–89a],

(ii) V_1 and V_2 need not be diffeomorphic [FarJ–89b].

There are examples showing (ii) in any dimension $n \geq 5$, and with sectional curvature satisfying $-1 \leq K \leq -1 + \epsilon$ for any $\epsilon > 0$.

53. Theorem. *Let V be a compact Riemannian manifold.*

(i, Preissmann). *Assume that the sectional curvature of V satisfies*

$$K < 0.$$

Then any non-trivial abelian subgroup of the fundamental group $\pi_1(V)$ is isomorphic to $\mathbb{Z}$.

(ii, Gromoll-Wolf and Lawson-Yau). *Assume that the sectional curvature of V satisfies*

$$K \leq 0$$

and let Γ be a soluble subgroup of $\pi_1(V)$. Then Γ is finitely generated; moreover there is a finite normal subgroup F_1 of Γ, an integer $k \geq 0$, a finite group F_2, and a short exact sequence

$$1 \longrightarrow \mathbb{Z}^k \longrightarrow \Gamma/F_1 \longrightarrow F_2 \longrightarrow 1.$$

In particular, any soluble subgroup of $\pi_1(V)$ is commensurable up to finite kernels with a free abelian group.

Preissmann's theorem appears in [Preis–42]. Statement (ii) was shown independently by Gromoll-Wolf and by Lawson-Yau in the early 70's; see Theorem 9.1 in [CheEb–75].

For Lie groups which have left-invariant Riemannian metrics with sectional curvature ≥ 0, or ≤ 0, see [Miln–76] and references there.

For a discussion of sign conditions on curvature, see [Gromo–91].

It seems too ambitious a project to try to define *precisely* the curvature for metric spaces much more general than Riemannian manifolds. But theorems such as the four previous ones make it natural to define curvature *inequality conditions* in a larger setting. The most successful recent attempt in this direction is certainly Gromov's definition of hyperbolic metric spaces, and in particular of hyperbolic finitely-generated groups [Gromo–87]. Among many possible equivalent definitions, we will state below one suggested by the following fact. If Y is a subspace of a metric space X and if δ is a positive number, we set $\mathcal{N}_\delta(Y) = \{\, x \in X \mid d(x, Y) \leq \delta \,\}$.

54. Triangles in negatively curved Riemannian manifolds. *For each constant $\kappa > 0$ there exists a constant $\delta > 0$ with the following property.*

Let V be a simply-connected complete Riemannian manifold with sectional curvature satisfying $K \leq -\kappa^2$. Let x, y, z be three points in V and let Δ be the triangle consisting of the three geodesic segments $[x, y], [y, z], [z, x]$. Then

$$[z, x] \subset \mathcal{N}_\delta\Big([x, y] \cup [y, z]\Big).$$

On the proof. If V has constant curvature, this is a simple consequence of the Gauss-Bonnet theorem for surfaces, which implies that there is a bound on the area of any triangle in V (and a fortiori a bound on the radius of the incircle of any triangle). In the general case, one has to use a comparison theorem à la Alexandrov-Toponogov; for a short reference on this, see M. Troyanov's Chapter 3 in [GhyHp–90]; for a comprehensive one, see [BriHa–99]. □

V.D. Complement on Gromov's hyperbolic groups

55. Hyperbolicity for metric spaces and for finitely-generated groups. A geodesic metric space X is defined to be *hyperbolic* if there exists a constant $\delta > 0$ such that the following holds: given three points $x, y, z \in X$ and three geodesic segments between them denoted respectively by $[x, y]$, $[y, z]$, $[x, z]$, we have

$$[x, z] \subset \mathcal{N}_\delta \Big([x, y] \cup [y, z] \Big).$$

The exact value of δ is irrelevant for most purposes.

Examples include metric trees (and real trees) and simply-connected complete Riemannian manifolds with sectional curvature satisfying $K \leq -\kappa^2$ for some constant $\kappa > 0$. (Spaces of finite diameter are also trivial examples.) For $n \geq 2$, the Euclidean space $\mathbb{R}^n$ is *not* hyperbolic.

A fundamental fact is that *a geodesic space quasi-isometric to a hyperbolic space is itself hyperbolic.* This is not trivial to show, and we refer to one of [Gromo–87, page 76] (Gromov's indications remain obscure to this writer), [CooDP–90, chap. 3], [GhyHp–90, chap. 5]. Thus it makes sense to define a *hyperbolic group* as a finitely-generated group whose Cayley graph is hyperbolic; the invariance of hyperbolicity by quasi-isometries implies that the system of generators used to define the Cayley graph is irrelevant.

There are obvious examples of hyperbolic groups, namely finite groups, infinite cyclic groups, and more generally finitely-generated groups which are quasi-isometric to $\mathbb{Z}$ (or equivalently which contain an infinite cyclic subgroup of finite index); they are called *elementary*. What follows provides examples of non-elementary hyperbolic groups.

It is a consequence of Theorem IV.23 and Item 54 that the fundamental group of a compact Riemannian manifold of scalar curvature $K < 0$ is a hyperbolic group. It is easy to check that the free product of two hyperbolic groups is again hyperbolic. Many other examples come from the "small cancellation theory" which we do not discuss in the present notes. For more examples, see also [Gromo–93], [Champ–94], [Delz–96a], and [Ol's–92].

For examples of pairs $\Gamma_0 \subset \Gamma$ with Γ hyperbolic and Γ_0 finitely generated *and not hyperbolic,* see [Rips–82] and the comments on Rips' construction in Lecture 3 of [Brids–99] (see also [Wise–98]). For pairs with Γ hyperbolic and Γ_0 finitely presented *and not hyperbolic,* see Lecture 4 in [Brids–99], and [Brady–99].

Let us mention without proof some properties of hyperbolic groups.

56. Hyperbolic groups are finitely presented in a very strong sense. Indeed, let Γ be a group generated by a finite set S, let $r > 0$ be a positive number and let $Rips_r(\Gamma, S)$ be the corresponding Rips complex. Recall from Item 3 that Γ is finitely presented if and only if $Rips_r(\Gamma, S)$ is simply-connected for r large enough.

If Γ is hyperbolic, it is a theorem of Rips that $Rips_r(\Gamma, S)$ is *contractible* for r large enough. An analysis of the canonical action of Γ on a contractible $Rips_r(\Gamma, S)$ shows that Γ is finitely presented, that Γ has finitely many conjugacy classes of elements of finite order, and that Γ is of finite cohomological dimension over $\mathbb{Q}$. (If Γ happens to be the fundamental group of a compact Riemannian manifold of negative curvature, the action of Γ on the universal covering of this manifold can be used instead of the action on the Rips complex for the same purpose.)

More on this in [Gromo–87] and, e.g., in chap. 4 of [GhyHp–90]. (Note that, although it is a fact that a hyperbolic group contains a finite number of conjugacy classes of elements of finite order, the proof of this in Proposition 13 of [GhyHp–90, chap. 4] is not correct; see rather [CooDP–90], Lemma 3.5 in Chapter 9.)

57. Linear isoperimetric inequalities. Let $\Gamma = \langle s_1, \ldots, s_n \mid r_1, \ldots, r_m \rangle$ be a finitely-presented group. The corresponding *Dehn function* $D : \mathbb{N} \to \mathbb{N}$ associates to an integer $k \geq 0$ the smallest integer $l \geq 0$ with the following property: each reduced word of length at most k in the free group on $\{s_1, \ldots, s_n\}$ which represents 1 in Γ can be written as a product of at most l conjugates of the relators r_j and of their inverses. If $\tilde{D}$ is the Dehn function for another presentation of the same group Γ, one can show that there exist constants c_1, c_2, c_3 such that

$$\tilde{D}(k) \leq c_1 D(c_2 k) + c_3 k \qquad \text{and} \qquad D(k) \leq c_1 \tilde{D}(c_2 k) + c_3 k$$

for all $k \geq 0$ (see for example [Gers–92b]). Consequently, it makes sense to say that a group Γ has a linear, or quadratic, or polynomial, or subexponential, or exponential, or superexponential (...) Dehn function, without reference to an actual finite presentation.

A finitely-presented group is known to be hyperbolic if and only if its Dehn function is linear (see [Gromo–87] and [CooDP–90, chap. 6]), if and only if its Dehn function is subquadratic in the sense that $\lim_{k \to \infty} D(k) k^{-2} = 0$ [Ol's–91b], [Bowdi–95].

For a list of examples of Dehn functions, with references, see [Pitte–97].

58. Boundaries of hyperbolic groups. Suppose that Γ is the fundamental group of a compact Riemannian manifold V of nonpositive curvature. Let $\partial \tilde{V}$ be the *ideal boundary* of the universal cover $\tilde{V}$ of V, namely the space of geodesic rays in $\tilde{V}$ modulo the relation of being at finite Hausdorff distance (N° I.3.2 in [BalGS–85]). Then $\partial \tilde{V}$ has a natural topology which makes it a compact space, and indeed a sphere of dimension $dim(V) - 1$.

In case of a hyperbolic group Γ, one defines similarly a *Gromov boundary* $\partial \Gamma$ which is a compact space on which Γ acts. If $\Gamma = \pi_1(V)$ with V as above and also of negative curvature (so that the fundamental group of V is indeed hyperbolic), $\partial \Gamma$ can be identified naturally with $\partial \tilde{V}$. If Γ is a non-abelian free group, $\partial \Gamma$ is a Cantor space.

By using the action of a hyperbolic group Γ on its Gromov boundary, one can show that, for any element $\gamma \in \Gamma$ of infinite order, the centralizer $Z_\Gamma(\gamma)$ contains an infinite cyclic group of finite index. Also, any infinite hyperbolic group which does not contain a cyclic subgroup of finite index does contain a non-abelian free group. Thus, groups such as $\mathbb{Z}^n$ for $n \geq 2$ or $SL(n, \mathbb{Z})$ for $n \geq 3$ are certainly *not* hyperbolic.

More on this in [Gromo–87] and, e.g., in chap. 6 through 8 of [GhyHp–90]. See also [Pauli] for a study of extra structures on boundaries of hyperbolic groups, and [CooPa–93] for the action of Γ on $\partial\Gamma$.

59. Rationality of growth series. Let Γ be a hyperbolic group. For *any finite set S of generators of Γ*, it is known that the *growth series* $\Sigma(\Gamma, S; z)$ introduced in Section VI.A below is rational (a particular case is proved in VI.8).

60. Expositions, open problems, extensions. After Gromov's original paper [Gromo–87], there have been several introductory notes on hyperbolic groups, including [Bowdi–91], [CooDP–90], [CooPa–93], [Ghys–90], [GhyHp–90], and [Shor8–91]. Let us mention two among several standard open problems:

- does every hyperbolic group contain a subgroup of finite index without torsion ? (compare with the result of Selberg quoted in V.20.iii),
- is every hyperbolic group residually finite ? (see Complement III.18.xiii).

(These questions are equivalent, by Theorem 5.1 of [KapWi–00].)

Work has also been done to find an appropriate definition of *semi-hyperbolic groups,* which should be to curvature ≤ 0 what hyperbolic groups are to curvature < 0. See [Epst6–92] and [AloBr–95].

CHAPTER VI

GROWTH OF FINITELY-GENERATED GROUPS

To a group Γ, given together with a finite generating set S, there corresponds a growth function; examples are given in Section VI.A. To obtain an invariant of the group Γ, and not only of the pair (Γ, S), one needs an appropriate equivalence relation for growth functions; Section VI.B presents two natural definitions of equivalence, which themselves turn out to be equivalent in the present context (at least for infinite groups: see Proposition 34). In Section VI.C, we observe that the growth function of a pair (Γ, S) is submultiplicative and we define the three types of groups, which are of

exponential growth,

polynomial growth,

intermediate growth.

We postpone to the next chapters the discussion of more examples and of applications to Riemannian geometry, which have been important motivations for the study of group growth.

VI.A. Growth functions and growth series of groups

1. Definitions. Let Γ be a finitely-generated group and let S be a finite set of generators of Γ. Recall from Section IV.A that the word length $\ell_S(\gamma)$ of an element $\gamma \in \Gamma$ is the smallest integer $n \geq 0$ for which there exist $s_1, \ldots, s_n \in S \cup S^{-1}$ such that $\gamma = s_1 \ldots s_n$.

The *growth function* of the pair (Γ, S) associates to an integer $k \geq 0$ the number $\beta(\Gamma, S; k)$ of elements $\gamma \in \Gamma$ such that $\ell_S(\gamma) \leq k$. The formal power series

$$B(\Gamma, S; z) = \sum_{k=0}^{\infty} \beta(\Gamma, S; k) z^k \in \mathbb{Z}[[z]]$$

is the corresponding *growth series*.

One also defines the *spherical growth function*

$$k \mapsto \sigma(\Gamma, S; k) = \beta(\Gamma, S; k) - \beta(\Gamma, S; k - 1),$$

with $\sigma(\Gamma, S; 0) = \beta(\Gamma, S; 0) = 1$, and the corresponding *spherical growth series*

$$\Sigma(\Gamma, S; z) = \sum_{\gamma \in \Gamma} z^{\ell_S(\gamma)} = \sum_{k=0}^{\infty} \sigma(\Gamma, S; k) z^k = (1 - z) B(\Gamma, S; z).$$

When there is no risk of confusion, we write $\beta(k), B(z), \sigma(k), \Sigma(z)$ instead of $\beta(\Gamma, S; k), \ldots, \Sigma(\Gamma, S; z)$.

2. Examples. (i) The cyclic group of order n, considered with one generator, provides the spherical growth series (or here polynomials)

$$1 + 2z + \ldots + 2z^{m-1} + z^m \qquad \text{if} \quad n = 2m \qquad \text{is even}$$

$$1 + 2z + \ldots + 2z^{m-1} + 2z^m \qquad \text{if} \quad n = 2m + 1 \quad \text{is odd}.$$

(ii) For the infinite cyclic group $\mathbb{Z}$ with the natural generator 1, the spherical and the plain growth series are

$$\Sigma(z) = 1 + \sum_{k=1}^{\infty} 2z^k = \frac{1+z}{1-z} \qquad \text{and} \qquad B(z) = \frac{1+z}{(1-z)^2}.$$

Consider the infinite dihedral group Γ, namely the free product of two groups $\{1, a\}$ and $\{1, b\}$ of order 2, with the generating set $\{a, b\}$. Then the spherical growth series is also

$$\Sigma(z) = \frac{1+z}{1-z}$$

because the Cayley graph of the free product $\{1, a\} * \{1, b\}$ with respect to $\{a, b\}$ is isomorphic to that of $\mathbb{Z}$ with respect to $\{1\}$. (Recall that in this book Cayley graphs do *not* have edges colored by generators; see Definition IV.2 and Remark IV.3.iv.)

(iii) The spherical growth series of $\mathbb{Z}$ with respect to $\{2, 3\}$ is

$$\Sigma(z) = 1 + 4z + 8z^2 + 6(z^3 + z^4 + \ldots) = \frac{1 + 3z + 4z^2 - 2z^3}{1-z};$$

indeed, one easily shows by induction on k that, for $k \geq 3$, there are 6 elements of length k, which are $\pm(3k - 2), \pm(3k - 1),$ and $\pm 3k$.

(iv) For the free abelian group $\mathbb{Z}^2$ with the usual basis of two generators, we have

$$\sigma(0) = 1 \qquad \text{and} \qquad \sigma(k) = 4k \quad \text{for} \quad k \geq 1$$

so that

$$\Sigma(z) = 1 + \sum_{k=1}^{\infty} 4kz^k = \left(\frac{1+z}{1-z}\right)^2$$

(see also Example 5.i below). For the free abelian group $\mathbb{Z}^n$ with the usual basis of n generators, it is straightforward to check that

$$\lim_{k \to \infty} \frac{\beta(k)}{k^n} = vol(K)$$

where $vol(K)$ denotes the volume of the convex hull of the generators and their inverses in $\mathbb{R}^n$.

This holds more generally for sets of generators of the form $S = K \cap \Gamma$, where Γ is a lattice in $\mathbb{R}^n$ and where K is a convex polytope in $\mathbb{R}^n$ with vertices in Γ. In this situation, the *Ehrhart polynomial* E_K counts for each integer $k \geq 0$ the number $E_K(k)$ of lattice points in kK. This has been investigated in [Ehrha–67] and later work; see the exposition of [Brion–96]; see also [BacHV–99].

One has $\beta(k) \leq E_K(k)$ for all $k \geq 0$, and $\lim_{k\to\infty} \frac{\beta(k)}{k^n} = \lim_{k\to\infty} \frac{E_K(k)}{k^n} = vol(K)$. (For the last equality, it is assumed that a fundamental domain for the lattice Γ in $\mathbb{R}^n$ has unit volume.) However, simple examples in dimension 3 with $-K = K$ show that one can have $\beta(2) < E_K(2)$ [BacHV–97].

(v) In each of the previous examples, as $\beta(k)$ is a polynomial in k, it makes sense to consider negative values of k. For example, in the "Ehrhart situation", it is known that $E_K(-k)$ counts the number of lattice points in the *interior of* kK; see again [Brion–96].

It would be nice to find further pairs (Γ, S) giving rise to growth functions β for which $\beta(-k)$ has an interpretation, at least for some $k \geq 1$.

(vi) Let Γ be a group generated by a finite set S, and let $\ell_S : \Gamma \longrightarrow \mathbb{N}$ denote the corresponding length function. For each $k \geq 0$, denote by $S(k)$ the set of elements $\gamma \in \Gamma$ of length $\ell_S(\gamma) = k$ and by A_k the free complex vector space over $S(k)$, with basis $(\delta_\gamma)_{\gamma \in S(k)}$. The direct sum $A = \bigoplus_{k=0}^{\infty} A_k$ is a complex associative algebra for the product defined by

$$\delta_\gamma \delta_{\gamma'} = \begin{cases} \delta_{\gamma\gamma'} & \text{if } \ell_S(\gamma\gamma') = \ell_S(\gamma) + \ell_S(\gamma') \\ 0 & \text{otherwise.} \end{cases}$$

It is also a *graded algebra*, satisfying $A_k A_{k'} \subset A_{k+k'}$, which is *finitely generated* by the subspace A_1 of elements of degree 1, a subspace of dimension $|S \cup S^{-1}|$ (if $1 \notin S$) or $|S \cup S^{-1}| - 1$ (if $1 \in S$). Now the spherical growth series $\Sigma(\Gamma, S; z) = \sum_{k=0}^{\infty} \sigma(\Gamma, S; k) z^k$ is precisely the Hilbert-Poincaré series $\sum_{k=0}^{\infty} dim_{\mathbb{C}}(A_k)$ of the graded algebra of finite type A.

Assume now that Γ is *abelian*, so that A is a commutative algebra. It is a classical result of Hilbert-Poincaré that such a series is rational. For Γ free abelian of rank m, one has more precisely

$$\Sigma(\Gamma, S; z) = \frac{P_S(z)}{(1 - z)^m}$$

where $P_S(z) \in \mathbb{Z}[z]$ is a polynomial such that $P_S(0) = 1$; see for example [KosMa–89], Chapter 3, Theorem 11.9 (and also [BacHV–99] for some specific polynomials P_S).

In other words, the growth series of a finitely-generated abelian group is rational. This method of proof is mentioned in [Billi–85]; there are other methods of proof of the same result in [Benso–87] and in [Klarn–81].

(vii) A straightforward generalization of Definition 1 applies to *connected rooted graphs*. For an integer $n \geq 2$, the spherical growth function of a regular tree of degree n is

$$\Sigma(z) \ = \ 1 + \sum_{k=1}^{\infty} n(n-1)^{k-1} z^k \ = \ \frac{1+z}{1-(n-1)z}.$$

This applies in particular to a free product of p copies of $\mathbb{Z}/2\mathbb{Z}$ and q copies of $\mathbb{Z}$, with $n = p + 2q$; see Example IV.6.

(viii) One does not understand well enough which properties of a group Γ are reflected in growth series of the form $\Sigma(\Gamma, S; z)$. A remarkable example appears in [FloPa–97]: there exist two groups Γ_1, Γ_2, respectively given with finite generating sets S_1, S_2, such that

$$\Sigma(\Gamma_1, S_1; z) \ = \ \Sigma(\Gamma_2, S_2; z) \ = \ \frac{1 + 88z + 28z^2}{1 - 20z + 28z^2}$$

with Γ_1 Gromov hyperbolic and Γ_2 not.

See also the examples of Item IV.9, showing non-isomorphic groups having isomorphic Cayley graphs.

3. Immediate properties. *Let Γ be a group generated by a finite set S of n generators.*

(i) The length function is symmetric and subadditive:

$$\ell_S\left(\gamma^{-1}\right) \ = \ \ell_S(\gamma)$$
$$\ell_S\left(\gamma_1\gamma_2\right) \ \leq \ \ell_S\left(\gamma_1\right) + \ell_S\left(\gamma_2\right)$$

for all $\gamma, \gamma_1, \gamma_2 \in \Gamma$.

(ii) The growth function is submultiplicative:

$$\beta(\Gamma, S; k_1 + k_2) \ \leq \ \beta(\Gamma, S; k_1)\beta(\Gamma, S; k_2)$$

for all $k_1, k_2 \in \mathbb{N}$.

(iii) The growth function is majorized by that of the free group of rank n with respect to a free generating set:

$$\beta(\Gamma, S; k) \ \leq \ 2n(2n-1)^{k-1}$$

for all $k \geq 1$.

(iv) If Γ is infinite, the growth function $\beta(\Gamma, S; k)$ is strictly increasing in k.

We will return to Property (iii) in Example 23 and in Remark 53.i.

The proof of the second part of the next proposition appears in [Johns–91].

4. Proposition. *For $j = 1, 2$, consider a group Γ_j generated by a finite set S_j, and the corresponding spherical growth series $\Sigma_j(z)$.*

(i) The spherical growth series of the direct product $\Gamma_1 \times \Gamma_2$ with respect to the generating set $S \doteq (S_1 \times \{1\}) \cup (\{1\} \times S_2)$ is given by

$$\Sigma_{\Gamma_1 \times \Gamma_2}(z) = \Sigma_1(z)\,\Sigma_2(z).$$

(ii) The spherical growth series of the free product $\Gamma_1 * \Gamma_2$ with respect to the generating set $S_1 \sqcup S_2$ is given by

$$\frac{1}{\Sigma_{\Gamma_1 * \Gamma_2}(z)} - 1 = \frac{1}{\Sigma_1(z)} - 1 + \frac{1}{\Sigma_2(z)} - 1.$$

Proof. (i) Any element in the direct product $\Gamma_1 \times \Gamma_2$ is of the form (γ_1, γ_2) with $\gamma_1 \in \Gamma_1$ and $\gamma_2 \in \Gamma_2$, and its S-length is given by

$$\ell_S(\gamma_1, \gamma_2) = \ell_{S_1}(\gamma_1) + \ell_{S_2}(\gamma_2).$$

Hence

$$\sigma(\Gamma, S; k) = \sum_{j=0}^{k} \sigma(\Gamma_1, S_1; j)\, \sigma(\Gamma_2, S_2; k - j)$$

for all $k \geq 0$ and the result follows.

(ii) Any element in the free product $\Gamma_1 * \Gamma_2$ is of the form

(*) $$b_0 a_1 b_1 a_2 b_2 \ldots a_n b_n a_{n+1}$$

where $n \geq 0$ and

$$a_1, \ldots, a_n \in \Gamma_1 \setminus \{1\} \quad , \quad a_{n+1} \in \Gamma_1$$
$$b_0 \in \Gamma_2 \quad , \quad b_1, \ldots, b_n \in \Gamma_2 \setminus \{1\}$$

(normal form), and its $(S_1 \sqcup S_2)$-length is

$$\ell_{S_2}(b_0) + \sum_{i=1}^{n} \ell_{S_1}(a_i) + \sum_{j=1}^{n} \ell_{S_2}(b_j) + \ell_{S_1}(a_{n+1}).$$

The set of words of the form (*) with a fixed value of n provides the growth series

$$\Sigma_2(z) \Big((\Sigma_1(z) - 1)(\Sigma_2(z) - 1) \Big)^n \Sigma_1(z).$$

Summing over all $n \geq 0$ we get

$$\Sigma_{\Gamma_1 * \Gamma_2}(z) = \frac{\Sigma_1(z)\Sigma_2(z)}{1 - (\Sigma_1(z) - 1)(\Sigma_2(z) - 1)}$$

and this is equivalent to the formula of (ii). $\square$

See also Exercise 17.

5. Examples. (i) For a free abelian group $\mathbb{Z}^d$ with the natural set of d generators, the spherical growth function is given by

$$\Sigma(z) = \left(\frac{1+z}{1-z}\right)^d$$

(but see also Exercise 16).

(ii) From the previous proposition, we again find the examples of growth series of free products given in 2.vii.

6. Heisenberg group. Consider the *Heisenberg group*

$$H_1 = \left\{ \begin{pmatrix} 1 & 0 & 0 \\ k & 1 & 0 \\ m & l & 1 \end{pmatrix} \middle| k, l, m \in \mathbb{Z} \right\}$$

of Example IV.8, generated by the set S of the two matrices

$$s = \begin{pmatrix} 1 & 0 & 0 \\ 1 & 1 & 0 \\ 0 & 0 & 1 \end{pmatrix} \qquad \text{and} \qquad t = \begin{pmatrix} 1 & 0 & 0 \\ 0 & 1 & 0 \\ 0 & 1 & 1 \end{pmatrix}.$$

It is easy to compute the equivalence class (in the sense of Section VI.B below) of the corresponding growth function:

$$\beta(H_1, S; k) \sim k^4$$

(see Lemma 4 in [Miln–68b] or Proposition VII.22). It is known that the growth series is

$$\Sigma(H_1, S; z) = \frac{1 + z + 4z^2 + 11z^3 + 8z^4 + 21z^5 + 6z^6 + 9z^7 + z^8}{(1-z)^4 (1+z+z^2)(1+z^2)};$$

see [Benso–87] (beware of a mistake in the formula for Σ !), [Shapi–89], and [Weber–89].

7. Modular group and Machì computations. Set $\mathbb{Z}/2\mathbb{Z} = \{1, a\}$ and $\mathbb{Z}/3\mathbb{Z} = \{1, b, b^2\}$. The modular group

$$\Gamma = (\mathbb{Z}/2\mathbb{Z}) * (\mathbb{Z}/3\mathbb{Z})$$

is generated by both

$$S = \{a, b\} \qquad \text{and} \qquad S' = \{a, t\}$$

where $t = ab$. The corresponding spherical growth series are

$$\Sigma(\Gamma, S; z) = \frac{1 + 3z + 2z^2}{1 - 2z^2} \qquad \text{and} \qquad \Sigma(\Gamma, S'; z) = \frac{1 + 2z + 2z^2 + z^3}{1 - z - z^2}$$

(this example was shown to me by A. Machì).

Let us first prove these formulas ((i) and (ii) below), before spelling out two observations by Machì ((iii) and (iv) below).

(i) Denote by $\sigma_a(k)$ [respectively $\sigma_b(k)$] the number of elements in the free product $(\mathbb{Z}/2\mathbb{Z}) * (\mathbb{Z}/3\mathbb{Z})$ which are represented by reduced words of length k in a and $b^{\pm 1}$ ending in a [respectively in b or b^{-1}]. We have $\sigma(\Gamma, S; k) = \sigma_a(k) + \sigma_b(k)$ as well as

$$\sigma_a(k+1) = \sigma_b(k) \qquad \text{and} \qquad \sigma_b(k+1) = 2\sigma_a(k)$$

for all $k \geq 1$; moreover $\sigma(\Gamma, S; 0) = 1$, $\sigma_a(1) = 1$ and $\sigma_b(1) = 2$. In terms of the series $\Sigma_a(z) \doteq \sum_{k=1}^{\infty} \sigma_a(k)z^k$ and $\Sigma_b(z) \doteq \sum_{k=1}^{\infty} \sigma_b(k)z^k$, the recurrence relations above become

$$\Sigma_a(z) = z + z\sum_{k=1}^{\infty} \sigma_a(k+1)z^k = z\left(1 + \Sigma_b(z)\right)$$

and

$$\Sigma_b(z) = 2z + z\sum_{k=1}^{\infty} \sigma_b(k+1)z^k = 2z\left(1 + \Sigma_a(z)\right).$$

Since the solution of this linear system is

$$\Sigma_a(z) = z\frac{1 + 2z}{1 - 2z^2} \qquad \text{and} \qquad \Sigma_b(z) = 2z\frac{1 + z}{1 - 2z^2},$$

we have

$$\Sigma(\Gamma, S; z) = 1 + \Sigma_a(z) + \Sigma_b(z) = \frac{1 + 3z + 2z^2}{1 - 2z^2} = 1 + 3z + 4z^2 + 6z^3 + 8z^4 + \dots.$$

(ii) Consider now the generating set $S' = \{a, t\}$; recall that $a^2 = 1$ and that $(at)^3 = 1$. Any element in Γ has a normal form which is a word on the alphabet $\{a, t, t^{-1}\}$ without subwords of the form a^2, tat, or $t^{-1}at^{-1}$ (because tat can be replaced by $at^{-1}a$ and $t^{-1}at^{-1}$ by ata). Let $\sigma'_a(k)$ [respectively $\sigma'_t(k)$] denote the number of words of this kind of length k ending in a [respectively in t or t^{-1}]. We have

$$\sigma'_a(k+1) = \sigma'_t(k) \qquad \text{and} \qquad \sigma'_t(k+1) = \sigma'_a(k) + \sigma'_t(k)$$

for all $k \geq 2$. Therefore, writing $\sigma'(k)$ for $\sigma(\Gamma, S'; k) = \sigma'_a(k) + \sigma'_t(k)$, we also have

$$\sigma'(k+1) = \sigma'(k) + \sigma'(k-1)$$

for all $k \geq 3$ [mind the 3, because this fails for $k = 2$], and $\sigma'(k)$ is a Fibonacci sequence. If one writes $\Sigma'(z)$ for $\Sigma(\Gamma, S'; z)$, the above recurrence relations become

$$\Sigma'(z) = 1 + 3z + 6z^2 + 10z^3 + \sum_{k=3}^{\infty} \sigma'(k+1)z^{k+1}$$

$$= 1 + 3z + 6z^2 + 10z^3 + z\left(\sum_{k=3}^{\infty} \sigma'(k)z^k\right) + z^2\left(\sum_{k=2}^{\infty} \sigma'(k)z^k\right)$$

$$= 1 + 3z + 6z^2 + 10z^3 + z\left(\Sigma'(z) - 1 - 3z - 6z^2\right) + z^2\left(\Sigma'(z) - 1 - 3z\right),$$

whence

$$\Sigma'(z) = \frac{1 + 2z + 2z^2 + z^3}{1 - z - z^2} = 1 + 3z + 6z^2 + 10z^3 + 16z^4 + \ldots.$$

(iii) From the formula for $\Sigma(\Gamma, S; z)$ and from calculus, we find

$$\beta(\Gamma, S; 2k) = 7 \cdot 2^k - 6$$
$$\beta(\Gamma, S; 2k + 1) = 10 \cdot 2^k - 6$$

for all $k \geq 1$. In particular, writing $\beta(k)$ for $\beta(\Gamma, S; k)$, we have

$$\lim_{k\to\infty} \frac{\beta(2k)}{\beta(2k-1)} = \frac{7}{5} < \lim_{n\to\infty} \sqrt[n]{\beta(n)} = \sqrt{2} < \lim_{k\to\infty} \frac{\beta(2k+1)}{\beta(2k)} = \frac{10}{7}$$

so that $\beta(k+1)/\beta(k)$ has no limit when $k \to \infty$; this is related to the fact that the series

$$\sum_{n=0}^{\infty} \beta(n)z^n = \frac{1 + 3z + 2z^2}{(1 - 2z^2)(1 - z)}$$

is a rational function which has *strictly more than one pole* on its circle of convergence. (The formula for the Taylor series about the origin of a rational function without a pole at the origin is discussed in Section 7.3 of [GraKP–91] — among other places.)

(iv) For the other generating set $S' = \{a, t\}$ of the same group and for the resulting growth function $\beta'(k) = \beta(\Gamma, S'; k)$, a similar computation shows that

$$\lim_{k\to\infty} \frac{\beta'(k+1)}{\beta'(k)} = \frac{\sqrt{5}+1}{2} = \lim_{k\to\infty} \sqrt[k]{\beta'(k)}.$$

This limit exists because the rational function

$$\sum_{k=0}^{\infty} \beta'(k)z^k = \frac{1 + 2z + 2z^2 + z^3}{(1 - z - z^2)(1 - z)}$$

has a *single pole* on its circle of convergence.

8. Surface groups, and rationality. Let us compute the growth series of
the fundamental group

$$\Gamma_g = \left\langle\, a_1, b_1, \ldots, a_g, b_g \mid a_1 b_1 a_1^{-1} b_1^{-1} \ldots a_g b_g a_g^{-1} b_g^{-1} = 1 \,\right\rangle$$

of an orientable closed surface of genus $g \geq 2$ with respect to the generating set

$$S_g = \{\, a_1, b_1, \ldots, a_g, b_g \,\}$$

described in Example V.46. The computation is due to Cannon [Canno–80]
(see also [CanWa–92]).

Consider a fundamental polygon P_g as in V.46 and all its images by the
elements of Γ_g; these form a tesselation $\mathcal{T}$ of the hyperbolic plane by isometric
copies of P_g. Consider also a point x_0 inside P_g; we may identify the group Γ_g
with the corresponding orbit $\Gamma_g x_0$. If one connects two points in this orbit by
an edge whenever they are in images of P_g which share a side, one obtains the
tesselation $\mathcal{T}'$ *dual* to $\mathcal{T}$. The growth function of the pair (Γ_g, S_g) is then the
growth function of the graph underlying the 1-skeleton of $\mathcal{T}'$.

In case Γ_g is a regular polygon, with its $4g$ sides of the same length and with
its $4g$ interior angles equal to $\frac{2\pi}{4g}$, the tesselations $\mathcal{T}$ and $\mathcal{T}'$ are isometric.

From now on, we assume for simplicity that $g = 2$. Define the *type* of a vertex
x of $\mathcal{T}'$ as follows:

x is of type 0 if it is the basis vertex x_0,
x is of type 1 if it has precisely one neighbour x_{pred} such that
$\quad d(x_0, x_{pred}) = d(x_0, x) - 1$,
x is of type 2 if it has precisely two neighbours x'_{pred} and x''_{pred} such that
$\quad d(x_0, x'_{pred}) = d(x_0, x''_{pred}) = d(x_0, x) - 1$.

Denote by $\sigma_1(k)$ [respectively $\sigma_2(k)$] the number of vertices of $\mathcal{T}'$ which are of
type 1 [respectively of type 2] and at combinatorial distance k from x_0. Count
the edges joining some vertex at distance k to some vertex at distance $k - 1$
from x_0 in two ways, summing on the vertices at distance k or $k - 1$, to get

$$\sigma_1(k) + 2\sigma_2(k) = 7\sigma_1(k-1) + 6\sigma_2(k-1) \qquad \text{for all} \quad k \geq 2.$$

Similarly, by counting in two ways the number of octagons joining some vertex
at distance k to some vertex at distance $k - 4$ from x_0, we get

$$\sigma_2(k) = 6\sigma_1(k-4) + 5\sigma_2(k-4) \qquad \text{for all} \quad k \geq 5.$$

We also have

$$\sigma_1(1) = 8 \qquad \sigma_2(1) = \sigma_2(2) = \sigma_2(3) = 0 \qquad \sigma_2(4) = 8.$$

In terms of the series $\Sigma_1(z) \doteq \sum_{k=1}^{\infty} \sigma_1(k) z^k$ and $\Sigma_2(z) \doteq \sum_{k=1}^{\infty} \sigma_2(k) z^k$, the
above recurrence relations become

$$\Sigma_1(z) - 8z + 2\Sigma_2(z) = 7z\Sigma_1(z) + 6z\Sigma_2(z)$$
$$\Sigma_2(z) - 8z^4 = 6z^4\Sigma_1(z) + 5z^4\Sigma_2(z).$$

Since the solution of this linear system is

$$\Sigma_1(z) = \frac{8z + 8z^2 + 8z^3 - 8z^4}{1 - 6z - 6z^2 - 6z^3 + z^4} \quad \text{and} \quad \Sigma_2(z) = \frac{8z^4}{1 - 6z - 6z^2 - 6z^3 + z^4}$$

we have

$$\Sigma(\Gamma_2, S_2; z) = 1 + \Sigma_1(z) + \Sigma_2(z) = \frac{1 + 2z + 2z^2 + 2z^3 + z^4}{1 - 6z - 6z^2 - 6z^3 + z^4}.$$

Similarly, we have

$$\begin{aligned}
\Sigma(\Gamma_g, S_g; z) &= \frac{1 + 2z + 2z^2 + \ldots + 2z^{2g-1} + z^{2g}}{1 - (4g-2)z - (4g-2)z^2 - \ldots - (4g-2)z^{2g-1} + z^{2g}} \\
&= \frac{(1+z)(1-z^{2g})}{1 - (4g-1)z + (4g-1)z^{2g} - z^{2g+1}}
\end{aligned}$$

for all $g \geq 2$. (This is Theorem 4 in [Canno–80]; observe that the genus of the surface, denoted here by g, is denoted by $2g$ in Cannon's typescript. For many other computations of this kind, see [BarCe–a].)

From the point of view of growth series, the choice of S_g is possibly not "the simplest one"; indeed, Alonso [Alons–91] has found a generating set S_g^{Alo} of Γ_g for which

$$\Sigma(\Gamma_g, S_g^{\text{Alo}}; z) = \frac{1 + 2z + z^2}{1 - (8g - 6)z + z^2}.$$

More generally, as already recalled in Item V.59, for Γ a Gromov hyperbolic group and for S an arbitrary finite set of generators, it is known that the series $\Sigma(\Gamma, S; z)$ is a *rational function*. (See 5.2 and 8.5 in [Gromo–87], as well as [Canno–80], [Canno–84], and Chap. 9 in [GhyHp–90].)

Observe that a power series with rational coefficients is rational [respectively algebraic] over $\mathbb{C}$ if and only if it is rational [respectively algebraic] over $\mathbb{Q}$. (There is for example a proof in Lemma 3.1 in [Stoll–96].)

Surprisingly, it is more difficult to find simple examples of pairs (Γ, S) with growth series which are *not* rational.

Let us first mention the following class of examples, from work of W. Parry [Parr–92b]. Let A be a free product of p copies of $\mathbb{Z}/2\mathbb{Z}$ and q copies of $\mathbb{Z}$, and set $n = p + 2q$; thus the natural generating subset S_A of A, with $S_A = S_A^{-1}$, has n elements and the corresponding Cayley graph of A is a regular tree of degree n (Example IV.6). Let B be a group generated by a finite subset S_B. Let $\Gamma = B \wr A$ denote the restricted wreath product of B by A, with B and A identified with subgroups of Γ in the usual way (see VIII.3). Let S denote the generating set $S_B \cup S_A$ of Γ. If $n \geq 3$ and if $B \neq \{1\}$, the growth series $\Sigma(\Gamma, S; z)$ is not a rational function; there are large classes of examples (in particular all examples with B a finite group) for which this series is an algebraic function.

There are other examples of pairs (Γ, S) with growth series which are not algebraic, and thus a fortiori not rational: finitely presented groups with non-solvable word problem (Theorem 9.1 in [Canno–80]), groups of intermediate growth (see Chapter VIII), and nilpotent examples worked out by M. Stoll [Stoll–96]. See also the discussion in Section D of [GriHa–97].

9. An example for which the spherical growth function is NOT monotonic. This example appears as Example 8.2 in [Canno–80] (the proof being left as an exercise), where Cannon observes that the corresponding spherical growth function σ never becomes monotonic. It was observed in [GriHa–98] that it also provides an answer to Problem 5.2 in [Kouro–95], because $\sigma(2k) > \frac{1}{2}\big(\sigma(2k-1) + \sigma(2k+1)\big)$ for all $k \geq 3$.

Consider an equilateral triangle of the Euclidean plane, the group generated by the three reflections defined by the three sides of the triangle, and its subgroup Γ of orientation-preserving isometries. Then Γ has the presentation

$$\Gamma = \big\langle\, s, t \mid s^3 = t^3 = (st)^3 = 1 \,\big\rangle,$$

as follows for example from the Poincaré Theorem of Section V.B.

Figure 1 shows part of the corresponding Cayley graph, embedded in the Euclidean plane. The vertices of this graph are the points with Cartesian coordinates

$$(4x, y) \quad \text{for} \quad x \in \mathbb{Z} \quad \text{and} \quad y = 4k\frac{\sqrt{3}}{2} \quad \text{with} \quad k \in \mathbb{Z}$$

$$(4x + 2, y) \quad \text{for} \quad x \in \mathbb{Z} \quad \text{and} \quad y = (4k + 2)\frac{\sqrt{3}}{2} \quad \text{with} \quad k \in \mathbb{Z}$$

$$(2x + 1, y) \quad \text{for} \quad x \in \mathbb{Z} \quad \text{and} \quad y = (2k + 1)\frac{\sqrt{3}}{2} \quad \text{with} \quad k \in \mathbb{Z}$$

and each vertex has been represented in Figure 1 by its combinatorial distance from the origin.

The two relations $s^3 = 1$ and $t^3 = 1$ correspond to triangles (for example the two triangles with vertices at combinatorial distances $0, 1, 1$ from the origin) and the relation $(st)^3 = 1$ corresponds to hexagons (for example two hexagons with vertices at combinatorial distances $0, 1, 2, 3, 2, 1$ from the origin and two with vertices at combinatorial distances $1, 2, 3, 3, 2, 1$ from the origin).

Computations show that the spherical growth function for Γ and the generating system $\{s, t\}$ is given by

$$\Sigma(z) = \sum_{n=0}^{\infty} \sigma(n)z^n = \frac{(1 + 2z)\left(1 + 2z + 2z^2 + 2z^3 - z^4\right)}{(1 - z^2)^2}$$

$$= 1 + 4z + 6z^2 + 14z^3 + 18z^4 + 22z^5 + 28z^6 + 30z^7 + 38z^8 + 38z^9$$

$$+ 48z^{10} + 46z^{11} + 58z^{12} + 54z^{13} + 68z^{14} + 62z^{15} + 78z^{16} + \ldots\ldots$$

(see, e.g., [GriHa–98]). Thus, for the function σ, we have $\sigma(0) = 1$, $\sigma(1) = 4$ and

$$\sigma(2k - 1) = 8k - 2 \qquad \text{for all} \qquad k \geq 2$$
$$\sigma(2k) = 10k - 2 \qquad \text{for all} \qquad k \geq 1.$$

In particular, and as stated above, we have

$$\sigma(2k) > \frac{\sigma(2k - 1) + \sigma(2k + 1)}{2} \qquad \text{for all} \qquad k \geq 3,$$
$$\sigma(2k + 1) < \sigma(2k) \qquad\qquad\qquad \text{for all} \qquad k \geq 5.$$

```
            6           6           6           6
        6   5   5   5   5   5   5   6
          6           4           4           4           6
        6   5   4   3   3   3   3   4   5   6
      6           4           2           2           4           6
        5   4   3   2   1   1   2   3   4   5
          5           3           0           3           5
        5   4   3   2   1   1   2   3   4   5
      6           4           2           2           4           6
        6   5   4   3   3   3   3   4   5   6
          6           4           4           4           6
        6   5   5   5   5   5   5   6
            6           6           6           6
```

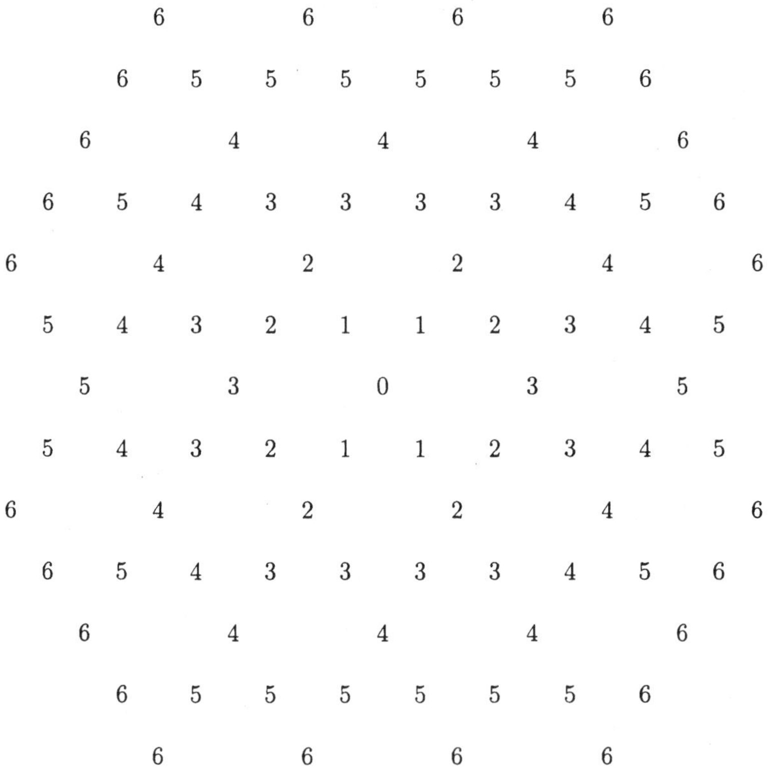

FIGURE 1. Cayley graph for the group of Example 9

10. Finite groups and Coxeter groups. Even though finite groups are not our main concern in these notes (they are quasi-isometric to the group reduced to one element!), they do provide interesting growth series, or rather growth polynomials.

For example, for the symmetric group W of $\{1, 2, \ldots, n\}$ and the generating set $S = \{(1,2), (2,3), \ldots, (n-1, n)\}$, one finds

$$\Sigma(W, S; z) = \prod_{i=1}^{n-1} (1 + z + z^2 + \ldots + z^i).$$

This formula, due to O. Rodrigues, goes back to 1838 ! In those days, it was of course stated in a different and purely combinatorial way; indeed, as L. Solomon puts it, "groups were unknown, except to Galois" [Solom–66].

More generally, one knows the growth series for any Coxeter system with finite generating set (the Coxeter *group* may be infinite); in particular, these series are all rational. The original paper is by L. Solomon [Solom–66]; see also *exercices* 15–26 in § IV.1 of [Bourb–68], as well as [Paris–91] and [Harpe–91]. As a sample result, the growth series of the affine Coxeter groups acting on the Euclidean plane are

$$\Sigma(W, S; t) = \frac{1 + t + \ldots + t^m}{1 - t^m} \frac{1 + t + \ldots + t^n}{1 - t^n}$$

where (m, n) is $(1, 2)$ [respectively $(1, 3)$, $(1, 5)$] for the group W generated by the set S of the reflections fixing the sides of a triangle with angles $(\pi/3, \pi/3, \pi/3)$ [respectively $(\pi/2, \pi/4, \pi/4)$, $(\pi/2, \pi/3, \pi/6)$]. The formulas for the other "affine Coxeter groups" are analogous, in terms of their "Coxeter exponents" (*exercice* VI.4 of [Bourb–68]).

Another example involving finite groups is that of the "Jacobian" of a finite graph (a finite group introduced, among other places, in [BacHN–97]). For an appropriate set of generators of this group, the resulting growth polynomial is, up to normalization, the value $T(1, z)$ of the Tutte polynomial of the graph [Biggs–99].

Exercises for VI.A

11. Exercise. For a finitely-generated group Γ, one can consider the set of all growth series of Γ, with respect to all possible finite sets of generators. Check that these sets are the same for the two groups of order 4.

12. Exercise. Generalize Definition 1 to monoids (see Section II.A), and compute the growth series of

- a free monoid of finite rank with respect to a "basis"
  [this is Exercise II.9 again],
- a free *abelian* monoid $\mathbb{N}^n$ with respect to the natural set of n generators,
- the monoid $\mathbb{N}$ with respect to $\{1, c, c^2, \ldots, c^n\}$ for positive integers
  c and n.

13. Exercise. For a finitely-generated group Γ generated *as a monoid* by a subset S, one has a growth series $B'(z)$, in which the coefficient of z^k is the number of elements in Γ which can be written as words of length at most k in letters of S (no letters in S^{-1} !), and a corresponding spherical growth series $\Sigma'(z) = (1 - z)B'(z)$.

Show that $\Sigma'(z) = 1 + z + z^2 + z^3$ occurs for one of the two groups of order 4 but not for the other. Compare with Exercise 11.

14. Exercise. Given a sequence $(a_k)_{k \geq 0}$ of complex numbers, one can consider, besides the formal power series $\Sigma(z) = \sum_{k=0}^{\infty} a_k z^k$, other expressions such as

$$\text{the exponential series} \quad \Sigma_{exp}(z) = \sum_{k=0}^{\infty} \frac{a_k}{k!} z^k,$$

$$\text{the Dirichlet series} \quad \Sigma_{Dir}(s) = \sum_{k=0}^{\infty} a_k k^{-s}$$

(see [GraKP–91], [Stanl–86], [Wilf–90]). In the particular case of the values $a_k = \sigma(\Gamma, S; k)$ of the spherical growth function of a pair (Γ, S), one finds corresponding variations on the theme of growth series.

(i) For $a_k = \sigma(\mathbb{Z}, \{1\}; k)$, check that

$$\Sigma_{exp}(z) = 2e^z - 1,$$
$$\Sigma_{Dir}(s) = 1 + 2\zeta(s),$$

where ζ denotes the Riemann zeta function.

(ii) Compute a few other examples.

Research question: are there other groups for which exponential series or Dirichlet series have interesting properties that standard growth series do not show ?

15. Exercise. Let Γ, S and $B(\Gamma, S; z)$ be as in Definition 1. For each integer $m \geq 2$, the set $S_m \doteq \{\gamma \in \Gamma \mid \ell_S(\gamma) \leq m\}$ is obviously a generating set of Γ. Show that

$$B(\Gamma, S_m; z^m) = \frac{1}{m} \sum_{j=0}^{m-1} B(\Gamma, S; \zeta^j z)$$

$$\Sigma(\Gamma, S_m; z^m) = \frac{1}{m} \sum_{j=0}^{m-1} \frac{1 - z^m}{1 - \zeta^j z} \Sigma(\Gamma, S; \zeta^j z)$$

where ζ denotes a primitive m^{th} root of 1. (This appears in several places, including [Parr–88]; see also (3.7) in [Wagre–82].)

In particular, if $B(\Gamma, S; z)$ (or, equivalently, $\Sigma(\Gamma, S; z)$) is an algebraic function, then $B(\Gamma, S_m; z^m)$ is also algebraic.

[Hint. With $\beta(k) = \beta(\Gamma, S; k)$, we have on the one hand

$$\frac{1}{m} \sum_{j=0}^{m-1} B(\Gamma, S; \zeta^j z) = \sum_{k=0}^{\infty} \beta(k) z^k \frac{1}{m} \sum_{j=0}^{m-1} \zeta^{kj} = \sum_{k=0}^{\infty} \beta(mk) z^{mk}$$

and on the other

$$B(\Gamma, S_m; z) = \sum_{k=0}^{\infty} \beta(mk) z^k.$$

The first formula follows. Next,

$$\Sigma(\Gamma, S_m; z^m) = (1 - z^m) B(\Gamma, S_m; z^m) = \frac{1}{m} \sum_{j=0}^{m-1} \frac{1 - z^m}{1 - \zeta^j z} (1 - \zeta^j z) B(\Gamma, S; \zeta^j z)$$

and the second formula follows.]

16. Exercise. Consider an integer N (thought of as being large), an integer $d \geq 2$, and the group $\mathbb{Z}$ together with the generating set $S_{N,d} = \{1, N, N^2, \ldots, N^d\}$. Compute the spherical growth series $\Sigma(\mathbb{Z}, S_{N,d}; z)$.

Compare the first terms of this series with those of the spherical growth series of $\mathbb{Z}^d$ with respect to a basis (Example 5.i).

17. Exercise. (i) With the notation of Proposition 4, show that the growth series of the direct product $\Gamma_1 \times \Gamma_2$ with respect to the generating set $(S_1 \cup \{1\}) \times (S_2 \cup \{1\})$ is given by

$$B_{\Gamma_1 \times \Gamma_2}(z) = B_{\Gamma_1}(z) \circ B_{\Gamma_2}(z)$$

where $\circ$ denotes the *Hadamard product* of series:

$$\sum_{k=0}^{\infty} \phi_k z^k \circ \sum_{k=0}^{\infty} \psi_k z^k = \sum_{k=0}^{\infty} (\phi_k \psi_k) z^k.$$

(Mind the fact that $S_1 \times S_2$ needs not generate $\Gamma_1 \times \Gamma_2$.)

(ii) Show that the Hadamard product of two rational power series is again a rational power series. (See Section II.4 of [SalSo–78], or Proposition 4.2.5 of [Stanl–86].)

Problems and complements for VI.A

18. Complete growth series. Let Γ be a finitely-generated group and let S be a finite set of generators. Let $\mathbb{Z}[\Gamma]$ denote the *group ring* of Γ, namely the ring of all functions $\Gamma \to \mathbb{Z}$ with finite support, for the *convolution product* defined by

$$\left(\phi * \psi\right)(\gamma) = \sum_{\gamma_1 \in \Gamma} \phi(\gamma_1) \psi(\gamma_1^{-1}\gamma)$$

for all $\phi, \psi \in \mathbb{Z}[\Gamma]$ and $\gamma \in \Gamma$. One identifies any $\gamma \in \Gamma$ with its characteristic function.

The *complete growth series* of (Γ, S) is the formal power series

$$\Sigma_{\text{comp}}(\Gamma, S; z) \;=\; \Sigma_{\text{comp}}(z) \;=\; \sum_{n=0}^{\infty} \left(\sum_{\substack{\gamma \in \Gamma \\ \ell_S(\gamma)=n}} \gamma \right) z^n \;=\; \sum_{\gamma \in \Gamma} \gamma z^{\ell_S(\gamma)} \;\in\; \mathbb{Z}[\Gamma][[z]]$$

with coefficients in the group ring. The *augmentation map*

$$\epsilon : \begin{cases} \mathbb{Z}[\Gamma] & \to & \mathbb{Z} \\ \sum a_\gamma \gamma & \mapsto & \sum a_\gamma \end{cases}$$

induces a morphism of rings $\mathbb{Z}[\Gamma][[z]] \to \mathbb{Z}[[z]]$, again denoted by ϵ, and we clearly have $\epsilon\left(\Sigma_{\text{comp}}(z)\right) = \Sigma(z) \in \mathbb{Z}[[z]]$.

Show that

$$\Sigma_{\text{comp}}(\mathbb{Z}, S; z) \;=\; 1 \;+\; \sum_{n=1}^{\infty} \left(\delta_n + \delta_{-n} \right) z^n \;=\; \frac{1 - z^2}{1 - (\delta_1 + \delta_{-1})\, z + z^2} \,,$$

and more generally that

$$\Sigma_{\text{comp}}(F_k, S_k; z) \;=\; \frac{1 - z^2}{1 - \left(\sum_{t \in S_k \cup S_k^{-1}} \delta_t \right) z + (2k-1)z^2}$$

for the free group F_k on a set S_k with k elements. (Compare with *exercice 3*, on Hecke operators, in N° II.1.1 of [Serr–77].)

Complete growth series were introduced by F. Liardet [Liard–96]. They have been studied by T. Smirnova-Nagnibeda (see [GriNa–97]), N. Changey [Chang–97], and L. Bartholdi (unpublished), among others; see also Section (F) of [GriHa–97]. They are known to be rational in the following cases: abelian groups (for any choice of generators), Gromov hyperbolic groups (for any choice of generators), and Coxeter groups (for generators of a Coxeter system).

19. Open problem on spheres becoming large. Do there exist an infinite group Γ generated by a finite set S and a positive constant C such that Γ does *not* contain a cyclic group of finite index and such that

$$\sigma(\Gamma, S; k) \le C \qquad \text{for infinitely many values of } k \ ?$$

(compare with Problem 65). If $\sigma(\Gamma, S; k) \le C$ for *all* values of k, then Γ either is finite or has two ends, so that Γ contains a cyclic subgroup of finite index. (This is Theorem 8.3 of [Canno–80], stated there for finitely-*presented* groups. See also Exercise 16 of Chapter 1 in [GhyHp–89].)

20. Computing more growth series. Computations of growth series are often interesting challenges, and we list a few open cases:

- braid groups on standard sets of generators
  (see [Charn–95], though this is *not* for the standard Artin generators),
- knot groups (see [JohKS–95]),
- the Richard Thompson group $\langle s, t \mid [st^{-1}, s^{-1}ts] = [st^{-1}, s^{-2}ts^2] = 1 \rangle$
  [CanFP–96],
- the Baumslag-Solitar group $\Gamma = \langle a, b \mid ab^2a^{-1} = b^3 \rangle$
  (see [BauSo–62], as well as our III.21 and IV.43),
- the Burnside groups [Adian–75],
- the Grigorchuk group of Chapter VIII.

VI.B. Generalities on growth types

There are several ways of defining comparison and equivalence for growth functions. We introduce two of these (Definitions 22 and 29), and the main result of this section is that they are equivalent as far as growth functions of infinite finitely-generated groups are concerned: see Proposition 34 below, which is our formulation of Proposition 3.1 in [Grigo–84]. (Observe however that the argument in [Grigo–84] relies on [Gromo–81] but that ours does not.)

The section ends with Proposition 36, which shows in particular that the growth of the fundamental group of a closed Riemannian manifold is equivalent to the growth of the volumes of Riemannian balls in its universal covering. As far as we know, this was the first historical motivation for the study of growth of groups by Efremovic [Efrem–53] and Milnor [Miln–68b].

We denote by $\mathbb{R}_+$ the closed interval $[0, \infty[$ of the real line. Remember that our set $\mathbb{N} = \{0, 1, 2, \ldots\}$ does contain 0.

21. Definition. A *growth function* is a non-decreasing function from $\mathbb{R}_+$ to $\mathbb{R}_+$.

To any non-decreasing function $\beta : \mathbb{N} \longrightarrow \mathbb{N}$ one can associate the growth function $\alpha(t) = \beta(\lceil t \rceil)$, where $\lceil t \rceil$ denotes the smallest integer such that $t \leq \lceil t \rceil$. On can also choose $\tilde{\alpha}(t) = \beta(k)^{1-\tau}\beta(k+1)^\tau$, where $t = k + \tau$ with $0 \leq \tau \leq 1$; observe $\tilde{\alpha}$ is continuous. If β is submultiplicative in the sense of Item 54, α and $\tilde{\alpha}$ are also submultiplicative. Consequently, all considerations below about growth functions apply also to non-decreasing functions $\mathbb{N} \longrightarrow \mathbb{N}$.

Both choices occur in the literature, e.g. α in Lemma 3.1 of [Grigo–84] and $\tilde{\alpha}$ in § 1 of [Shubi–86]. A third frequent choice is $\alpha'(t) = \beta[t]$, with $[t]$ is the integral part of t. However, β may be submultiplicative and α' not: if β is the growth function of the free group of rank 2 with respect to a basis, then $\frac{\alpha'(p+q+1)}{\alpha'(p+0.5)\alpha'(q+0.5)} = \frac{4 \cdot 3^{p+q}}{4 \cdot 3^{p-1} \cdot 4 \cdot 3^{q-1}} = \frac{9}{4}$ is strictly larger than 1.

Most growth functions which occur below are unbounded, but some *are* bounded, such as growth functions of finite groups.

22. Definition. A growth function α_2 *weakly dominates* a growth function α_1, and this is denoted by

$$\alpha_1 \overset{w}{\prec} \alpha_2 \qquad \text{or} \qquad \alpha_1(t) \overset{w}{\prec} \alpha_2(t),$$

if there exist constants $\lambda \geq 1$ and $C \geq 0$ such that

$$\alpha_1(t) \leq \lambda \alpha_2(\lambda t + C) + C$$

for all $t \in \mathbb{R}_+$.

For strictly positive growth functions eventually bounded below by 1, an equivalent definition of the relation $\overset{w}{\prec}$ appears in Proposition 25 below.

Two growth functions α_1, α_2 are *weakly equivalent*, and this is denoted by

$$\alpha_1 \overset{w}{\sim} \alpha_2 \qquad \text{or} \qquad \alpha_1(t) \overset{w}{\sim} \alpha_2(t),$$

if each weakly dominates the other. We denote by $[\alpha]_w$ or $[\alpha(t)]_w$ the equivalence class modulo $\overset{w}{\sim}$ of a growth function α.

When we wish to emphasize the values of growth functions on integers, we write also $\alpha_1(k) \overset{w}{\prec} \alpha_2(k)$ and $\alpha_1(k) \overset{w}{\sim} \alpha_2(k)$.

23. Examples. For $a, b \in {]}0, \infty[$, we have

$$t^a \overset{w}{\prec} t^b \quad \Longleftrightarrow \quad a \leq b$$

and

$$e^{at} \overset{w}{\sim} e^{bt}.$$

If a growth function α is polynomial of degree d, we have $\alpha(t) \overset{w}{\prec} t^a$ if and only if $d \leq a$ and $\alpha(t) \overset{w}{\sim} t^a$ if and only if $d = a$.

Let Γ be a group generated by a finite set S. Then

$$\beta(\Gamma, S; k) \overset{w}{\prec} e^k$$

by Property 3.iii.

In the notation of Theorem IV.23 (Γ acting properly and cocompactly by isometries on a metric space X which is geodesic and proper), the growth functions

$$k \longmapsto \beta(\Gamma, S; k) = \left| \{ \gamma \in \Gamma \mid d_S(1, \gamma) \leq k \} \right|$$

and

$$k \longmapsto \left| \{ \gamma \in \Gamma \mid d(x_0, \gamma x_0) \leq k \} \right|$$

are weakly equivalent (the second of these is a weight as defined in Complement 42 below). If moreover X is a Riemannian manifold, they are also equivalent to the function whose value at k is the Riemannian volume of a ball in X of radius k about the base point x_0 (Proposition 36).

24. Lemma. *Let α_1, α_2 be two growth functions. Assume that $\alpha_2(t) \geq 1$ for t large enough. Let t_0 be such that $\alpha_2(t_0) > 0$. If $\alpha_1 \overset{w}{\prec} \alpha_2$, there exists a constant $\rho \geq 1$ such that*

$$\alpha_1(t) \leq \rho\alpha_2(\rho t)$$

for all $t \geq t_0$.

Proof. By hypothesis, there exist two constants $\lambda \geq 1$ and $C \geq 0$ such that

$$\alpha_1(t) \leq \lambda\alpha_2(\lambda t + C) + C$$

for all $t \in \mathbb{R}_+$. Let $t_1 \geq 1$ be such that $\alpha_2(\lambda t_1 + C) \geq 1$; then

$$\alpha_1(t) \leq (\lambda + C)\alpha_2((\lambda + C)t)$$

for all $t \geq t_1$. If $t_1 \leq t_0$, there is nothing more to prove.

Otherwise, set

$$\rho = \max\left\{\frac{\alpha_1(t_1)}{\alpha_2(t_0)}, \lambda + C\right\}.$$

Then

$$\alpha_1(t) \leq \alpha_1(t_1) \leq \rho\alpha_2(t_0) \leq \rho\alpha_2(\rho t)$$

for all $t \in [t_0, t_1]$, and this concludes the proof. $\square$

The following proposition is a particular case of Lemma 24.

25. Proposition. *Let α_1, α_2 be two growth functions; assume that $\alpha_2(0) > 0$ and that $\alpha_2(t) \geq 1$ for t large enough.*

Then $\alpha_1 \overset{w}{\prec} \alpha_2$ if and only if there exists a constant ρ such that

$$\alpha_1(t) \leq \rho\alpha_2(\rho t)$$

for all $t \in \mathbb{R}_+$.

26. Growth for uniformly quasi-locally bounded pseudo-metric spaces. A pseudo-metric space X with pseudo-metric d is *uniformly quasi-locally bounded* if

$$\sup_{x \in X} |B(x,t)| < \infty \qquad \text{for all} \qquad t \in \mathbb{R}_+$$

where $B(x,t) = \{y \in X \mid d(x,y) \leq t\}$ denotes the (closed!) ball of radius t about x. This is a property which plays an important role in the study of various "pseudo-difference operators" on X, such as the Laplace operator and the Green's function [Shubi–86]. The terminology is that of Gromov. (See page 12 of [Gromo–93]; see also Exercise IV.37.)

The *growth function* of such a space X with respect to a base point x_0 is the function β whose value at $t \geq 0$ is the cardinality of the ball $B(x_0, t)$.

The metric space underlying a finitely-generated group endowed with some word metric is clearly uniformly quasi-locally bounded, and in this case the present definition of the growth function coincides with that of Section VI.A. Recall from Example 9 that *spherical* growth functions need not be monotonic (even for infinite groups), and thus are not necessarily growth functions in the sense of Definition 21.

27. Proposition. *For $j = 1, 2$, let X_j be a pseudo-metric space which is uniformly quasi-locally bounded and let x_j be a base point in X_j. Let β_j denote the corresponding growth function.*

If there exists a quasi-isometric embedding of X_1 into X_2, then

$$\beta_1 \overset{w}{\prec} \beta_2.$$

In particular, if X_1 and X_2 are quasi-isometric, then

$$\beta_1 \overset{w}{\sim} \beta_2.$$

Proof. By hypothesis, there exist a mapping $\phi : X_1 \longrightarrow X_2$ and constants $\lambda \geq 1, C \geq 0$ such that

(*) $$\lambda^{-1} d_1(x, y) - C \leq d_2(\phi(x), \phi(y)) \leq \lambda d_1(x, y) + C$$

for all $x, y \in X_1$. Setting $D = C + d_2(x_2, \phi(x_1))$, we have

$$\phi\big(B(x_1, t)\big) \subset B(x_2, \lambda t + D)$$

for all $t \in \mathbb{R}_+$ by the second inequality in (*). Thus

(**) $$\big|\phi\big(B(x_1, t)\big)\big| \leq \beta_2(\lambda t + D)$$

for all $t \in \mathbb{R}_+$.

If $x, y \in X_1$ have the same image by ϕ, we have $d_1(x, y) \leq \lambda C$ by the first inequality in (*). It follows from the uniformly quasi-locally boundedness that there exists a constant $E \geq 1$ such that

$$\sup_{z \in X_2} \big|\phi^{-1}(z)\big| \leq E.$$

Thus

(***) $$\beta_1(t) = |B(x_1, t)| \leq E\big|\phi\big(B(x_1, t)\big)\big|$$

for all $t \in \mathbb{R}_+$.

The proposition follows from (**) and (***). $\square$

28. Consequences. Let Γ be a finitely-generated group. If S, T are two finite sets of generators of Γ, the corresponding metric spaces (Γ, d_S) and (Γ, d_T) are quasi-isometric (Item IV.22), so that the corresponding growth functions are weakly equivalent:

$$\beta(\Gamma, S; k) \overset{w}{\sim} \beta(\Gamma, T; k).$$

This equivalence class, denoted by $[\beta(\Gamma; k)]_w$, depends only on the group, and not on any generating set; it is called the *w-growth type* of the group. [This definition will be slightly amended below! see Definition 35.]

Two groups Γ_1, Γ_2 which are quasi-isometric have the same w-growth type.

In particular, note the following consequence of Theorem IV.23. Consider a metric space X which is geodesic and proper, and two groups Γ_1, Γ_2 acting on X by isometries; assume that the two actions are proper with compact quotients. Then the finitely-generated groups Γ_1 and Γ_2 have the same w-growth type.

$$* * * * * * * \qquad * * * * * * * \qquad * * * * * * *$$

29. Definition. A growth function α_2 *strongly dominates* a growth function α_1, and this is denoted by

$$\alpha_1 \overset{s}{\prec} \alpha_2 \qquad \text{or} \qquad \alpha_1(t) \overset{s}{\prec} \alpha_2(t),$$

if there exists a constant $\lambda \geq 1$ such that

$$\alpha_1(t) \leq \alpha_2(\lambda t)$$

for all $t \in \mathbb{R}_+$. Two growth functions α_1, α_2 are *strongly equivalent*, and this is denoted by

$$\alpha_1 \overset{s}{\sim} \alpha_2 \qquad \text{or} \qquad \alpha_1(t) \overset{s}{\sim} \alpha_2(t),$$

if each strongly dominates the other. We denote by $[\alpha]_s$ or $[\alpha(t)]_s$ the equivalence class modulo $\overset{s}{\sim}$ of a growth function α.

30. Example. For two growth functions α_1, α_2, one obviously has

$$\alpha_1 \overset{s}{\prec} \alpha_2 \qquad \Longrightarrow \qquad \alpha_1 \overset{w}{\prec} \alpha_2,$$

$$\alpha_1 \overset{s}{\sim} \alpha_2 \qquad \Longrightarrow \qquad \alpha_1 \overset{w}{\sim} \alpha_2.$$

Set $\alpha(t) = \max\{1, \log\log t\}$ and $\beta(t) = 2\alpha(2t)$ for all $t \in \mathbb{R}_+$. Then $\beta \overset{w}{\prec} \alpha$ by definition, but it is *not true* that $\beta \overset{s}{\prec} \alpha$ (Exercise 37).

31. Example. Let X be a pseudo-metric space which is uniformly quasi-locally bounded (see Item 26) and let x_1, x_2 be two base points in X. Then the two corresponding growth functions $\beta_j : t \longmapsto |B(x_j, t)|$ are strongly equivalent.

Indeed, if $C = d(x_1, x_2)$, then $\beta_1(t) \leq \beta_2(t + C)$ for all $t \in \mathbb{R}_+$. A fortiori, $\beta_1(t) \leq \beta_2(\lambda t)$ for $\lambda = 1 + \frac{C}{t_{min}}$, where t_{min} is the minimal distance separating x_2 from another point of X.

32. Consequences. Let Γ be a finitely generated group. If S, T are two finite sets of generators of Γ, the corresponding growth functions are strongly equivalent:

$$\beta(\Gamma, S; k) \overset{s}{\sim} \beta(\Gamma, T; k).$$

We will see below (Definition 35) that there is no reason to introduce a special notation for the strong equivalence class of the growth functions of finitely-generated groups.

$$* * * * * * * \qquad * * * * * * * \qquad * * * * * * *$$

33. Lemma. *Let Γ be an infinite finitely-generated group given together with a finite set S of generators, and let $\beta(k) = \beta(\Gamma, S; k)$ denote the corresponding growth function. For any real number $\rho \geq 1$, there exists an integer $K \geq 1$ such that*

$$\rho\beta(\rho k) \leq \beta(Kk)$$

for all $k \geq 1$.

Proof. Consider an integer $k \geq 1$ and the ball $B(1, 3k)$ of radius $3k$ centred at the group unit. The argument indicated for Exercise IV.12 shows that there exists a geodesic segment $(x_{-3k}, \ldots, x_{-1}, x_0 = 1, x_1, \ldots, x_{3k})$ centred at the group unit and contained in $B(1, 3k)$. As the balls of radius k about x_{-k-1} and x_{k+1} are disjoint,

$$2\beta(k) \leq \beta(3k).$$

By iteration, we also have

$$2^m\beta(k) \leq \beta(3^m k)$$

for all $m \geq 0$ and $k \geq 1$. Without loss of generality, we may assume that ρ is an integer; choosing m such that $2^m \geq \rho$ and setting $K = 3^m \rho$, we have finally

$$\rho\beta(\rho k) \leq 2^m\beta(\rho k) \leq \beta(3^m \rho k) = \beta(Kk)$$

for all $k \geq 1$. $\square$

34. Proposition. *For $j = 1, 2$, let Γ_j be an infinite finitely-generated group given together with a finite set S_j of generators, and let $\beta_j(k) = \beta(\Gamma_j, S_j; k)$ denote the corresponding growth function. Then,*

$$\beta_1 \overset{w}{\prec} \beta_2 \quad \Longleftrightarrow \quad \beta_1 \overset{s}{\prec} \beta_2.$$

Proof. The non-trivial implication is a straightforward consequence of the previous lemma and of Proposition 25. $\square$

Observe that, if Γ_1, Γ_2 in the proposition were *finite* groups with $|\Gamma_1| > |\Gamma_2|$, we would have $\beta_1 \overset{w}{\prec} \beta_2$ but *not* $\beta_1 \overset{s}{\prec} \beta_2$!

35. Definition. The *growth type* of an infinite finitely generated group Γ is the equivalence class (according to one of Definitions 22 and 29) of its growth functions, as defined in Section VI.A; it is denoted by $[\beta(\Gamma; k)]$. Observe that there is no need to specify a generating set (see Item 28) or an index (w or s).

It is convenient to decide that all finite groups have the same growth type.

It follows from Proposition 27 that two finitely-generated groups which are quasi-isometric have the same growth type. For example, the fundamental groups of two compact Riemannian manifolds with isometric universal coverings have the same growth type; and groups with different growth types, such as $\mathbb{Z}^m$ and $\mathbb{Z}^n$ for $m \neq n$, are not quasi-isometric (see Examples 2.iv and 23).

It is straightforward to construct two abstract growth functions (in the sense of Definition 21), say α_1, α_2, such that neither $\alpha_1 \overset{w}{\prec} \alpha_2$ nor $\alpha_2 \overset{w}{\prec} \alpha_1$ holds. It is less straightforward to construct two finitely generated groups Γ_1, Γ_2, with corresponding growth functions β_1, β_2, such that neither $\beta_1 \prec \beta_2$ nor $\beta_2 \prec \beta_1$ holds; but it is indeed a result of Grigorchuk that there exist uncountably many finitely-generated groups of which the growth functions are pairwise noncomparable with respect to $\prec$ (Theorem 7.2 in [Grigo–84], and Theorem 1 in [Grigo–91]).

36. Proposition (Schwarzc, Milnor). *Let X be a complete Riemannian manifold with Riemannian distance d. Choose a point $x_0 \in X$ and, for each real number $t \geq 0$, let $v(t)$ denote the Riemannian volume of the closed ball of centre t and radius x_0 in X.*

Let Γ be a group acting properly by isometries on X in such a way that the quotient $\Gamma \setminus X$ is compact. Let $\beta(k)$ denote the growth function of Γ with respect to some finite set of generators.

Then v and β are weakly equivalent growth functions.

Proof. For $x \in X$ and $t \geq 0$, let $B(x,t)$ denote the closed ball of centre x and radius t in X. As in the proof of Theorem IV.23, consider

the diameter R of the quotient space $\Gamma \setminus X$,
the covering $(\gamma B(x_0, R))_{\gamma \in \Gamma}$ of X,
the finite set of generators $S = \{s \in \Gamma \mid s \neq 1 \text{ and } sB \cap B \neq \emptyset\}$ of Γ,
the separating distance $r = \min \{d(B, sB) \mid \gamma \in \Gamma, \gamma \notin S \cup \{1\}\}$,
the distance $\lambda = \max \{d(x_0, sx_0) \mid s \in S\}$

as well as

the order a of the isotropy subgroup $\Gamma_0 = \{\gamma \in \Gamma \mid \gamma x_0 = x_0\}$

(observe that Γ_0 is finite and contained in $S \cup \{1\}$).

Let k be an integer, $k \geq 0$. As the closed balls $B(x, \frac{1}{3}r)$, for x in the orbit Γx_0, are pairwise disjoint, we have

$$\frac{1}{a} \beta(\Gamma, S; k) \, v \left(\frac{1}{3}r \right) \leq v \left(k\lambda + \frac{1}{3}r \right).$$

It follows that $\beta(\Gamma, S; k) \overset{w}{\prec} v(k)$. [Note that $\overset{w}{\prec}$ may not be replaced by $\overset{s}{\prec}$, since $\beta(\Gamma, S; 0) = 1$ and $v(0) = 0$.]

Let $x \in B(x_0, k)$. There exists $\gamma \in \Gamma$ such that $x \in B(\gamma x_0, R)$. Recall from the proof of Theorem IV.23 that the word length $\ell_S(\gamma)$ of γ satisfies

$$\ell_S(\gamma) \leq \frac{1}{r} d(x_0, \gamma x_0) + 1 \leq \frac{1}{r} d(x_0, x) + c$$

where c denotes the constant $\frac{R}{r} + 1$. Consequently the ball $B(x_0, k)$ is covered by the balls $\gamma B(x_0, R)$, for $\gamma \in \Gamma$ with $\ell_S(\gamma)$ bounded as above, and

$$v(k) \leq \beta\left(\Gamma, S; \frac{k}{r} + c\right) v(R).$$

It follows that $v(k) \overset{w}{\prec} \beta(\Gamma, S; k)$. $\square$

Proposition 36 applies to the universal covering manifold and the fundamental group of a compact manifold. For the generalization to other Galois coverings and the corresponding coset spaces of the fundamental group, we refer to Section 1 of [MouPe–74].

Consequences of Proposition 36 appear in Items VII.6 and VII.7.

Exercises for VI.B

37. Exercise. Check the details of Example 30.

38. Exercise. Let (Γ, S) and β be as in Lemma 33. Show that

$$\beta(5k) \leq \frac{\left(\beta(4k)\right)^2}{\beta(k)}$$

for all $k \geq 1$.

[Sketch. For $x \in \Gamma$ and $r \geq 0$, set $B(x, r) = \{\gamma \in \Gamma \mid d(x, \gamma) \leq r\}$. Let N be a subset of $B(1, 3k)$ such that

$$\gamma, \gamma \in N, \quad \gamma \neq \gamma' \quad \Longrightarrow \quad d(\gamma, \gamma') \geq 2k + 1$$

and which is maximal for this property. First, we have

$$|N|\beta(k) \leq \beta(4k).$$

Second, we have

$$B(1, 5k) \subset \bigcup_{y \in N} B(y, 4k)$$

so that

$$\beta(5k) \leq |N|\beta(4k)$$

for all $k \geq 1$.]

39. Exercise. Deduce from the previous exercise that, when $k \geq 1$ is a multiple of 4, we have successively

$$\log \beta(6k) \leq 4 \log \beta(4k) - 3 \log \beta(k)$$

[write $6k \leq 5\left(k + \frac{k}{4}\right)$], then

$$\log \beta(7k) \leq 8 \log \beta(4k) - 7 \log \beta(k)$$

[write $7k \leq 5\left(k + \frac{2k}{4}\right)$], and finally

$$\log \beta(8k) \leq 16 \log \beta(4k) - 15 \log \beta(k)$$

[write $8k \leq 5\left(k + \frac{3k}{4}\right)$]. Deduce that

$$\log \beta\left(2^j k\right) \leq 16^j \left[\log \beta\left(k\right) - \log \beta\left(\frac{k}{4}\right)\right] + \log \beta\left(\frac{k}{4}\right)$$

for all $k \equiv 0 \pmod{16}$ and for all $j \geq 0$.

The content of Exercises 38 and 39 appears in Section 3 of [Gromo–81].

Problems and complements for VI.B

40.   Which functions are growth functions of finitely-generated groups ? Given a growth function $\alpha : \mathbb{R}_+ \longrightarrow \mathbb{R}_+$, little is known about when there exist pairs (Γ, S) such that $\alpha(k)$ and $\beta(\Gamma, S; k)$ are equal, or sufficiently "similar"; a particular case of this question appears in Item VII.29. Rather than speculate on possible partial answers, let us quote two results from other subjects.

Let $\alpha : \mathbb{R}_+ \longrightarrow \mathbb{R}_+$ be a function which is non-decreasing (as in Definition 21), unbounded, and continuous. Then there exists an entire function f such that

$$\sup \left\{ |f(z)| \,\big|\, |z| \leq t \right\} \geq \alpha(t)$$

for all $t \in \mathbb{R}_+$. See [Rubel–96], Theorem 10.3; *there*, a growth function is defined to be non-decreasing, continuous, and unbounded (Definition 13.1.7); but the last two requirements are not relevant to *our* subject.

Let M be a non-compact smooth manifold of dimension $n \geq 2$ given together with a base-point x_0. Let $\alpha : \mathbb{R}_+ \longrightarrow \mathbb{R}_+$ be a function of class C^1 such that $\alpha(0) = 0$, $\alpha'(t) > 0$ for all $t > 0$ and $\lim_{t \to 0} \frac{\alpha'(t)}{\alpha_0'(t)} = 1$, where $\alpha_0(t)$ denotes the volume of a ball of radius t in $\mathbb{R}^n$. Then there exists on M a continuous Riemannian metric for which the growth of balls centred in x_0 is precisely α [GriPa–94].

41. Another interpretation of (the class of) a growth function. Let Γ be a finitely-generated group, let S be a finite set of generators, let F_S denote the free group of base S, let $\pi : F_S \longrightarrow \Gamma$ denote the corresponding projection, and set $N = Ker(\pi)$.

For each integer $k \geq 0$, let N_k denote the subgroup of N generated by $\{\gamma \in N \mid \ell_S(\gamma) \leq k\}$ and let $r_N(k)$ denote the smallest number n for which there exists a system of n generators of N_k. It is known that there exist constants $C_1, C_2, \lambda_1, \lambda_2 > 0$ and an integer $K \geq 0$ such that

$$C_1 \beta(\Gamma, S; \lambda_1 k) \leq r_N(k) \leq C_2 \beta(\Gamma, S; \lambda_2 k)$$

for all $k \geq K$ [Osin–99a]. As a particular case, this shows again the claim of Example III.3: N is finitely generated as a group if and only if Γ is finite.

More generally [Osin–99b], for a group Γ generated by a finite set S and a subgroup Δ (the free group F_S and the subgroup N above), define the *general rank growth function* of H by

$$r_\Delta(k) =$$

$$\min \left\{ n \geq 0 \; \middle| \; \begin{array}{c} \text{there exists a } n\text{-generator subgroup } \Delta_0 \text{ of } \Delta \text{ containing} \\ \text{the } \ell_S\text{-ball of radius } k \text{ about the origin} \end{array} \right\}.$$

D.V. Osin has shown that the weak equivalence class of r_Δ (in the sense of Definition 22) is independent of the choice of S.

Given any non-decreasing function $f : \mathbb{N} \longrightarrow \mathbb{N}$ such that $f(k) \leq a^k$ for some $a \geq 1$ and all $k \in \mathbb{N}$, Osin has also announced that there is a subgroup Δ of the free group on two generators such that $r_\Delta(k) \overset{w}{\sim} f(k)$. He has moreover results on the set of functions appearing as general rank growth functions for subgroups of various classes of groups Γ, e.g. for subgroups of finitely-generated solvable groups, or linear groups.

42. Weights: definition and examples. A *weight* on a group Γ is a function

$$\lambda : \Gamma \longrightarrow \mathbb{R}_+$$

such that

$$\lambda(\gamma_1 \gamma_2) \leq \lambda(\gamma_1) + \lambda(\gamma_2)$$

for all $\gamma_1, \gamma_2 \in \Gamma$. To such a weight there corresponds the left-invariant pseudo-metric d_λ on Γ defined by

$$d_\lambda(\gamma_1, \gamma_2) = \lambda(\gamma_1^{-1} \gamma_2)$$

(see IV.19 for "pseudo-metric"). In case the function λ is *proper*, namely in case $\lambda^{-1}[0, t]$ is a finite subset of Γ for all $t \in \mathbb{R}_+$, it is natural to define the function

$$t \longmapsto \beta(\Gamma, \lambda; t) \doteq \left| \left\{ \gamma \in \Gamma \mid \lambda(\gamma) \leq t \right\} \right|.$$

Weights on groups generalize word lengths. Here are three kinds of examples.

(i) If Γ is a subgroup of a finitely generated group Δ, the restriction to Γ of a word length on Δ is a proper weight, sometimes called a *relative growth function*. (See Item 43 below.)

(ii) If Γ is a group given together with a finite set S of generators and a function $\sigma : S \longrightarrow \mathbb{R}_+$ (say for simplicity with $S^{-1} = S$ and $\sigma(s^{-1}) = \sigma(s)$ for all $s \in S$), one defines a weight λ on Γ by

$$\lambda(\gamma) \; = \; \min \left\{ t \in \mathbb{R}_+ \; \middle| \; \begin{array}{c} \text{there exist } s_1, \ldots, s_n \in S \\ \text{with } \gamma = s_1 \ldots s_n \text{ and } t = \sum_{j=1}^{n} \sigma(s_j) \end{array} \right\}.$$

A weight of this kind has been used quite efficiently in [Barth–98].

(iii) For a group Γ which acts properly by isometries on a metric space (X, d) with base point x_0, the formula $d_\Gamma(\gamma_1, \gamma_2) = d(\gamma_1 x_0, \gamma_2 x_0)$ defines a pseudo-metric on Γ, and the formula

$$\lambda(\gamma) \; = \; d(x_0, \gamma x_0)$$

defines a proper weight on Γ.

A subexample of (iii) is given by the weight of Section I.A, namely by the "Euclidean" proper weight on $\mathbb{Z}^2$:

$$\lambda(a, b) \; = \; \sqrt{a^2 + b^2}$$

for all $(a, b) \in \mathbb{Z}^2$.

43. Distortion functions for subgroups. Let Γ be a subgroup of a finitely-generated group Δ and let $\lambda : \Gamma \to \mathbb{N}$ be the restriction to Γ of some word length on Δ, as in Complement 42.i. It is obvious that

$$\begin{array}{lll} \lambda(\gamma^{-1}) = \lambda(\gamma) & \text{for all} & \gamma \in \Gamma, \\ \lambda(\gamma) = 0 & \text{if and only if} & \gamma = 1, \\ \lambda(\gamma_1 \gamma_2) \leq \lambda(\gamma_1) + \lambda(\gamma_2) & \text{for all} & \gamma_1, \gamma_2 \in \Gamma, \\ |\{\gamma \in \Gamma \mid \lambda(\gamma) \leq k\}| \leq a^k & \text{for some} \quad a \geq 1 \quad \text{and for all} & k \geq 0. \end{array}$$

Conversely, given a weight $\lambda : \Gamma \to \mathbb{N}$ satisfying these conditions, there exist a finitely-generated overgroup Δ of Γ and a word length on Δ whose restriction λ' to Γ satisfies

$$c_1 \lambda(\gamma) \leq \lambda'(\gamma) \leq c_2 \lambda(\gamma)$$

for all $\gamma \in \Gamma$ and for appropriate constants $c_1, c_2 > 0$ [Ol's–99b]; see [Ol's–97] for the case of finitely-*presented* overgroups.

More on distortion in Chapter 3 of [Gromo–93], [LubMR–93], [Farb–94], [Varop–99], and in [Osin].

44. Weights which do not extend. The following example, which is Remark 2.1.7 in [Jolis–90], shows that proper weights *do not always extend to over-groups.*

Consider the free group F_2 on two generators x and y, the automorphism α of F_2 of order 2 defined by $\alpha(x) = xy$ and $\alpha(y) = y^{-1}$, and the corresponding semi-direct product

$$\Gamma = F_2 \rtimes_\alpha C_2$$

where $C_2 = \{1, -1\}$ denotes the group of order two.

Suppose there existed a weight $\lambda : \Gamma \to \mathbb{R}_+$ which extends the usual word length $\ell = \ell_{\{x,y\}}$ on F_2. One would have

$$2k = \ell\big(\alpha(x^k)\big) = \lambda\big((\alpha(x^k), 1)\big) = \lambda\big((1, -1)(x^k, 1)(1, -1)\big)$$
$$\leq \ell(x^k) + 2\lambda\big((1, -1)\big) = k + 2\lambda\big((1, -1)\big)$$

for all $k \geq 1$, and this is absurd.

Similarly, considering the Baumslag-Solitar group $BS(p, q)$ with 2 generators s, t and one relation $st^p s^{-1} = t^q$ (with $p \neq q$), we leave it to the reader to check that the weight $t^k \longmapsto |k|$ on the subgroup generated by t does not extend to a weight on $BS(p, q)$.

45. Comparison between a word metric and a geometrical metric.

Let G be a semi-simple Lie group which is connected and with finite centre, and let Γ be an irreducible lattice in G. Let S be a finite set of generators of Γ and let d_S denote the corresponding *word metric* on Γ. (See Complements IV.42 and V.20.)

Let d_R denote the *geometrical metric,* given by the restriction to Γ of a left-invariant Riemannian distance function on G. (Observe that $\gamma \mapsto d_R(1, \gamma)$ is a proper weight in the sense of Complement 42.)

If $\Gamma \backslash G$ is compact, the proof of Theorem IV.23 shows that there exist two constants $\lambda \geq 1, C \geq 0$ such that

(*)
$$\frac{1}{\lambda} d_S(\gamma, \gamma') - C \leq d_R(\gamma, \gamma') \leq \lambda d_S(\gamma, \gamma')$$

for all $\gamma, \gamma' \in \Gamma$. In particular, the metric spaces (Γ, d_R) and (Γ, d_S) are quasi-isometric.

If $G = SL(2, \mathbb{R})$ and $\Gamma = SL(2, \mathbb{Z})$, this does not hold. Indeed, if $u = \begin{pmatrix} 1 & 1 \\ 0 & 1 \end{pmatrix}$, then it is easy to check that $d_R(1, u^k) = O(\log k)$ whereas $d_S(1, u^k)$ grows linearly with k.

It is a theorem that (*) holds for any lattice in $SL(n, \mathbb{R})$ when $n \geq 3$, and more generally if the semi-simple group G has "real rank" at least 2 [LubMR–93].

46. Another comparison between a word metric and a geometrical metric.

Let Γ be a discrete group of isometries of the n-dimensional hyperbolic space $\mathbb{H}^n$. Let $\lambda_S : \Gamma \to \mathbb{N}$ be some word length on Γ and let $\lambda_{\mathbb{H}} : \Gamma \to \mathbb{N}$ be defined by $\lambda_{\mathbb{H}}(\gamma) = d(x_0, \gamma x_0)$ for some $x_0 \in \mathbb{H}^n$, as in Complement 42.iii. We assume that $\Gamma \backslash \mathbb{H}^n$ is compact and that $\gamma x_0 \neq x_0$ for $\gamma \in \Gamma$ with $\gamma \neq 1$. It follows from Claim 2 in the proof of Theorem IV.23 that there exist constants $A, B > 0$ such that

$$A \leq \frac{\lambda_S(\gamma)}{\lambda_{\mathbb{H}}(\gamma)} \leq B$$

for all $\gamma \in \Gamma$.

This has been refined by Pollicott and Sharp: the quantity

$$\frac{\sum_{\gamma \in \Gamma, \lambda_{\mathbb{H}}(\gamma) \leq T} \left(\frac{\lambda_S(\gamma)}{\lambda_{\mathbb{H}}(\gamma)} \right)}{\sharp \{ \gamma \in \Gamma \mid \lambda_{\mathbb{H}}(\gamma) \leq T \}}$$

has a limit when $T \to \infty$ and there exists a constant $C > 0$ such that

$$\sum_{\gamma \in \Gamma, \lambda_{\mathbb{H}}(\gamma) \leq T} \frac{\lambda_S(\gamma)}{\lambda_{\mathbb{H}}(\gamma)} \sim C e^{(n-1)T}$$

as $T \to \infty$ (i.e., the quotient of the left-hand side by the right-hand side tends to 1 when $T \to \infty$); see [PolSh–98].

47. An exercise on maps of degree $\neq 0$, and comparison of fundamental groups. Let X_1, X_2 be two connected closed manifolds of the same dimension and let Γ_1, Γ_2 denote their fundamental groups. Assume that there exists a continuous mapping

$$f : X_1 \longrightarrow X_2$$

of degree different from zero. Show that

$$[\beta(\Gamma_2; k)] \prec [\beta(\Gamma_1; k)]$$

(the notation $[\beta(\Gamma; k)]$ was defined in Definition 35).

Sketch. It is enough to check that the canonical image of Γ_1 by f_* is of finite index in Γ_2. If this were not the case, there would exist an infinite covering Y_2 of X_2 and a lift $\tilde{f} : X_1 \longrightarrow Y_2$ of f corresponding to the subgroup $f_*(\Gamma_1)$ of Γ_2; hence f would induce the zero map $H_n(X_1) \longrightarrow H_n(X_2)$ in the top-dimensional homology (where $n = dim(X_1) = dim(X_2)$), in contradiction with the hypothesis that the degree of f is not zero. (This exercise and its solution were shown to me by Marc Troyanov.)

48. Other orderings of growth functions. There are other interesting orderings on the set of growth functions, besides $\overset{w}{\prec}$ and $\overset{s}{\prec}$ of Definitions 22 and 29. For example one can define

$$\alpha_1 \overset{D}{\prec} \alpha_2$$

if there exist constants $A, B, C > 0$ such that

$$\alpha_1(t) \leq A\alpha_2(Bt) + Ct$$

for all $t \in \mathbb{R}_+$; this is indeed "the" natural definition in the study of the so-called *Dehn functions* on finitely-presented groups (see Item V.57). One can also define

$$\alpha_1 \overset{pol}{\prec} \alpha_2$$

if there exist integers $p, q \geq 1$ such that

$$\alpha_1(t) \leq (\alpha_2(t^p))^q$$

for all $t \in \mathbb{R}_+$; this is indeed "the" natural definition in some study of relative growth functions, in work of D.V. Osin (see Complements 42 and 43).

VI.C. Exponential growth rate and entropy

49. Definition. Let $(a_k)_{k \geq 0}$ be a non-decreasing sequence of positive real numbers. Its *exponential growth rate* is the upper limit

$$\omega = \limsup_{k \to \infty} \sqrt[k]{a_k} \in [0, \infty].$$

50. Observations. Let $(a_k)_{k \geq 0}$, $(a'_k)_{k \geq 0}$ be two sequences as above and let ω, ω' be the corresponding exponential growth rates.
  (i) ω^{-1} is the radius of convergence of the series $\sum_{k=0}^{\infty} a_k z^k$.
  (ii)   If $a_k \overset{w}{\prec} a'_k$ in the sense of Definition 22, then

$$\omega > 1 \quad \Longrightarrow \quad \omega' > 1.$$

In general, there is no more precise relationship between ω and ω'. Consider the example $a_k = \beta(k), a'_k = \beta'(k)$, with the notation of (iii) and (iv) in Item 7. We have $a_k \sim a'_k$, and also $\omega > 1$ and $\omega' > 1$, in accordance with Observation (ii) above. But note that

$$\sqrt{2} = \omega \neq \omega' = \frac{1}{2}(\sqrt{5} + 1).$$

51. Definition. Let Γ be a group generated by a finite set S and let $\beta(\Gamma, S; k)$ denote the corresponding growth function. The *exponential growth rate* of the pair (Γ, S) is the upper limit

$$\omega(\Gamma, S) = \limsup_{k \to \infty} \sqrt[k]{\beta(\Gamma, S; k)}$$

and its *entropy* is

$$h(\Gamma, S) = \log \omega(\Gamma, S).$$

(It follows from Proposition 56 below that the upper limit is in fact a limit.)
The group Γ is said to be of

exponential growth if $\omega(\Gamma, S) > 1$,

subexponential growth if $\omega(\Gamma, S) = 1$,

polynomial growth if $\beta(\Gamma, S; k) \prec k^d$ for some $d \geq 0$,

and of *intermediate growth* if it is of subexponential growth and *not* of polynomial growth.

These definitions will be reformulated in Item 58. There is in [DeuSS–95] an interesting notion of "exponential growth rate" for a discrete metric space; but this is definitely not (a generalization of) the notion of exponential growth discussed here; see the discussion in [CecGH–99].

52. Observations. Let (Γ, S) be as in Definition 51.

Let Γ' be a subgroup of Γ. If Γ' is generated by $S' = S \cap \Gamma'$, then $\omega(\Gamma', S') \leq \omega(\Gamma, S)$.

Let Γ'' be a quotient of Γ and let $\pi : \Gamma \to \Gamma''$ denote the canonical projection. If $S'' = \pi(S)$, then $\omega(\Gamma'', S'') \leq \omega(\Gamma, S)$.

In particular, if a group Γ has either a subgroup or a quotient group of exponential growth, then Γ itself is of exponential growth. (See also Proposition VII.1.)

53. Remarks. (i) For (Γ, S) as above, we have

$$1 \leq \omega(\Gamma, S) \leq 2|S| - 1$$

because, with $n = |S|$, one has $1 \leq \beta(\Gamma, S; k) \leq 2n(2n - 1)^{k-1}$ for all $k \geq 1$ by Property 3.iii.

(ii) The property of being of exponential growth [respectively subexponential growth, polynomial growth, intermediate growth] depends only on the group Γ, and not on the choice of S (see Items 28 and 50 above). Indeed, *these properties are invariant by quasi-isometry.*

(iii) A finitely-generated group is necessarily of one (and only one) of the three types: exponential growth, intermediate growth, polynomial growth.

(iv) Among the groups appearing in Section VI.A, observe that finite groups and free abelian groups are of polynomial growth, whereas non-abelian free groups are of exponential growth.

The discovery of finitely-generated groups of intermediate growth is more recent, and due to Grigorchuk ([Grigo–84], [Grigo–91]), answering in this manner a question of Milnor [Miln–68a]. See Chapter VIII.

The existence of *semi-groups* of intermediate growth is easier to establish: for a recent short proof, see [Natha–99]. The same paper recalls the following related open problem, which has Number 10.12 in [Kouro–95].

Let Γ be a finitely-generated group and let M be a sub-*semi*-group of Γ; assume that M is finitely-generated and generates Γ. Can Γ be of exponential growth and M be of intermediate growth ?

(v) Let Γ, S be as in Definition 51 and let $\sigma(\Gamma, S; k)$ denote the corresponding spherical growth function. The upper limit

$$\limsup_{k \to \infty} \sqrt[k]{\sigma(\Gamma, S; k)}$$

is the inverse of the radius of convergence R' of the spherical growth series $\Sigma(z) = \sum_{k=0}^{\infty} \sigma(\Gamma, S; k) z^k$. If Γ is an infinite group, $\sigma_k(\Gamma, S; k) \geq 1$ for all $k \geq 0$ and $R' \leq 1$, so that $\Sigma(z)$ has the same radius of convergence as the growth series $\sum_{k=0}^{\infty} \beta_k(\Gamma, S; k) z^k = \frac{1}{1-z} \Sigma(z)$. It follows that

$$\omega(\Gamma, S) = \limsup_{k \to \infty} \sqrt[k]{\sigma(\Gamma, S; k)}$$

(and the upper limit is in fact a limit, again by Proposition 56 below.)

(vi) The term *entropy* for $h(\Gamma, S) = \log \omega(\Gamma, S)$ is used, for example, in [GroLP–81]. One reason is that, if Γ is the fundamental group of a compact Riemannian manifold of unit diameter, and if S is an appropriate generating set (given by the geometry), then $h(\Gamma, S)$ is a lower bound for the topological entropy of the geodesic flow of the manifold. See [Manni–79], strengthening a previous result of [Dinab–71], and the exposition in Chapter 5 of [Pater–99].

(The same word "entropy" has a different meaning in [Avez–72] and [Avez–76], where it applies to a *pair* (Γ, μ) of a group Γ and a symmetric probability measure μ on this group.)

Here is another lower bound result. Consider a compact manifold M, its fundamental group Γ, generated by some finite set S, and the corresponding word length ℓ_S; consider also a continuous mapping $f : M \longrightarrow M$ and the endomorphism f_* of Γ induced by f. The *growth rate* of f_* is defined as

$$\omega(f_*) = \sup_{\gamma \in \Gamma} \left(\limsup_{k \to \infty} \sqrt[k]{\ell_S\left(f_*^k(\gamma)\right)} \right) = \sup_{s \in S} \left(\limsup_{k \to \infty} \sqrt[k]{\ell_S\left(f_*^k(s)\right)} \right)$$

(the proof of the second equality is left as an exercise to the reader). Then we have a lower bound

$$h(f) \geq \omega(f_*)$$

for the topological entropy $h(f)$ of the mapping f [Bowen–78].

(vii) The number $\omega(\Gamma, S)$ appears in the study of random walks on the Cayley graph of a pair (Γ, S). For example, if $\nu(\Gamma, S)$ denotes the spectral radius of the simple random walk on this graph (see Item VII.37), then $\nu(\Gamma, S) \geq \omega(\Gamma, S)^{-1/2}$ by a simple application of the Cauchy-Schwarz inequality (see Section (F) in [GriHa–97]).

Lyons [Lyons–95] has studied a one-parameter family of nearest-neighbour random walks $(RW_\lambda)_{\lambda \geq 1}$ on a Cayley graph which are transient [respectively positive recurrent] if and only if $\lambda < \omega(\Gamma, S)$ [respectively if and only if $\lambda > \omega(\Gamma, S)$]. See also [LyoPP–96] for the interesting particular case of lamplighter groups.

(viii) Our next object is to show that the lim sup above is in fact a limit.

54. Definition. A function $\alpha : \mathbb{N} \longrightarrow \mathbb{R}$ is *subadditive* if

$$\alpha(p + q) \leq \alpha(p) + \alpha(q)$$

for all $p, q \in \mathbb{N}$. A function $\beta : \mathbb{N} \longrightarrow \mathbb{R}_{>0}$ is *submultiplicative* if

$$\beta(p + q) \leq \beta(q)\beta(q)$$

for all $p, q \in \mathbb{N}$.

55. Examples. Let Γ be a group generated by a finite set S. Then $\beta(\Gamma, S; k)$ and $\sigma(\Gamma, S; k)$ are submultiplicative functions.

The notion of subadditive function also makes sense for real-valued functions defined on groups and semi-groups. The word length $\ell_S : \Gamma \to \mathbb{N}$ is then subadditive; this means that $\ell_S(\gamma_1 \gamma_2) \leq \ell_S(\gamma_1) + \ell_S(\gamma_2)$ for all $\gamma_1, \gamma_2 \in \Gamma$. (This was already observed in VI.3.)

The next proposition is classical, and appears for example in [PólSz–76, Problem 98 of Part I, page 23], as well as in standard books on ergodic theory [Walte–82, Theorem 4.9]. It is essentially the so-called *Fekete lemma*, going back at least to 1923, given as Number 11.6 in [VaLWi–92].

56. Proposition. *(i) For a subadditive function $\alpha : \mathbb{N} \longrightarrow \mathbb{R}$, the sequence $\left(\frac{\alpha(k)}{k}\right)_{k \geq 1}$ converges and*

$$\lim_{k \to \infty} \frac{\alpha(k)}{k} = \inf_{k \geq 1} \frac{\alpha(k)}{k}.$$

Moreover, if L denotes this limit, then $kL \leq \alpha(k) \leq k\alpha(1)$ for all $k \geq 1$.

(ii) For a submultiplicative function $\beta : \mathbb{N} \longrightarrow \mathbb{R}_{>0}$, the sequence of kth term $\sqrt[k]{\beta(k)}$ converges and

$$\lim_{k \to \infty} \sqrt[k]{\beta(k)} = \inf_{k \geq 1} \sqrt[k]{\beta(k)}.$$

Moreover, if λ denotes this limit, then $\lambda^k \leq \beta(k) \leq \beta(1)^k$ for all $k \geq 1$.

Proof. (i) The argument below applies to a slightly more general case, where $\alpha(k)$ is defined only for $k \geq N$, for some $N \geq 1$.

Choose an integer $a \geq N$. Write each $k \geq N$ as $k = qa + r$ with $q \geq 0$ and $r \in \{N, N + 1, \ldots, N + a - 1\}$. One has

$$\frac{\alpha(k)}{k} \leq \frac{q\alpha(a) + \alpha(r)}{qa + r} \leq \frac{q\alpha(a)}{qa} + \frac{\alpha(r)}{qa}.$$

It follows that $\limsup_{k\to\infty} \frac{\alpha(k)}{k} \leq \frac{\alpha(a)}{a}$. As this holds for all $a \geq N$, this shows (i).

(ii) This follows from (i) applied to the subadditive function defined by $\alpha(k) = \log \beta(k)$. □

Observe that the sequence $\left(\frac{\alpha(k)}{k}\right)_{k\geq 1}$ need *not* be decreasing (as the sequence with $\alpha(k) = 1$ for $k \leq 10$ and $\alpha(k) = 2$ for $k \geq 11$ shows).

57. Corollary. *Let* $\beta : \mathbb{N} \longrightarrow \mathbb{R}_{>0}$ *be a submultiplicative function, and set*

$$\lambda = \lim_{k\to\infty} \sqrt[k]{\beta(k)}.$$

(i) $\beta(k) \overset{s}{\prec} e^k$.

(ii) $\beta(k) \overset{s}{\sim} e^k$ *if and only if* $\beta(k) \overset{w}{\sim} e^k$, *if and only if* $\lambda > 1$.

(iii) *For* $v \in \mathbb{R}, v > 1$, *the following are equivalent:*

    there exists $c > 0$ *such that* $\beta(k) \geq cv^k$ *for all* $k \geq 1$;

    $\beta(k) \geq v^k$ *for all* $k \geq 1$.

Proof. Claims (i) and (ii) are straightforward consequences of Proposition 56. For (iii), if $\beta(k) \geq cv^k$ for all $k \geq 1$, then $\beta(k)^a \geq \beta(ak) \geq cv^{ak}$ for all $a \geq 1$ and $k \geq 1$, so that $\beta(k) \geq \left(\inf_{a\geq 1} c^{1/a}\right) v^k = v^k$ for all $k \geq 1$. □

58. Reformulation of some definitions of Item 51. Let Γ be a group generated by a finite set S and let $\beta(k) = \beta(\Gamma, S; k)$ denote the corresponding growth function. The group Γ is said to be of

exponential growth if $\beta(k) > cv^k$ *for some* $c > 0$ *and* $v > 1$,

polynomial growth if $\beta(k) \leq ck^d$ *for some* $c > 0$ *and* $d \geq 1$,

intermediate growth if   *it is neither of exponential nor of polynomial growth.*

Exercises for VI.C

59. Exercise. For $j = 1, 2$, let Γ_j be a group generated by a finite set S_j and let $\omega(\Gamma_j, S_j)$ be the corresponding exponential growth rate, as in Definition 51. Let Γ be the direct product $\Gamma_1 \times \Gamma_2$ and set $S \doteq (S_1 \times \{1\}) \cup (\{1\} \times S_2)$. Show that

$$\omega(\Gamma, S) = \max\left\{\omega(\Gamma_1, S_1), \omega(\Gamma_2, S_2)\right\}.$$

[Hint: use Proposition 4.i.]

60. Exercise. Show that, for any integer $n \geq 2$ and for any $\epsilon > 0$, there exists a group Γ generated by a set S of n generators, such that

$$2n - 1 - \epsilon < \omega(\Gamma, S) < 2n - 1.$$

[Sketch. Consider an integer $N \geq 1$ and the group with presentation

$$\Gamma = \langle s_1, \ldots, s_n \mid s_n^{2N} = 1 \rangle.$$

Proposition 4.ii shows that, for $S = \{s_1, \ldots, s_n\}$, we have

$$\Sigma(\Gamma, S; z) = \frac{(1+z)(1-z^N)}{1 - (2n-1)z + (2n-2)z^{N+1}}.$$

The smallest pole of this rational function on the positive real half-axis is strictly smaller than, but almost equal to, $\frac{1}{2n-1}$, and this proves the claim.]

Problems and complements for VI.C

61. Research problem. For an integer $n \geq 2$, denote by Ω_n the set of those numbers of the form $\omega(\Gamma, S)$, where Γ is a group generated by a set S of at most n elements. It is known that

(i) the set Ω_n is a subset of $[1, 2n-1]$,
(ii) the point $2n-1$ is an accumulation point of Ω_n,
(iii) the point 1 is an accumulation point of Ω_n.

(See respectively Remark 53.i, Exercise 60, and [GriHa].)

It follows from (iii) and Exercise 17 that $\cup_{n \geq 1} \Omega_n$ is dense in $[1, \infty[$. I do not know whether this set is all of $[1, \infty[$, let alone whether it is uncountable. An obvious project would be to compute the exponential growth rates of the Neumann groups described in Section III.B (Problem III.35.v).

62. Research problem on growth and non-amenability. For an integer $n \geq 2$, does there exist a constant $c_n > 1$, with $c_n < 2n - 1$, such that the following holds ?

Let Γ be a group generated by a set S of n generators (so that the corresponding exponential growth rate satisfies $\omega(\Gamma, S) \in [1, 2n-1]$). If $\omega \geq c_n$, then Γ is not amenable.

(If $\omega(\Gamma, S) = 2n - 1$, it is easy to check that Γ is free of rank n, and in particular non-amenable. The proof appears in Section V of [GriHa–97].)

63. Research problem. Does there exist a finitely-*presented* group of intermediate growth ? (Chapter VIII is an exposition of *the* canonical example of a finitely-*generated* group of intermediate growth.)

64. Research problem. Is it true that a finitely-generated group of subexponential growth is residually finite ?

This is the last question in [AdeSr–57]. Note that it is a standard result that finitely-generated nilpotent groups are residually finite; hence it follows from Gromov's theorem VII.29 that groups of polynomial growth are residually finite.

65. Research problem. For $j = 1, 2$, let Γ_j be a group generated by a finite set S_j, and let $\sigma_j(k) = \sigma(\Gamma_j, S_j; k)$ denote the corresponding spherical growth function. Assume that the groups Γ_1 and Γ_2 are quasi-isometric, so that the growth functions $\beta(\Gamma_1, S_1; k)$ and $\beta(\Gamma_2, S_2; k)$ are strongly equivalent by Items 34 and 28.

When is it true that $\sigma_1 \overset{s}{\sim} \sigma_2$? This question is related to that of Item 19.

Observe that this *is* true when the groups are of exponential growth; indeed, Remark 53.v and Proposition 56 show that in this case $\sigma_j \overset{s}{\sim} \beta_j$.

66. Complement on admissible groups. A finitely-generated group Γ is *admissible* if there exists a nested sequence of subgroups

$$\Gamma_0 = \{1\} < \Gamma_1 < \ldots < \Gamma_k = \Gamma$$

with Γ_{j-1} normal in Γ_j and Γ_j/Γ_{j-1} of subexponential growth for any j with $1 \leq j \leq k$. Such groups, which are amenable, play some role in analysis (see Definition 7.1 in [Anton–88]) and deserve further study.

CHAPTER VII

GROUPS OF EXPONENTIAL OR POLYNOMIAL GROWTH

This chapter provides further examples of groups of exponential or polynomial growth. It also mentions the motivating connection between, on the one hand, the growth of the fundamental group of a closed manifold and, on the other hand, curvature properties of Riemannian metrics on the manifold. Historically, these go back to the early 1950s in the former Soviet Union ([Schwa–55], [Efrem–53]), and also (independently!) to the late 1960s in the West [Miln–68b].

One should also mention work which may well have been under-estimated [Dixmi–60]. There, in order to prove a property of the group algebra $L^1(G)$, Dixmier shows that a nilpotent connected real Lie group G has polynomial growth.

VII.A. On groups of exponential growth

1. Groups with free sub-semi-groups. Let SF_2 denote the free monoid on two generators, say here on $A = \{a_1, a_2\}$, so that $SF_2 = W(\{a_1, a_2\})$ in the notation of Section II.A. The growth function $\beta(k) = \beta(SF_2, A; k)$ is given by

$$\beta(k) = 1 + 2 + \ldots + 2^k = 2^{k+1} - 1$$

and satisfies $\beta(k) \sim 2^k$ in the sense of Definitions VI.22 and VI.29. If Γ is a finitely-generated group containing SF_2 as a sub-semi-group, one may choose a finite set S of generators for Γ containing a free set of generators of this free sub-semi-group; consequently

$$\beta(\Gamma, S; k) \geq 2^{k+1} - 1$$

for all $k \geq 1$. This proves the following result (compare with Observation VI.52).

Proposition. *A finitely-generated group which contains a free semi-group on two generators is of exponential growth.*

This proposition is a very efficient way of recognizing groups of exponential growth: see Example 3 for a finitely-generated subgroup "of $ax + b$ type", Theorem 28 for finitely-generated solvable groups in general, or Corollary 4.7 in [CanFP–96], where it is shown that Thompson's group F is of exponential growth. However, there are groups of exponential growth without free semi-groups — such as the appropriate Burnside groups [Adian–75].

To recognize free monoids, one can use an analogue of the Table-Tennis Lemma of Section II.B.

2. Proposition. *Let Γ be a group acting on a set X. Assume that there exist $\gamma_1, \gamma_2 \in \Gamma$ and $X_1, X_2 \subset X$ such that*

$$X_1 \cap X_2 = \emptyset, \qquad \gamma_1(X_1 \cup X_2) \subset X_1, \qquad \gamma_2(X_1 \cup X_2) \subset X_2.$$

Then the semi-group generated in Γ by γ_1 and γ_2 is free, and in particular Γ is of exponential growth.

Proof. Let $\phi : SF_2 \longrightarrow \Gamma$ denote the map which sends a word $w \in SF_2$, considered as a word in the two letters a_1 and a_2 (positive powers only), onto the element of Γ obtained by substituting γ_1 for a_1 and γ_2 for a_2. Let $w, w' \in SF_2$ be two words such that $\phi(w) = \phi(w') \in \Gamma$; we have to show that $w = w' \in SF_2$.

If w is empty, the word w' may begin neither with a_1 (this would imply $\phi(w')(X_2) \subset X_1$, in contradiction with $\phi(w)(X_2) = X_2$) nor with a_2 (for an analogous reason); thus w' is also empty.

Hence we may assume that both w and w' are non-empty, without restricting generality. Thus they are of the form

$$w = sv \qquad w' = s'v' \qquad s, s' \in \{a_1, a_2\}.$$

The action of $\phi(w) = \phi(w')$ on X_1 and X_2 shows that $s = s'$. Now, because $\phi(w), \phi(w')$ are elements of a *group* Γ, we have

$$\phi(w) = \phi(w') \quad \Longrightarrow \quad \phi(v) = \phi(v').$$

The claim follows by induction on the smaller of the lengths of w and w'. $\square$

3. Example. The subgroup Γ of $GL(2, \mathbb{R})$ generated by

$$h = \begin{pmatrix} 2 & 0 \\ 0 & 1 \end{pmatrix} \qquad \text{and} \qquad t = \begin{pmatrix} 1 & 1 \\ 0 & 1 \end{pmatrix}$$

is of exponential growth.

Proof. The group, which is

$$\Gamma = \left\{ \begin{pmatrix} a & b \\ 0 & 1 \end{pmatrix} \mid a = 2^n \text{ with } n \in \mathbb{Z}, b \in \mathbb{Z}\left[\frac{1}{2}\right] \right\},$$

acts on $\mathbb{R}$ by

$$\begin{pmatrix} a & b \\ 0 & 1 \end{pmatrix} x = ax + b.$$

If $a \neq 1$, the element $\begin{pmatrix} a & b \\ 0 & 1 \end{pmatrix}$ has a unique fixed point which is $\frac{b}{1-a}$.

For $j = 1, 2$, choose an element $s_j = \begin{pmatrix} a_j & b_j \\ 0 & 1 \end{pmatrix} \in \Gamma$ with

$$a_1 < 1, \quad a_2 < 1, \quad \frac{b_1}{1 - a_1} \neq \frac{b_2}{1 - a_2}$$

and choose an open interval I_j of $\mathbb{R}$ containing $\frac{b_j}{1-a_j}$, with $I_1 \cap I_2 = \emptyset$; choose also an interval I in $\mathbb{R}$ containing $I_1 \cup I_2$.

Upon replacing s_1, s_2 by large enough powers of themselves, we may assume that

$$s_1(I) \subset I_1 \quad \text{and} \quad s_2(I) \subset I_2.$$

It follows from the two previous propositions that Γ is of exponential growth. $\square$

One shows similarly that other Baumslag-Solitar groups are of exponential growth (Complement 8).

The next proposition appears in the first publication of Schwarzc on group growth [Schwa–55]. We follow essentially Lemma 6.2 of [Tits–81b].

4. Lemma. *Let $\beta \in GL(d, \mathbb{Z})$ be a matrix which has an eigenvalue $\lambda \in \mathbb{C}$ such that $|\lambda| \geq 2$. Then there exists a vector $v \in \mathbb{Z}^d$ such that, for any $k \geq 0$, the 2^{k+1} vectors*

$$\epsilon_0 v + \epsilon_1 \beta(v) + \ldots + \epsilon_k \beta^k(v)$$

are all distinct, where $\epsilon_0, \epsilon_1, \ldots, \epsilon_k$ are in $\{0, 1\}$.

Proof. Let $L : \mathbb{C}^d \longrightarrow \mathbb{C}$ be a linear form such that $L \circ \beta = \lambda L$ and let $v \in \mathbb{Z}^d$ be a vector such that $L(v) \neq 0$. Then

$$L\left(\epsilon_0 v + \epsilon_1 \beta(v) + \ldots + \epsilon_k \beta^k(v)\right) = \left(\sum_{j=0}^{k} \epsilon_j \lambda^j\right) L(v)$$

and the numbers $\sum_{j=0}^{k} \epsilon_j \lambda^j$ are pairwise distinct for $\epsilon_0, \epsilon_1, \ldots, \epsilon_k \in \{0, 1\}$. $\square$

The lemma and its proof hold more generally for matrices having an eigenvalue outside the set of zeros of polynomials with $0, \pm 1$ coefficients. This subset of $\mathbb{C}$ is smaller than the open disc $\{|\lambda| < 2\}$: see [OdlPo–93], and results of T. Bousch quoted there.

5. Proposition. *Let $\alpha \in GL(d, \mathbb{Z})$ and let $\Gamma = \mathbb{Z}^d \rtimes_\alpha \mathbb{Z}$ denote the corresponding semi-direct product. If α has an eigenvalue $\lambda \in \mathbb{C}$ such that $|\lambda| \neq 1$, then Γ is of exponential growth.*

Proof. There exists $r \in \mathbb{Z}$ such that $\beta \doteq \alpha^r$ satisfies the hypothesis of the lemma. By a straightforward induction argument, one shows that, for any integer $k \geq 0$ and for any $\epsilon_0, \epsilon_1, \ldots, \epsilon_k \in \{0, 1\}$,

$$(\epsilon_0 v, r)(\epsilon_1 v, r) \ldots (\epsilon_k v, r) = \left(\epsilon_0 v + \epsilon_1 \beta(v) + \ldots + \epsilon_k \beta^k(v), (k+1)r\right)$$

and these 2^{k+1} elements of Γ are all distinct by Lemma 4.

Thus Γ contains the free semi-group on the two generators $(v, 0)$ and $(0, r)$, so that Γ is of exponential growth by Proposition 1. $\square$

The following result is Theorem 2 of [Miln–68b].

6.   Theorem. *Let X be a connected compact Riemannian manifold of strictly negative sectional curvature. The fundamental group of X is of exponential growth.*

Proof. Let x_0 be a base point in the universal covering $\tilde{X}$ of X and let $\beta(r)$ denote the Riemannian volume of a ball of radius r centred at x_0 in $\tilde{X}$. If X (hence also $\tilde{X}$) is of constant curvature $-\alpha^2 < 0$, we have

$$\beta(r) = c_0 \int_0^r \left(\alpha^{-1} \sinh(\alpha t)\right)^{n-1} dt \quad \overset{r \to \infty}{\sim} \quad \frac{c_0}{(n-1)(2\alpha)^n} \, exp\big((n-1)\alpha r\big)$$

for some constant c_0. In the general case of curvature satisfying $\kappa \leq -\alpha^2 < 0$, we have

$$\beta(r) \geq c_1 exp(c_2 r)$$

for constants c_1, c_2 and for all r large enough, by results of Günther [Gunth–60]. (See for example Section III.H in [GalHL–87].)

It follows from Theorem IV.23, Example VI.23, and Proposition VI.36 that the fundamental group of X is a finitely-generated group with growth function weakly equivalent to $\beta(r)$, and is in particular a group of exponential growth. □

Here is an improvement due to Avez. *Let X be a connected compact Riemannian manifold of dimension $n \geq 3$ and of curvature ≤ 0. Then either $\pi_1(X)$ is of exponential growth, or X is flat* [Avez–70].

7.   Consequences. Theorem 6 implies that a torus does not carry any Riemannian manifold of strictly negative sectional curvature.

As another application, consider the quotient $M \doteq H_{\mathbb{R}}/H_{\mathbb{Z}}$ of the Heisenberg group over $\mathbb{R}$ by the Heisenberg group over $\mathbb{Z}$. Then M does not carry any Riemannian metric with strictly negative curvature, because $H_{\mathbb{Z}}$ has polynomial growth (Example VI.6 and Proposition 22 below). This carries over to any compact *nilmanifold,* namely to any quotient of a connected and simply-connected nilpotent real Lie group by a lattice. (For $H_{\mathbb{R}}$-invariant Riemannian metrics on M, see [Wolf–64], and the sharper Theorem 2.4 of [Miln–76].)

However, as the following two examples show, a compact manifold V may have a fundamental group of exponential growth even if it is not of negative curvature!

(i) Let $V = \mathbb{S}^1 \times \Sigma_g$ be the direct product of a circle and a closed orientable surface of genus $g \geq 2$. Then $\pi_1(V) = \pi_1(\Sigma_g) \times \mathbb{Z}$ is of exponential growth by Observation VI.52, but V does not admit any Riemannian structure with sectional curvature < 0 by Preissmann's Theorem V.53.i.

(ii) Let V be a closed manifold with fundamental group a soluble group of exponential growth, for example a Baumslag-Solitar group. Then V does not admit any Riemannian structure with sectional curvature ≤ 0 by Gromoll-Wolf and Lawson-Yau, Theorem V.53.ii.

Problems and complements for VII.A

8. Growth of Baumslag-Solitar groups. *The Baumslag-Solitar group*

$$\Gamma_{p,q} = \langle s,t \mid st^p s^{-1} = t^q \rangle$$

of Complements III.21 and IV.43 is of exponential growth for any pair of integers (p,q) with $1 \leq p \leq q$ and $(p,q) \neq (1,1)$.

The following proof shows more, namely, using a notion defined later in Section VII.B, that $\Gamma_{p,q}$ *has uniformly exponential growth.*

Proof. Suppose first that $p = q$. Then $\Gamma_{p,p}$ has a quotient isomorphic to the free product Γ of an infinite cyclic group (generated by the image of s) and a cyclic group of order p (generated by the image of t). The kernel of the natural homomorphism from Γ to the cyclic group of order p is free by Kurosh's Theorem II.22. Thus $\Gamma_{p,p}$ has a subgroup of finite index which has a non-abelian free quotient, hence $\Gamma_{p,p}$ has uniformly exponential growth.

Suppose now that $p < q$ with p, q coprime. Consider the subgroup $\Gamma_{1,q/p}$ of $GL(2,\mathbb{R})$ generated by $\begin{pmatrix} q/p & 0 \\ 0 & 1 \end{pmatrix}$ and $\begin{pmatrix} 1 & 1 \\ 0 & 1 \end{pmatrix}$, and the subgroup A of $GL(2,\mathbb{R})$ consisting of matrices $\begin{pmatrix} 1 & a \\ 0 & 1 \end{pmatrix}$ with $a \in \mathbb{Z}\left[\frac{1}{p}\right]$. Then A is a subgroup of $\Gamma_{1,q/p}$; indeed, for any $n \geq 2$, there exist $x_n, y_n \in \mathbb{Z}$ such that $x_n p^n + y_n q^n = 1$, and

$$\begin{pmatrix} 1 & 1 \\ 0 & 1 \end{pmatrix}^{x_n} \begin{pmatrix} q/p & 0 \\ 0 & 1 \end{pmatrix}^{n} \begin{pmatrix} 1 & 1 \\ 0 & 1 \end{pmatrix}^{y_n} \begin{pmatrix} q/p & 1 \\ 0 & 1 \end{pmatrix}^{-n} = \begin{pmatrix} 1 & p^{-n} \\ 0 & 1 \end{pmatrix} \in \Gamma_{1,q/p}\,.$$

There is a short exact sequence

$$1 \longrightarrow A \longrightarrow \Gamma_{1,q/p} \overset{\pi}{\longrightarrow} \mathbb{Z} \longrightarrow 1$$

with $\pi\begin{pmatrix} q/p & 0 \\ 0 & 1 \end{pmatrix} = 1$ and $\pi(A) = 0$, and the proof of Example 3 [respectively the proof of Proposition 2.6 in [CecGr–97], relying itself on Lemma 4.8 in [Rosbl–74]] shows that $\Gamma_{1,q/p}$ is of exponential growth [respectively of uniformly exponential growth]; as $\Gamma_{1,q/p}$ is a quotient of $\Gamma_{p,q}$, with $\begin{pmatrix} q/p & 0 \\ 0 & 1 \end{pmatrix}$ the image of s and $\begin{pmatrix} 1 & 1 \\ 0 & 1 \end{pmatrix}$ that of t, it follows that $\Gamma_{p,q}$ itself is of exponential growth [respectively of uniformly exponential growth].

Suppose finally that $p = rp' < q = rq'$, with $r > 1$ and with p', q' coprime. Then $\Gamma_{p',q'}$ has uniformly exponential growth by the previous step, and is clearly a quotient of $\Gamma_{p,q}$, so that the proof is complete. $\square$

9. Exponential growth of free products with amalgamation. Let $\Gamma = \Gamma_1 * \Gamma_2$ be a free product of two finitely-generated groups. If Γ_1 and Γ_2 are

both of order 2, then Γ is the infinite dihedral group of Exercise II.30, which has a subgroup of index 2 isomorphic to $\mathbb{Z}$ and is consequently of linear growth. If $(|\Gamma_1| - 1)(|\Gamma_2| - 1) \geq 2$, then Γ contains a non-abelian free subgroup (see for example Complement II.16), so that Γ is of exponential growth by Observation VI.52.

Now let $\Gamma = \Gamma_1 *_A \Gamma_2$ be a free product with amalgamation (III.14), where Γ_1, Γ_2 are finitely generated and where A is isomorphic to a *proper* subgroup of both Γ_1 and Γ_2. There is a well-known normal form theorem for elements of Γ, for which we refer to N° I.1.2 in [Serr-77].

If $([\Gamma_1 : A] - 1)([\Gamma_2 : A] - 1) \geq 2$, it is easy to deduce from the normal form theorem that Γ contains a non-abelian free group. It follows (again by Observation VI.52) that Γ has exponential growth.

If $[\Gamma_1 : A] = [\Gamma_2 : A] = 2$, the group Γ may of course be of polynomial growth, as already observed above (in case $A = \{1\}$). But Γ may also be of exponential growth even if A is of polynomial growth, as the next example shows. (It is taken from an unpublished typescript of M.R. Bridson, dated April 1997, and entitled *A remark about the growth of groups.*)

Note the following fact, of which the proof is an easy exercise (and Lemma 3 in [BauSh–90]): a group Γ can be written as $\Gamma_1 *_A \Gamma_2$ with $[\Gamma_1 : A] = [\Gamma_2 : A] = 2$ if and only if there exists a homomorphism of Γ onto the infinite dihedral group $(\mathbb{Z}/2\mathbb{Z}) * (\mathbb{Z}/2\mathbb{Z})$.

10. An example of M.R. Bridson. Consider the two matrices

$$b = \begin{pmatrix} 0 & 1 \\ 1 & 0 \end{pmatrix} \quad \text{and} \quad c = \begin{pmatrix} 2 & 1 \\ 1 & 1 \end{pmatrix} \begin{pmatrix} 0 & 1 \\ 1 & 0 \end{pmatrix} \begin{pmatrix} 2 & 1 \\ 1 & 1 \end{pmatrix}^{-1} = \begin{pmatrix} -1 & 3 \\ 0 & 1 \end{pmatrix}$$

in $GL(2, \mathbb{Z})$. The properties of b, c used below are that b, c are both of order 2 and that their product bc is hyperbolic, namely has a real eigenvalue strictly larger than 1 (see Exercise II.32). To the matrix b there correspond an action of the group $J = \{1, j\}$ of order 2 on $\mathbb{Z}^2$ and a semi-direct product $\Gamma_b = \mathbb{Z}^2 \rtimes_b J$; each element of Γ_b can be written uniquely either as $(x, 1)$ or as (x, j), with $x \in \mathbb{Z}^2$. Similarly, we have a semi-direct product $\Gamma_c = \mathbb{Z}^2 \rtimes_c K$ with $K = \{1, k\}$ another copy of the group of order 2. We identify $A = \mathbb{Z}^2$ with a subgroup of index 2 in both Γ_b and Γ_c by the injective homomorphisms

$$\phi : \begin{cases} A \longrightarrow \Gamma_b \\ x \longmapsto (b(x), 1) \end{cases} \quad \text{and} \quad \psi : \begin{cases} A \longrightarrow \Gamma_c \\ x \longmapsto (c(x), 1) \end{cases}.$$

We thus have a free product with amalgamation

$$\Gamma = \Gamma_b *_A \Gamma_c$$

which obviously satisfies

$$[\Gamma_b : A] = [\Gamma_c : A] = 2.$$

Proposition. *The group* $\Gamma = \Gamma_b *_A \Gamma_c$ *defined above is of exponential growth.*

Proof. We have short exact sequences

$$1 \longrightarrow A \xrightarrow{\phi} \Gamma_b \longrightarrow J \longrightarrow 1$$

$$1 \longrightarrow A \xrightarrow{\psi} \Gamma_c \longrightarrow K \longrightarrow 1.$$

As $\phi(A), \psi(A)$ are of index 2, and in particular are normal, in Γ_b, Γ_c respectively, the canonical image of A is a normal subgroup of Γ and we also have a short exact sequence

$$1 \longrightarrow A \longrightarrow \Gamma \longrightarrow D_\infty \longrightarrow 1$$

where $D_\infty = J * K$ is the infinite dihedral group generated by j and k. Indeed, Γ may be seen as a semi-direct product

$$\Gamma = \mathbb{Z}^2 \rtimes_\rho D_\infty$$

where the action ρ of D_∞ on $\mathbb{Z}^2$ is given by

$$\rho(j)(x) = b(x) \qquad \text{for all} \qquad x \in \mathbb{Z}^2$$
$$\rho(k)(x) = c(x) \qquad \text{for all} \qquad x \in \mathbb{Z}^2.$$

Thus, if $\mathbb{Z}$ is identified with the subgroup of D_∞ generated by jk, the group Γ has a subgroup of index 2 of the form

$$\Gamma_0 = \mathbb{Z}^2 \rtimes_{bc} \mathbb{Z},$$

where the notation indicates that the generator jk of $\mathbb{Z}$ acts on $\mathbb{Z}^2$ by the hyperbolic element $bc \in GL(2, \mathbb{Z})$. As Γ_0 is of exponential growth by Proposition 5, this concludes the proof. $\square$

11. On the regularity of the growth functions for groups of exponential growth. In general, for a pair (Γ, S) with $\beta(k) = \beta(\Gamma, S; k)$ of exponential growth, little is known about when the quotients $\beta(k+1)/\beta(k)$ converge towards the exponential growth rate ω, or when the quotients $\omega^{-k}\beta(k)$ converge towards some constant, for $k \to \infty$ (see also Problem 34.B below).

If Γ is hyperbolic, Coornaert [Coorn–93] has shown that there exist constants $c_1, c_2 > 0$ such that $c_1\omega^k \le \beta(k) \le c_2\omega^k$ for all $k \ge 0$. There are some interesting cases for which there exist constants $c, d > 0$ such that $c\omega^k - d \le \beta(k) \le c\omega^k + d$ for all $k \ge 0$ (see [Canno–80], [CanWa–92], various papers by W. Floyd, and [BarCe–b]). But Machì's example VI.7 shows that this does not hold in general.

VII.B. On uniformly exponential growth

12. Definition. Let Γ be a finitely-generated group of exponential growth. Recall from Section VI.C that, to each finite set S of generators of Γ, one associates the exponential growth rate

$$\omega(\Gamma, S) = \lim_{k \to \infty} \sqrt[k]{\beta(\Gamma, S; k)}$$

such that $1 < \omega(\Gamma, S) \leq 2|S| - 1$. The *minimal growth rate* of Γ is defined to be

$$\omega(\Gamma) = \inf \omega(\Gamma, S),$$

where the infimum is taken over all S as above; clearly, $1 \leq \omega(\Gamma) \leq 2|S| - 1$.

The group Γ is said to have *uniformly exponential growth* if $\omega(\Gamma) > 1$. This notion appears in [GroLP–81] (Definition 5.11 and Proposition 5.13). See also [ShaWa–92], [GriHa–97], [CecGr–97], and [Shalo–a].

13. Proposition. *The free group F_k on $k \geq 2$ generators has uniformly exponential growth, and $\omega(F_k) = 2k - 1$.*

Proof. If S_k is a free generating set of F_k, it follows from Example VI.2.vii that

$$\omega(F_k, S_k) = 2k - 1.$$

Now let S be any finite generating set of F_k. The canonical image $\underline{S}$ of S in the abelianized group $(F_k)^{\mathrm{ab}} = \mathbb{Z}^k$ generates $\mathbb{Z}^k$. Thus $\underline{S}$ contains a subset $\underline{R}$ of k elements generating a subgroup of finite index in $\mathbb{Z}^k$. Let R be a subset of S projecting bijectively onto $\underline{R}$. The subgroup $\langle R \rangle$ of F_k generated by R is free (as a subgroup of a free group), of rank at most k (because $|R| = k$) and of rank at least k (because the abelianized group of $\langle R \rangle$ is of rank k). Hence R is a free basis of $\langle R \rangle \approx F_k$, and it follows that $\omega(F_k, S) \geq \omega(\langle R \rangle, R) = \omega(F_k, S_k)$.

(This argument already appears in Example 5.13 of [GroLP–81].) $\square$

Let Γ be a group with the following properties: it is non-abelian and all its 2-generated subgroups are free. It is an obvious consequence of Proposition 13 that Γ has uniform exponential growth, indeed that $\omega(\Gamma) \geq 3$. It is a result of G. Arzhantseva and A. Ol'shanskii [ArzOl–96], already mentioned in Complement II.43, that "almost all finitely-generated groups" have these properties, so that *almost all finitely-generated groups have uniformly exponential growth.*

14. Proposition. *Let Γ be a finitely generated group, let Γ' be a subgroup of finite index in Γ, and let Γ'' be a quotient of Γ'.*

(i) If $\omega(\Gamma'') > 1$, then $\omega(\Gamma) > 1$.
(ii) If $\omega(\Gamma') > 1$, then $\omega(\Gamma) > 1$.

(iii) If Γ *has a subgroup of finite index which has a non-abelian free quotient, then* Γ *has uniformly exponential growth.*

Proof. Claim (i) is straightforward modulo Claim (ii). Claim (ii) follows from the elementary (and smart!) Proposition 3.3 of [ShaWa–92], which shows more precisely that

$$\omega(\Gamma)^{2[\Gamma:\Gamma']-1} \geq \omega(\Gamma').$$

Claim (iii) follows from Proposition 13 and from Claims (i)–(ii). $\square$

There are many examples of groups which can be shown to have uniformly exponential growth by applying Claim (iii) of Proposition 14. For instance, it applies to Coxeter groups which are not virtually abelian, by a result due independently to Gonciulea [Gonci] and to Margulis-Vinberg [MarVi].

Other cases need other techniques. For example, the Baumslag-Solitar group presented with two generators s, t and one relation $st^2s^{-1} = t^3$ has uniformly exponential growth by Complement 8; it has *no* subgroup of finite index which has a non-abelian free quotient (proof of Theorem 3.4 in [Pride–80]). Also, the groups constructed by Golod to answer Burnside's problem [Golod–64] have uniformly exponential growth; the argument of Section 3 in [BarGr–a] involves the growth of an appropriate graded algebra.

Proposition 14 applies in particular to the groups considered in the next proposition, and shows that $\omega(\Gamma_g) \geq 2g - 1$; however, Proposition 15 gives a more precise estimate.

15. Proposition. *The fundamental group* Γ_g *of a closed orientable surface of genus* $g \geq 2$ *has uniformly exponential growth, and* $\omega(\Gamma_g) \geq 4g - 3$.

Proof. Let S be an *arbitrary* system of generators of Γ_g. Observe that S contains some subset R of $2g$ elements which generates a subgroup of finite index in the abelianized group $\mathbb{Z}^{2g}$ of Γ_g. If R_0 is the complement of one (arbitrary) element in R, then R_0 generates a subgroup $\langle R_0 \rangle$ of infinite index in Γ_g. Such a group is free (since it is the fundamental group of a non-compact surface) of rank exactly $2g - 1$ (because its abelianization is isomorphic to $\mathbb{Z}^{2g-1}$). Hence $\omega(\Gamma_g, S) \geq \omega(\langle R_0 \rangle, R_0) = 4g - 3$. $\square$

16. Open problem. What is the minimal growth rate of Γ_g ? Is it the growth rate of Γ_g with respect to the standard set of $2g$ generators (see Item VI.8) ?

We refer to [GriHa–97] for more examples of groups of exponential growth which are known to have uniformly exponential growth; see also § 8 of [Shalo–a]. However, we wish to quote a result of M. Koubi which was not entirely clear at the time of [GriHa–97]:

17. Theorem [Koubi–98]. *A non-elementary hyperbolic group has uniformly exponential growth.*

18. Proposition. *Let* $\Gamma = \Gamma_1 * \Gamma_2$ *be a free product of two finitely-generated groups, with* $|\Gamma_1| \geq 2$ *and* $|\Gamma_2| \geq 3$. *Then* Γ *has uniformly exponential growth and*

$$\omega(\Gamma) \geq \sqrt[6]{3} \sim 1.2.$$

Proof. Let X be the bipartite graph with set of vertices $X^0 = \Gamma/\Gamma_1 \sqcup \Gamma/\Gamma_2$, and with set of edges identified with Γ, the edge γ having origin $\gamma\Gamma_1$ and end $\gamma\Gamma_2$. Thus X is the canonical tree on which Γ acts by automorphisms and without inversion. (See Theorem 7 in § I.4 of [Serr–77].) Observe that Γ acts freely on the edge set of X, namely acts in a 0-acylindrical way with the terminology of Problem II.36.

Observation: as the free product is non-trivial, Γ does not contain any normal subgroup isomorphic to $\mathbb{Z}$. This is a consequence of a result already mentioned in Item III.3, which is Theorem 3.11 in [ScoWa–79]: *In a free product* $\Gamma = \Gamma_1 * \Gamma_2$ *where* Γ_1, Γ_2 *are non-trivial, a finitely-generated subgroup of* Γ *which is normal is either reduced to* $\{1\}$ *or of finite index.*

Suppose first that there exists an element $s \in S$ which is hyperbolic on X, say of amplitude a. It follows from the above observation that there exists $u \in S$ such that $usu^{-1} \notin \{s, s^{-1}\}$. By Problem II.36, the axis of s and that of usu^{-1} have at most a edges in common, and the group $\langle s^2, us^2u^{-1} \rangle$ is free on s^2 and us^2u^{-1}. As these two elements are in the ball $B_S(4)$ of radius 4 with respect to S, the growth rate $\omega(\Gamma, B_S(4))$ is bounded below by the growth rate of the free group of rank 2 with respect to a basis, which is 3. Consequently

$$\omega(\Gamma, S) = \sqrt[4]{\omega(\Gamma, B_S(4))} \geq \sqrt[4]{3} \sim 1.316$$

and we have $\omega(\Gamma) \geq \sqrt[4]{3}$.

Suppose now that any element of S has a fixed point on X. Then there exist $s, t \in S$ such that st is hyperbolic on X (see Corollary 2 of Proposition 26 in n° I.6.5 of [Serr–77]). One checks as in the previous case that there exists $r \in S$ such that $rstr^{-1} \notin \{st, (st)^{-1}\}$, that

$$\omega(\Gamma, S) = \sqrt[6]{\omega(\Gamma, B_S(6))} \geq \sqrt[6]{3} \sim 1.2$$

and finally that $\omega(\Gamma) \geq \sqrt[6]{3}$. □

Michelle Bucher has shown that the constant in Proposition 18 can be improved from $\sqrt[6]{3} \sim 1.20$ to $\sqrt{2} \sim 1.41$.

A finitely-generated free product with amalgamation $\Gamma = \Gamma_1 *_A \Gamma_2$, satisfying the condition $([\Gamma_1 : A] - 1)([\Gamma_2 : A] - 1) \geq 2$, is of uniformly exponential growth; similarly, an HNN-extension Γ defined by a finitely-generated group Γ' and an isomorphism $\theta : A \longrightarrow B$ between two subgroups A, B of Γ', satisfying the condition $[\Gamma' : A] + [\Gamma' : B] \geq 3$, is of uniformly exponential growth. More pecisely, in these cases, we have $\omega(\Gamma) \geq \sqrt[4]{2}$; see [BucHa]; see also Lemmas 2.2 and 2.4 of [Lubo–96] for some special cases.

19. Main open problem. Does there exist a finitely-generated group of exponential growth which is not of uniformly exponential growth ? (This appears as Remark 5.12 in [GroLP–81].)

Related problems. Is it true that a soluble group of exponential growth has uniformly exponential growth ? Is it true that a lattice in a simple Lie group of higher rank has uniformly exponential growth ? (The question stands for Lie groups over any local field.)

Does there exist an infinite group with Kazhdan Property (T) which is not of uniformly exponential growth ?

What are the finitely-generated groups Γ for which there exist finite generating sets S such that $\omega(\Gamma, S) = \omega(\Gamma)$?

A. Sambusetti has shown that $\omega(\Gamma_1 * \Gamma_2, S) > \omega(\Gamma_1 * \Gamma_2)$ if Γ_1 is a non-Hopfian group, Γ_2 a group not reduced to one element, and S *any* finite generating set of $\Gamma_1 * \Gamma_2$ [Sambu–99].

In [GriMa–97], there are speculations on a type of construction which could lead to finitely-generated groups of exponential growth, not of uniformly exponential growth.

20. Open problem. For two finitely-generated groups Γ_1 and Γ_2 of exponential growth which are quasi-isometric, does $\omega(\Gamma_1) > 1$ imply $\omega(\Gamma_2) > 1$?

VII.C. On groups of polynomial growth

21. Computations with the Heisenberg group. Let Γ denote the Heisenberg group of Examples IV.8 and VI.6. The three matrices

$$s = \begin{pmatrix} 1 & 0 & 0 \\ 1 & 1 & 0 \\ 0 & 0 & 1 \end{pmatrix} \qquad t = \begin{pmatrix} 1 & 0 & 0 \\ 0 & 1 & 0 \\ 0 & 1 & 1 \end{pmatrix} \qquad u = \begin{pmatrix} 1 & 0 & 0 \\ 0 & 1 & 0 \\ 1 & 0 & 1 \end{pmatrix}$$

generate Γ and

$$s^k t^l u^m = \begin{pmatrix} 1 & 0 & 0 \\ k & 1 & 0 \\ m & l & 1 \end{pmatrix}$$

for all $k, l, m \in \mathbb{Z}$. In what follows, we consider the word length $\gamma \longmapsto |\gamma|$ defined by this generating set.

Lemma. *With the above notation,*

(*i*) $|s^k t^l u^m| \le |k| + |l| + 6\sqrt{|m|}$ *for all* $k, l, m \in \mathbb{Z}$,

(*ii*) $|s^k t^l u^m| \le r \implies \begin{cases} |k| + |l| \le r \\ |m| \le r^2, \end{cases}$

(*iii*) $\dfrac{1}{2}\left(|k| + |l| + \sqrt{|m|}\right) \le |s^k t^l u^m|$ *for all* $k, l, m \in \mathbb{Z}$.

Proof. Observe first that $s^k t^{-l} s^{-k} t^l = u^{kl}$. Consider then an integer $m \geq 0$, the integral part i of $\sqrt{m}$, and the difference $j = m - i^2$, which satisfies $j \leq 2\sqrt{m}$. As $s^i t^{-i} s^{-i} t^i u^j = u^m$, we have $|u^m| \leq 6\sqrt{m}$. The case of an integer $m \leq 0$ is similar, and we have

$$\left| s^k t^l u^m \right| \leq |s^k| + |t^l| + |u^m| \leq |k| + |l| + 6\sqrt{|m|}$$

for all $s^k t^l u^m \in \Gamma$.

From the formulas

$$\left(s^k t^l u^m \right) s = s^{k+1} t^l u^{m+l}$$
$$\left(s^k t^l u^m \right) t = s^k t^{l+1} u^m$$
$$\left(s^k t^l u^m \right) u = s^k t^l u^{m+1}$$

of Example IV.8, it is straightforward to check Claim (ii) by induction on $|s^k t^l u^m|$, and Claim (iii) follows. $\square$

22. Proposition. *The Heisenberg group is of polynomial growth. More precisely, if $\beta(r)$ denotes its growth function, there exist constants $c_1, c_2 > 0$ such that*

$$c_1 r^4 \leq \beta(r) \leq c_2 r^4$$

for all $r \geq 1$.

Proof. Let $r \geq 1$.

For $|k| \leq \frac{r}{8}$, $|l| \leq \frac{r}{8}$ and $|m| \leq \left(\frac{r}{8}\right)^2$, Claim (i) of the previous lemma shows that $|s^k t^l u^m| \leq r$. It follows that $\beta(r) \geq \left(2\left[\frac{r}{8}\right] + 1\right)^2 \left(2\left[\frac{r^2}{64}\right] + 1\right)$, namely that $\beta(r) \geq c_1 r^4$ for an appropriate constant c_1 and for all $r \geq 1$.

On the other hand, Claim (ii) of the previous lemma shows that

$$\beta(r) \leq (r+1)^2 (2r^2 + 1) \leq 12 r^4$$

for all $r \geq 1$. $\square$

Gromov's main result on groups of polynomial growth (our Theorem 29) has a "geometric corollary":

any expanding self-map of a compact manifold is topologically conjugate to an infra-nil-endomorphism.

The proof of this corollary involves a lemma of Franks which we state here for finitely-generated groups. See [Gromo–81], and Section 8 in [Frank–70].

23. Definition. Let Γ_1, Γ_2 be two finitely-generated groups, each given with some length function; denote by $|\gamma|$ the length of an element γ in either of these groups. A homomorphism $\phi : \Gamma_1 \longrightarrow \Gamma_2$ is *expanding* if there exist constants $\lambda > 1$ and $C \geq 0$ such that

$$|\phi(\gamma)| \geq \lambda |\gamma|$$

for all $\gamma \in \Gamma_1$ with $|\gamma| \geq C$.

For an example with $\Gamma_1 = \Gamma_2$, consider the Heisenberg group Γ with the same word length as in Item 21, an integer $a \geq 2$, and the endomorphism ϕ_a of Γ defined by

$$\phi_a \begin{pmatrix} 1 & 0 & 0 \\ k & 1 & 0 \\ m & l & 1 \end{pmatrix} = \begin{pmatrix} 1 & 0 & 0 \\ ak & 1 & 0 \\ a^2 m & al & 1 \end{pmatrix}$$

for all $k, l, m \in \mathbb{Z}$. Lemma 21 implies that

$$|\phi_a(s^k t^l u^m)| \geq \frac{a}{2}\left(|k| + |l| + \sqrt{|m|}\right) \geq \frac{a}{12}\left(|k| + |l| + 6\sqrt{|m|}\right) \geq 2|s^k t^l u^m|$$

and that ϕ_a is expanding, if $a \geq 24$.

We leave it to the reader to check that $\phi_a(\Gamma)$ is of finite index in Γ.

24. Lemma (Franks). *Let Γ be a finitely-generated group given with some word length $\gamma \longmapsto |\gamma|$, and let Γ_0 be a subgroup of finite index in Γ, endowed with the length induced by the word length on Γ.*

Assume that there exists a homomorphism $\phi : \Gamma_0 \longrightarrow \Gamma$ which is expanding, with an image of finite index. Then Γ has polynomial growth.

Proof. Let N denote the index $[\Gamma : \phi(\Gamma_0)]$, choose representatives $\gamma_1, \ldots, \gamma_N$ such that $\Gamma = \bigcup_{1 \leq i \leq N} \phi(\Gamma_0)\gamma_i$ and set $D = \max_{1 \leq i \leq N} |\gamma_i|$. Observe that the balls of radius D about the points of $\phi(\Gamma_0)$ cover Γ. For any real number $t \geq 0$, set

$$\beta(t) = |\{x \in \Gamma \mid |x| \leq \lceil t \rceil\}| \qquad \text{and} \qquad \beta_0(t) = |\{x \in \Gamma_0 \mid |x| \leq \lceil t \rceil\}|.$$

The functions $\beta, \beta_0 : \mathbb{R}_+ \longrightarrow \mathbb{R}_+$ are submultiplicative (see VI.21). Finally, let λ and C be constants for which ϕ satisfies the inequalities of Definition 23.

Consider a number

$$t \geq \max\{\lambda C, \, 1 + \lambda \max_{\gamma \in \Gamma, |\gamma| \leq C} |\phi(\gamma)|, \, D\}.$$

For any element $y \in \Gamma$ such that $|y| \leq t - D$, there exists $x \in \Gamma_0$ such that

$$d(\phi(x), y) = |\phi(x)^{-1} y| \leq D$$

and $|\phi(x)| \leq t$; moreover $|x| \leq \frac{t}{\lambda}$ by the hypothesis on ϕ. It follows that

$$\beta(t - D) \leq \beta(D)\beta_0\left(\frac{t}{\lambda}\right).$$

We will also use the rough estimates $\beta(t) \leq \beta(D)\beta(t - D)$ and $\beta_0(t) \leq \beta(t)$. Thus

$$(*) \qquad\qquad \beta(t) \leq \beta\left(\frac{t}{\lambda}\right) K$$

where K is the constant $\beta(D)^2$.

Now let k be the integral part of the logarithm of t/C in base λ; then $\log_\lambda(t/C) < k + 1$ and thus $t\lambda^{-k} \leq \lambda C$. Iterating (*) k times one obtains

$$\beta(t) \leq \beta\left(\frac{t}{\lambda^k}\right) K^k \leq \beta(\lambda C) K^{\log_\lambda(t/C)}.$$

As $K^{\log_\lambda(t/C)} = K^{\log_\lambda(K)\log_K(t/C)} = (t/C)^M$, with $M = \log_\lambda(K)$, we have

$$\beta(t) \leq L t^M$$

where L is the constant $\beta(\lambda C)C^{-M}$. In other words, $\beta(t)$ is bounded by a polynomial in t, for all $t \geq 0$. $\square$

We use the following terminology: a *virtual isomorphism* from a group Γ_1 to a group Γ_2 is an injective homomorphism from a subgroup of finite index in Γ_1 onto a subgroup of finite index in Γ_2. Then Franks' lemma can be reformulated as:

a group which has an expanding virtual automorphism has polynomial growth.

The following "higher-order Franks' lemma" appears with a slightly different formulation as Lemma 3.5 in [Grigo–84]. It is an important step in proving that the "first group of Grigorchuk" is of intermediate growth (see Proposition VIII.63).

Lemma 24'. *Let Γ be a finitely-generated group given with some word length $\gamma \longmapsto |\gamma|$. Let m be an integer, $m \geq 2$, and consider the direct product Γ^m together with the length function $(\gamma_1, \ldots, \gamma_m) \longrightarrow \sum_{j=1}^{m} |\gamma_j|$.*

If there exists an expanding virtual isomorphism from Γ^m to Γ, then there exists a constant σ with $0 < \sigma \leq 1$ such that the growth function $\beta(k)$ of Γ satisfies

$$e^{k^\sigma} \prec \beta(k).$$

In particular Γ is of intermediate or exponential growth.

25. Notation. For a group Γ, let $(\gamma_n(\Gamma))_{n\geq 1}$ denote the *lower central series,* defined inductively by

$$\gamma_1(\Gamma) = \Gamma \qquad \gamma_{n+1}(\Gamma) = [\Gamma, \gamma_n(\Gamma)].$$

For each $n \geq 1$, denote by d_n the dimension of the vector space

$$\left(\gamma_n(\Gamma)/\gamma_{n+1}(\Gamma)\right) \otimes_{\mathbb{Z}} \mathbb{Q}.$$

Assume now that Γ is finitely generated and *nilpotent* (i.e. $\gamma_n(\Gamma) = \{1\}$ for n large enough); the *homogeneous dimension* of Γ is the integer

$$d(\Gamma) = \sum_{j=1}^{n} j d_j$$

where n is such that $\gamma_{n+1}(\Gamma) = \{1\}$.

For example, the homogeneous dimension of $\mathbb{Z}^k$ is k and that of the Heisenberg group is 4.

26. Theorem (Dixmier, Wolf, Guivarc'h, Bass, and others). *Let* Γ *be a finitely-generated nilpotent group. Then* Γ *is of polynomial growth. More precisely its degree of polynomial growth and its homogeneous dimension coincide:*

$$\beta_\Gamma(k) \sim k^{d(\Gamma)}.$$

The equality of this theorem is due to Yves Guivarc'h. The fact that the degree of polynomial growth is at most $d(\Gamma)$ appears as a particular case of Theorem 2 in [Guiva–70]; the equality is Statement 5 of the main result in [Guiva–71]. See also the thesis [Guiva–73], and in particular Theorem II.1, which is a more general theorem applying to other groups than just finitely-generated nilpotent groups. (Part of the results of [Guiva–73] were found independently by Jenkins [Jenki–73].)

The same formula is often attributed to H. Bass, who published it later [Bass–72]. It has probably been discovered independently by several mathematicians (besides Guivarc'h and Bass), including Brian Hartley (unpublished) as I have been told.

Neither Guivarc'h nor Jenkins nor Bass quote a previous paper where Jacques Dixmier studied properties of unitary representations and group algebras of nilpotent Lie groups. In the first three lemmas of this paper [Dixmi–60], Dixmier shows that a connected nilpotent Lie group G has polynomial growth in the following sense: there exists an integer d, depending only on G, such that, given a Haar measure μ on G and a compact subset H of G, one has $\mu(H^k) = O(k^d)$ when $k \to \infty$. From this and from Mal'cev's results, it follows that any finitely-generated nilpotent group is of polynomial growth. We find it remarkable that this paper of Dixmier, preceding by 8 years the re-discovery of group growth in the West, has not attracted more attention.

It is appropriate to recall here results of Mal'cev [Mal'c–49]; see also Chapter II in [Raghu–72], Chapter 5 in [CorGr–89], or *exercices* 29-33 in § III.9 of [Bourb–72]. Recall that a *rational form* of a real Lie algebra $\underline{g}$ is a $\mathbb{Q}$-subalgebra $\underline{g}_\mathbb{Q}$ of $\underline{g}$ such that the natural mapping $\underline{g}_\mathbb{Q} \otimes_\mathbb{Q} \mathbb{R} \longrightarrow \underline{g}$ is an isomorphism. We say that two subgroups Γ_1, Γ_2 of a group G are *strictly commensurable* if $\Gamma_1 \cap \Gamma_2$ is of finite index in both Γ_1 and Γ_2. (Thus Γ_1, Γ_2 are commensurable in the sense of Item II.18 if and only if Γ_1 and $g\Gamma_2 g^{-1}$ are strictly commensurable for some $g \in G$.) For Assertion (ii), see Theorem 5.1.12 of [CorGr–89].

(i) *A necessary and sufficient condition for a group* Γ *to occur as a discrete subgroup of a connected simply-connected nilpotent Lie group* G *with* G/Γ *compact is that* Γ *be finitely generated, nilpotent, and without element* $\neq 1$ *of finite order.*

(ii) *Let* G *be a connected simply-connected nilpotent Lie group with Lie algebra* $\underline{g}$*. A necessary and sufficient condition for* G *to contain a discrete subgroup* Γ *with* G/Γ *compact is that* $\underline{g}$ *has a rational form. More precisely, there is a natural bijection between the set of strict commensurability classes of discrete cocompact subgroups of* G *and the set of rational forms of* $\underline{g}$*.*

(iii) Let Γ_j be a discrete subgroup of a simply-connected nilpotent Lie group G_j with G_j/Γ_j compact, for $j = 1, 2$; then any isomorphism from Γ_1 onto Γ_2 extends uniquely to an isomorphism from G_1 onto G_2.

The following is also a standard result; see, e.g., Corollary 9.18 in [MacDo–88].

In a nilpotent group N, the elements of finite order form a normal subgroup N_f such that N/N_f is torsion free.

Observe that if, moreover, N is finitely generated, then N_f is a finite group.

Some of the previous results carry over from nilpotent to solvable groups; see in particular [Witte–95].

Let us quote without proof a few basic results about groups of polynomial growth.

27.   Theorem (Milnor and Wolf). *A finitely-generated solvable group is of polynomial growth if and only if it contains a nilpotent subgroup of finite index, and is of exponential growth otherwise.*

See [Wolf–68], for a polycyclic group, and [Miln–68c], for the complement giving the theorem as stated here; see also [Tits–81a]. Lemma 1 of [Miln–68c] has been exploited in [Rosse–76] for showing the following theorem.

Let N be a normal subgroup of a finitely-generated group Γ of subexponential growth; if Γ/N is solvable, then N is finitely generated.

The alternative of Theorem 27, of being either of polynomial growth or of exponential growth, holds equally for finitely-generated linear groups, by Tits' theorem II.42 [Tits–72]. (See also [Shalo–98], and Complement II.42.)

Here is an interesting complement from [Rosbl–74]. (For further investigation on linear *semi*-groups, see [OknSa–95].)

28.   Theorem (Rosenblatt). *A finitely-generated solvable group is of exponential growth if and only if it contains a free sub-semi-group on two generators.*

More generally, Theorems 27 and 28 hold for all "elementary amenable groups", by Section 3 of [Chou–80].

29.   Theorem (Gromov). *A finitely-generated group is of polynomial growth if and only if it has a nilpotent subgroup of finite index.*

This theorem is a landmark in geometric group theory, and Gromov's proof [Gromo–81], which is beyond the scope of this book, has been an inspiration for much further work. Here, we wish to quote [VdDW–84a], [VdDW–84b], and [Tits–81b], on Gromov's proof, and [Loser–87], on the extension of Gromov's theorem to locally compact groups.

Here is a different formulation of Theorem 29. Consider a finitely-generated group Γ and a field $\mathbb{K}$; *then the group algebra $\mathbb{K}[\Gamma]$ has finite Gelfand-Kirillov dimension if and only if Γ has a nilpotent subgroup of finite index* [KraLe–85].

Gromov's proof shows more than what has just been stated; for example: *if Γ is a finitely-generated group such that $\liminf_{k\to\infty} k^{-d}\beta(k) < \infty$ for some $d \geq 1$, then Γ has a nilpotent subgroup of finite index.*

Recall that, given a prime p, a group Γ is said to be *residually-p* if, for any $\gamma \in \Gamma$ with $\gamma \neq 1$, there exist a finite p-group G and a homomorphism $\pi : \Gamma \to G$ such that $\pi(\gamma) \neq 1$. The following theorem has been proved by Grigorchuk [Grigo–89] (see also the end of Section 1 in [LubMa–91], Interlude E in [DiDMS–91], and [BarGr–a]):

a finitely-generated group which is residually-p for some prime p or which is residually nilpotent is of subradical growth if and only if it has a nilpotent subgroup of finite index.

(A finitely-generated group is of *subradical growth* if its growth function is weakly dominated by the function $e^{\sqrt{k}}$, in the sense of Definition VI.22.)

Research problems. Is it true that the growth function β_Γ of any finitely-generated group Γ which is not of polynomial growth satisfies

$$e^{\sqrt{k}} \prec \beta(k) \quad ?$$

Does there exist a finitely-generated group Γ of which the growth function β_Γ satisfes

$$e^{\sqrt{k}} \sim \beta(k) \quad ?$$

30. A consequence and a question. For a finitely-generated group Γ, the *degree of polynomial growth* is defined by

$$d(\Gamma) = \limsup_{k\to\infty} \frac{\log \beta_\Gamma(k)}{\log k} \in [0, \infty]$$

and is independent of the set of generators used to define the growth function β_Γ. One has $d(\Gamma) < \infty$ if and only if Γ is of polynomial growth.

It follows from Theorems 26 and 29 that $d(\Gamma)$ is an integer if $d(\Gamma) < \infty$. Can one show this without using the full strength of Gromov's theorem ?

Observe also that, as a consequence of Theorem 29, a finitely-generated group which is of polynomial growth is finitely presented.

There is an elementary *splitting lemma* in [Gromo–81]:

if Δ is a finitely-generated subgroup of infinite index of a finitely-generated group Γ, then $d(\Delta) \leq d(\Gamma) - 1$.

There is also the following lemma from [VdDW–84b] (compare with our Lemma VIII.60):

let Γ be a finitely-generated group of subexponential growth and let Γ_0 be a finitely-generated subgroup; let S be a finite set of generators of Γ containing a finite set S_0 of generators of Γ_0; assume that there exists a constant C such that $\beta(\Gamma, S; k) \leq C\beta(\Gamma_0, S_0; k)$ for all k large enough; then Γ_0 is of finite index $\leq C$ in Γ.

31.   Theorem (Milnor). *The fundamental group* Γ *of a compact Riemannian manifold* M *with mean curvature* ≥ 0 *is of polynomial growth. More precisely one has* $\beta_\Gamma(k) \prec k^n$, *where* n *is the dimension of* M.

This appears in [Miln–68b].

Group growth can serve as a tool for various problems of differential geometry. For example, it has been used for the classification of homogeneous Riemannian manifolds with zero Ricci curvature [AleKi–75].

32.   Degree of polynomial growth and isoperimetric inequalities.
For a group Γ generated by a finite set S, one defines the *inverse growth function*
$\Phi : \mathbb{R}_+ \longrightarrow \mathbb{N}$ by

$$\Phi(\lambda) = \min\{k \in \mathbb{N} \mid \beta(\Gamma, S; k) > \lambda\}.$$

For a subset Ω of Γ, define the *S-boundary* $\partial_S(\Omega)$ to be the subset of those $\gamma \in \Omega$ for which there exists $s \in S \cup S^{-1}$ with $\gamma s \notin \Omega$. It is then a result of [CouSC–93] that

$$\frac{|\partial_S(\Omega)|}{|\Omega|} \geq \frac{1}{8|S|\Phi(2|\Omega|)}$$

for all finite subsets Ω of Γ. It is remarkable that the function Φ, defined only in terms of the sizes of *balls* in Γ and their boundaries, controls the isoperimetry of *arbitrary finite subsets* in Γ.

In particular, if Γ has polynomial growth of degree d, there exists a constant $c > 0$ such that

$$\frac{|\partial_S(\Omega)|}{|\Omega|} \geq \frac{c}{|\Omega|^{1/d}}$$

for all finite subsets Ω of Γ (a result of Varopoulos).

Given any group Γ generated by a finite set S, define the *isoperimetric profile*

$$I_\Gamma(k) = \max_{m \leq k} \min_{|\Omega|=m} |\partial_S \Omega|$$

(this function does not depend on the choice of S up to the equivalence of Definition VI.22). As a consequence of results of Gromov (Theorem 29) and others, the following are then equivalent, for an integer d:

$I_\Gamma(k) \overset{w}{\sim} k^{(d-1)/d}$,

Γ has polynomial growth of degree d,

Γ contains a nilpotent subgroup of finite index Γ_0 for which $d(\Gamma_0) = d$

(the notation $d(\Gamma_0)$ is that of Item 30); see [PitSa–99].

33.   Asymptotics of the growth function for nilpotent groups. Let Γ be a finitely-generated group which has a nilpotent subgroup of finite index and let d be the index of polynomial growth of Γ. Pansu has shown that the limit

$$c_1 = \lim_{k \to \infty} \frac{\beta(k)}{k^d}$$

exists [Pansu–83]. (Compare with Item VI.2.iv in case Γ is abelian.) M. Stoll has shown that $\beta(k) - c_1 k^d = O\left(k^{d-1}\right)$ for the so-called 2-step nilpotent groups [Stoll–98]. It is an open problem to determine when the quantity

$$\frac{\beta(k) - c_1 k^d}{k^{d-1}}$$

has a limit for $k \to \infty$.

34. Growth and Følner sets. Let Γ be a group generated by a finite subset S and let $\omega = \omega(\Gamma, S) = \lim_{k \to \infty} \sqrt[k]{\beta(\Gamma, S; k)}$ denote the corresponding exponential growth rate. The *S-boundary* of a subset A of Γ is *now* the subset

$$\partial_S A = \left\{\gamma \in \Gamma \mid \gamma \notin A \quad \text{and} \quad \gamma s \in A \quad \text{for some} \quad s \in S \cup S^{-1}\right\}$$

of $\Gamma \setminus A$; for example, if $B(k) = \{\gamma \in \Gamma \mid \ell_S(\gamma) \leq k\}$, then $\partial_S B(k)$ is the sphere $B(k+1) \setminus B(k)$. A *Følner sequence* in Γ is a sequence $(F_k)_{k \geq 1}$ of finite subsets of Γ such that

$$\lim_{k \to \infty} \frac{|F_k \cup \partial_S F_k|}{|F_k|} = 1$$

(it is well-known and easy to check that this condition does not depend on the choice of the generating set S). The group Γ is *amenable* if it has a Følner sequence.

Proposition. *Let (Γ, S), ω and $B(k)$ be as above.*

(i) If Γ is of polynomial growth, the sequence $(B_k)_{k \geq 0}$ of balls is a Følner sequence in Γ.

(ii) If Γ is of subexponential growth, there exists a sub-sequence of $(B_k)_{k \geq 0}$ which is a Følner sequence in Γ.

(iii) If Γ is of exponential growth, no Følner sequence in Γ is a sub-sequence of $(B_k)_{k \geq 0}$.

Proof. Claim (i) follows from the result of Pansu [Pansu–83] already quoted in Item 33.

Recall from Calculus that

$$\liminf_{k \to \infty} \frac{\beta(\Gamma, S; k+1)}{\beta(\Gamma, S; k)} \leq \omega \leq \limsup_{k \to \infty} \frac{\beta(\Gamma, S; k+1)}{\beta(\Gamma, S; k)}$$

(Theorem 3.37 in [Rudin–53]). It follows that if Γ is of subexponential growth, i.e. if $\omega = 1$, then there exists a sub-sequence of $(B_k)_{k \geq 0}$ which is a Følner sequence.

Claim (iii) is an observation for which we refer to [Pitte]. □

Finitely-generated groups of subexponential growth are amenable.

This consequence of Claim (ii) has been observed on several occasions, apparently for the first time in [AdVSr–57], and later for example in [HirTh–75].

With the same notation as above, consider an infinite finitely-generated group Γ, and let

$$F_1 \subsetneq F_2 \subsetneq F_3 \subsetneq \cdots$$

be a strictly increasing sequence such that

$$F_{\beta(\Gamma, S; k)} = B(k)$$

for all $k \geq 0$. Thus $|F_j| = j$ for all $j \geq 1$, and the sequence $(F_j)_{j \geq 1}$ has precisely *linear growth*.

Problem 34.A. *When is it true that* $(F_j)_{j \geq 1}$ *is a Følner sequence ?*

Observe that, for j such that $\beta(\Gamma, S; k) \leq j < \beta(\Gamma, S; k+1)$, one has $B(k) \subset F_j$ and $F_j \cup \partial_S F_j \subset B(k+2)$, so that

$$\frac{|F_j \cup \partial_S F_j|}{|F_j|} \leq \frac{\beta(\Gamma, S; k+2)}{\beta(\Gamma, S; k)}.$$

Thus, if the growth function of (Γ, S) is regular enough for

$$(*) \qquad\qquad \lim_{k \to \infty} \frac{\beta(\Gamma, S; k+1)}{\beta(\Gamma, S; k)} = 1,$$

then $(F_j)_{j \geq 1}$ is a Følner sequence of linear growth. Thus, one may reformulate Problem 34.A as follows.

Problem 34.B. *Does there exist a group* Γ *of intermediate growth and a finite generating subset* S *of* Γ *such that (*) holds ? Does this hold for the group* Γ *and the generating set* $S = \{a, b, c, d\}$ *of Chapter VIII ?*

I am grateful to Slava Grigorchuk, Christophe Pittet, and Aryeh Vaillant for discussions about this item. As well as to Andrzej Zuk, who has shown me why any amenable group has a Følner sequence $(F_j)_{j \geq 1}$ such that $|F_j| = j$ (private communication); observe however that Problem 34.A refers to sequences with one more property, namely $F_{\beta(\Gamma, S; k)} = B(k)$.

VII.D. Complement on other kinds of growth

35. Relative growth and cogrowth. Consider a group Π generated by a finite set S, the word length $\ell_S : \Pi \longrightarrow \mathbb{N}$, and a subgroup Ξ of Π. The corresponding *relative spherical growth function* is the function defined by

$$\sigma(\Xi \text{ rel } \Pi, S; k) = |\{\xi \in \Xi \mid \ell_S(\xi) = k\}|$$

and is the spherical growth function of Ξ relative to the weight $\ell_S|\Xi$ (compare with Complements VI.42 and 43). For example, computations have been made for cases where Ξ is the commutator subgroup of Π (see [GriHa–97] for Π a free group and [Sharp–98] for Π a hyperbolic group).

The most important case is the following: for a group Γ given as a quotient $\pi : F_n \longrightarrow \Gamma$ of a free group F_n on a free set S_n of n generators $(n < \infty)$, set $(\Pi, S) = (F_n, S_n)$ and $\Xi = Ker(\pi)$. The relative spherical growth function $\sigma(Ker(\pi)$ rel $F_n, S_n; k)$ is the *cogrowth function* and the growth rate

$$\alpha = \limsup_{k \to \infty} \sqrt[k]{\sigma(Ker(\pi) \text{ rel } F_n, S_n; k)}$$

the *cogrowth exponent* of $\pi : F_n \longrightarrow \Gamma$. It is easy to see that either $\alpha = 1$ (this occurs if and only if $Ker(\pi) = \{1\}$) or $\sqrt{2n-1} \leq \alpha \leq 2n-1$. It is also true that, for $\alpha \neq 1$, one has

$$\sqrt{2n-1} < \alpha \leq 2n-1$$

with $\alpha = 2n - 1$ if and only if Γ is amenable [Grigo–78].

This last criterion of amenability has been used on several occasions, for example by Ol'shanskii [Ol's–80] to show the existence of a non-amenable group without non-abelian free subgroups, and also by Adian [Adian–83] to show that Burnside groups of large odd exponents are not amenable.

36. Open problem. Consider more generally a group Ξ generated by a finite set S and a surjective homomorphism $\pi : \Xi \longrightarrow \Gamma$; set

$$\alpha = \limsup_{k \to \infty} \sqrt[k]{\sigma(Ker(\pi) \text{ rel } \Xi, S; k)}.$$

Assuming Ξ hyperbolic and non-elementary in the sense of Gromov, and with the exponential growth rate $\omega(\Xi, S)$ as in Definition VI.51, is it true that

$$\sqrt{\omega(\Xi, S)} < \alpha \leq \omega(\Xi, S) \quad ?$$

(this already appears in [GriHa–97]).

37. Green's functions of simple random walks. Let X be a connected graph, say here regular of some degree n for simplicity, and let x, y be two vertices of X. For each integer $k \geq 1$, let g_k denote the number of all paths from x to y of combinatorial length k. The normalized growth rate

$$\nu = \frac{1}{n} \limsup_{k \to \infty} \sqrt[k]{g_k}$$

is independent of the choice of x and y, and is called the *spectral radius* of X. For the importance of this notion, see [Keste–59], as well as [Woess–94] and Chapter II of [Woess–00].

Assume moreover that X is the Schreier graph (see IV.15) corresponding to the free group F_S on a set S of m generators and a subgroup Γ_0 of F_S; thus the degree n of X is now $2m$. For $k \geq 1$, let σ_k denote the number of elements in Γ_0 of S-word length k; thus σ_k would be $\sigma(F_S \text{ rel } \Gamma_0, S; k)$ in the notation of Item 35; set $\alpha = \limsup_{k\to\infty} \sqrt[k]{\sigma_k}$. One may view σ_k as the number of paths in X, from the base point to itself, of combinatorial length k, in which an edge is never followed by the same edge backwards (this condition corresponds to words representing elements of F_S being reduced). This implies of course that $\sigma_k \le g_k$ for all $k \ge 0$ (with g_k counting paths from the base point to itself), but also suggests a more precise relationship between the σ_k's and the g_k's. Indeed, Grigorchuk has shown that

$$(*) \quad \nu = \begin{cases} \dfrac{\sqrt{2m-1}}{m} & \text{if } 1 \le \alpha \le \sqrt{2m-1} \\[2ex] \dfrac{\sqrt{2m-1}}{2m}\left(\dfrac{\sqrt{2m-1}}{\alpha} + \dfrac{\alpha}{\sqrt{2m-1}}\right) & \text{if } \sqrt{2m-1} \le \alpha \le 2m-1 \end{cases}$$

[Grigo–78]. (Moreover, if Γ_0 is normal in F_S and not reduced to 1, then $\alpha > \sqrt{2m-1}$.) More recently, the relationship $\sigma_k \rightsquigarrow g_k$ has been studied further, and we will quote a particular case of results of [Barth–99]: for the formal power series

$$\Sigma(t) = \sum_{k=0}^{\infty} \sigma_k t^k \quad \text{and} \quad G(t) = \sum_{k=0}^{\infty} g_k t^k,$$

so that G is the *Green's function* of the simple random walk on the Cayley graph of (Γ, S), we have

$$\frac{\Sigma(t)}{1-t^2} = \frac{G\left(\frac{t}{1+(2m-1)t^2}\right)}{1+(2m-1)t^2}.$$

Formula (*), relating the radius of convergence $\frac{1}{\alpha}$ of F to the radius of convergence $\frac{1}{2m\nu}$ of G, follows easily.

38. Remark on subgroup growth. Given a group Γ [respectively a profinite group G], consider for all $k \ge 1$ the number N_k of subgroups of Γ [respectively of closed subgroups of G] of index k. The study of the growth of $k \longmapsto N_k$ goes back to M. Hall's study of free groups [HallM–49], and is consequently older than the study of the growth of groups in the sense of our Chapters VI and VII. We refer to Problem II.21 and to more recent work quoted there.

39. Miscellanea. Given a group Γ generated by a finite set S, there are several kinds of objects with interesting growth properties; in particular, the following problems have been (and probably will be) studied.

• Growth of the number of subgraphs isomorphic to a given finite graph in the successive balls of a Cayley graph of a group [EpsIZ–96].

• Growth of conjugacy classes, with various weights; see in particular Items 5.2 and 8.5 in [Gromo–87]. This is related to the number of closed geodesics on a compact manifold. See also [Rivin].

- Growth of coset spaces $\Gamma_0 \setminus \Gamma$ with respect to subgroups; see Section (E.2) in [GriHa–97].
- Growth of double coset spaces $\Gamma_1 \setminus \Gamma/\Gamma_2$; this theme has not yet received much attention, but probably should.
- Growth of geodesic segments starting from 1; see e.g. [BarCe–a] and [Shapi–97]. Information on the growth of *pairs* of geodesic segments from 1 to the same group element provides estimates from below on the spectral radius [BarCe].
- Growth of graded associative algebras defined as follows. For a group Γ and a prime p, denote by Δ the augmentation ideal of the group algebra $\mathbb{F}_p[\Gamma]$. The corresponding graded algebra is defined by $\mathcal{A}_\Gamma = \bigoplus_{n \geq 0} \left(\Delta^n / \Delta^{n+1} \right)$. If Γ is residually-p (see Item VII.29), there are interesting relations between the growth of this graded algebra and the growth of Γ. See [Grigo–89], and Remark VIII.66.iii below; see also [BarGr–a].
- Growth of the ranks of the subquotients of the lower central series, and of related ranks [AlpPe–90]. (See also Remark VIII.31.e.)
- Growth of the maximal orders of finite subgroups contained in balls about $1 \in \Gamma$ [Ol's–99a] (say for a torsion group, or even better for a p-group Γ).
- Growth of the number of irreducible unitary representations as a function of their dimensions. See the discussions in [HarRV–93] and [Rapin–99], as well as [PasTe–96].

40. Infinitely-generated groups and infinite families of finite groups. There has been some work on the extension of group growth to countably generated groups.

A group Γ has *uniform polynomial growth* if there exists a sequence $(P_n)_{n \geq 1}$ of polynomials such that $\beta(\langle S \rangle, S; k) \leq P_n(k)$ for all n-element subsets S of Γ, where $\langle S \rangle$ denotes the subgroup of Γ generated by S. Nilpotent groups are known to have uniform polynomial growth [Bozej–80]. For example, if $\Gamma_j = \mathbb{Z}$ for each $j \in \mathbb{Z}$, then $\oplus_{j \in \mathbb{Z}} \Gamma_j$ has uniform polynomial growth; this shows that, in general, the degrees of the polynomials P_n's cannot be bounded.

A group Γ has *weak exponential growth* if there exists an integer n such that

$$\limsup_{k \to \infty} \sup_{\substack{S \subset \Gamma \\ |S| = n}} \sqrt[k]{\beta(\langle S \rangle, S; k)} > 1.$$

For example, the locally finite infinite group $Sym_f(\mathbb{N})$ of Item III.4 has weak exponential growth [KaiVe–83].

Another subject of study is provided by a family $\mathcal{G} = (\Gamma_i)_{i \in I}$ of finite groups, each Γ_i being given together with a set S_i of generators, with $\sup_{i \in I} |S_i| < \infty$. The growth function of $\mathcal{G}$ with respect to $\mathcal{S} = (S_i)_{i \in I}$ is defined by

$$\beta(k) = \sup_{i \in I} \beta(\Gamma_i, S_i; k)$$

(the sup is in fact a max). S. Black has obtained results in this context analogous to various results of Milnor, Wolf, and Gromov quoted above [Black–98].

CHAPTER VIII

THE FIRST GRIGORCHUK GROUP

The group Γ described in this chapter appears in [Grigo-80]. The word "first" refers to the chronological order of discovery; other groups, discovered later and sharing various properties with this one, will not be discussed here. Our exposition borrows from [Grigo-98]; for another exposition, see [CecMS].

After generalities on full automorphism groups of rooted d-ary trees in Section A, we introduce the group Γ in Section B. The main results we prove are the following:

Γ is a finitely-generated infinite 2-group (Corollary 15 and Theorem 17),

Γ does not have any finite-dimensional linear representation which is faithful (Item 19),

Γ and $\Gamma \times \Gamma$ are commensurable (Theorem 28),

for any $n \geq 1$ there exists $\gamma \in \Gamma$ such that $\gamma^{2^n} \neq 1$ (Theorem 32),

any normal subgroup of Γ distinct from $\{1\}$ is a congruence subgroup (Theorem 42),

the word problem in Γ is solvable (Algorithm 47 and Corollary 49),

Γ is not finitely presentable (Theorem 55),

Γ is of intermediate growth (Theorem 65),

any finite group of order a power of 2 is isomorphic to a subgroup of Γ (Exercise 84).

The reader mostly interested in the growth of Γ can read Section F (together with the beginning of Section E) immediately after Section B.

This chapter, the longest of this book, is devoted to just *one* group! We claim that this special treatment is justified by the importance of the Grigorchuk group for many problems, such as growth [Grigo-91], amenability [Grigo-96], or "just infinite" groups [Grigo-99]. There are other mathematical objects which play a really central role in their subject: the hyperbolic plane for spaces of negative curvature, $SL(2, \mathbb{R})$ for representations of semi-simple groups, or the hyperfinite II_1 factor for von Neumann algebras, to mention only a few examples. In our opinion, the Grigorchuk group is a good candidate for membership in this club.

VIII.A. Rooted d-ary trees and their automorphisms

1. The d-ary tree $T^{(d)}$. For an integer $d \geq 2$, we denote by $T^{(d)}$ the *infinite rooted d-ary tree*. The set $T^{(d)}$ of *vertices* of $T^{(d)}$ is the set of finite sequences of elements in $\{0, 1, \ldots, d-1\}$. The empty sequence defines the root vertex denoted by $x_\emptyset$. We sometimes write a typical sequence as $(j_1, j_2, \ldots, j_k)$ and sometimes as $x_{j_1, j_2, \ldots, j_k}$, so that for example $x_{0,0,1,0,1} = (0, 0, 1, 0, 1)$. Two vertices are connected by an *edge* in $T^{(d)}$ if their lengths (as sequences) differ by 1 and if the shorter can be obtained by erasing the last term of the longer. Thus $x_\emptyset$ has d neighbours which are $x_0, x_1, \ldots, x_{d-1}$, and any other vertex has $d + 1$ neighbours, one "ancestor" and d "descendants".

The d-ary tree has *levels* with vertex sets

$$L^{(d)}(k) = \{\text{sequences of length } k\}$$

for each $k \geq 0$, so that

$$T^{(d)} = \coprod_{k=0}^{\infty} L^{(d)}(k)$$

(where $\coprod$ stands for "disjoint union"). We think of $L^{(d)}(k+1)$ as being "below" $L^{(d)}(k)$.

For each $x \in T^{(d)}$, we denote by $T_x^{(d)}$ the subtree of $T^{(d)}$ spanned by vertices beginning by x (as sequences). Observe that there is a canonical isomorphism

$$\delta_x : T^{(d)} \longrightarrow T_x^{(d)}$$

mapping a vertex y to the vertex xy obtained by concatenation of the sequences x and y.

For groups acting on $T^{(d)}$ when $d \geq 3$, see for example [GupS–83a,b], [Gupta–89], Section II.2 in [Baums–93], [Bass4–96] and [BarGr–c]. In this chapter, we will concentrate on the case $d = 2$, and occasionally write T, T, and G instead of $T^{(2)}$, $T^{(2)}$, and $G^{(2)}$.

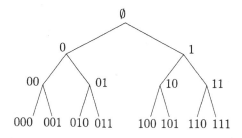

FIGURE 1. A neighbourhood of the root in $T^{(2)}$

2. The group $G^{(d)}$ of automorphisms of $T^{(d)}$. Any automorphism g of $T^{(d)}$ fixes the root $x_\emptyset$ and permutes the d subtrees $T_0^{(d)}, \ldots, T_{d-1}^{(d)}$. More generally, for any $k \geq 1$, the automorphism g permutes the d^k subtrees $T_x^{(d)}$ with $x \in L^{(d)}(k)$. It is easy to check that the automorphism group $G^{(d)}$ of $T^{(d)}$ is uncountable. (For example because, together with the topology of pointwise convergence, $G^{(d)}$ is a topological group which is compact, and homeomorphic to the Cantor set. See also Item 7 below.)

For $k \geq 1$ and $g \in G^{(d)}$, let $\chi_k(g)$ denote the signature of the permutation induced by g on the level $L^{(d)}(k)$. The direct product $\prod_{k \geq 1} \chi_k$ provides a homomorphism from $G^{(d)}$ to the direct product of infinitely many groups of order two. Thus the quotient of $G^{(d)}$ by the closure of its commutator subgroup is a compact abelian group which does not contain any finitely-generated dense subgroups. A fortiori, $G^{(d)}$ does not contain finitely-generated dense subgroups.

Given d automorphisms $g_0, g_1, \ldots, g_{d-1}$ of $T^{(d)}$, there is a well-defined automorphism g of $T^{(d)}$, which fixes the d vertices $x_0, x_1, \ldots, x_{d-1}$ of $L^{(d)}(1)$ and which acts on $T_j^{(d)}$ as $\delta_j g_j \delta_j^{-1}$ for each $j \in \{0, 1, \ldots, d-1\}$, with δ_j as in Item 1. Conversely, any automorphism of $T^{(d)}$ fixing these d vertices is of this form. It follows that, if

$$St_{G^{(d)}}(1) = \left\{ g \in G^{(d)} \mid g(x_j) = x_j \quad \text{for} \quad j = 0, 1, \ldots, d-1 \right\}$$

denotes the subgroup of $G^{(d)}$ fixing all vertices of level 1, we have an isomorphism

$$\psi = (\phi_0, \ldots, \phi_{d-1}) : \begin{cases} St_{G^{(d)}}(1) \longrightarrow G^{(d)} \times G^{(d)} \times \ldots \times G^{(d)} \\ g \longmapsto (g_0, g_1, \ldots, g_{d-1}) \end{cases}$$

where each ϕ_j is the composition of ψ and the appropriate projection. We will write improperly

$$g = (g_0, g_1, \ldots, g_{d-1})$$

instead of $\psi(g) = (g_0, g_1, \ldots, g_{d-1})$ for $g \in St_{G^{(d)}}(1)$. Observe also that $G^{(d)}$ acts transitively on $L^{(d)}(1)$ and that the corresponding homomorphism from $G^{(d)}$ to the symmetric group $Sym(d)$ of $\{0, 1, \ldots, d-1\}$ is onto. It follows that $St_{G^{(d)}}(1)$ is a normal subgroup of index $d!$ in $G^{(d)}$.

More generally, for any $k \geq 1$, one defines the subgroup

$$St_{G^{(d)}}(k) = \left\{ g \in G^{(d)} \mid g(x) = x \quad \text{for all} \quad x \in L^{(d)}(k) \right\}$$

fixing vertices of level $\leq k$, which is a normal subgroup of finite index in $G^{(d)}$. We have an isomorphism

$$\psi_k = (\phi_{0,\ldots,0}, \ldots, \phi_{d-1,\ldots,d-1}) : \begin{cases} St_{G^{(d)}}(k) \longrightarrow G^{(d)} \times \ldots \times G^{(d)} \\ g \longmapsto (g_{0,\ldots,0}, \ldots \ldots, g_{d-1,\ldots,d-1})_k \end{cases}$$

where the last subscript k indicates that there are d^k copies of $G^{(d)}$, where d^k-uplets are indexed by the $\{0, \ldots, d-1\}$-sequences of length k, and where $\phi_{i_1,\ldots,i_k}$ is in fact the composition $\phi_{i_k} \circ \ldots \circ \phi_{i_1}$. If $T^{(d)}(k)$ is the finite subtree of $T^{(d)}$ spanned by the vertices of level at most k, we have a short exact sequence

$$1 \longrightarrow St_{G^{(d)}}(k) \longrightarrow G^{(d)} \longrightarrow Aut\left(T^{(d)}(k)\right) \longrightarrow 1$$

which splits, and we will identify the quotient $G^{(d)}/St_{G^{(d)}}(k)$ with the automorphism group of $T^{(d)}(k)$. One also has natural projections $G^{(d)}/St_{G^{(d)}}(k+1) \longrightarrow G^{(d)}/St_{G^{(d)}}(k)$ and one could identify the corresponding inverse limit with the group $G^{(d)}$ itself.

The finitely-generated group Γ to be defined in Section B will share with $G^{(2)}$ this property of being commensurable with its square. See Definition IV.27 for "commensurable", and Theorem 28 below for the fact itself. For a finitely-generated group $\Gamma \neq 1$ *isomorphic* to $\Gamma \times \Gamma$, see [Tyrer–74].

3. Wreath products (a reminder). Let A, B be two groups and let L be a set on which A acts. Denote by B^L the group of functions from L to B, with pointwise multiplication. Then A acts on B^L by

$$({}^a f)(x) = f\left(a^{-1}x\right) \qquad \text{for all} \quad a \in A,\, f \in B^L \quad \text{and} \quad x \in L.$$

The corresponding semi-direct product

$$B \wr_L A = B^L \rtimes A$$

$$(f, a)(f', a') = (f\, {}^a f', aa') \quad \text{for all} \quad f, f' \in B^L \quad \text{and} \quad a, a' \in A$$

is called the *wreath product* of B by A relative to L. If $A, B,$ and L are finite, $B \wr_L A$ is also finite, and of order

$$|B \wr_L A| = |B|^{|L|} |A|.$$

In case $L = A$, with A acting on itself by left multiplication, one obtains the wreath product $B \wr A = B^A \rtimes A$.

For more on wreath products, see e.g. Section 2.6 of [DixMo–96], § 6.2 of [KarMe–85], or Chapter 7 of [Rotma–95]; see also § 25 of [Huppe–98] on linear representations.

4. Proposition. *Let $A, B, L,$ and $B \wr_L A$ be as above. Assume that the action of A on L is faithful and transitive.*

(i) The centre of $B \wr_L A$ is isomorphic to the centre of B.

(ii) Suppose that L is finite. The abelianization of $B \wr_L A$ is isomorphic to the direct product of the abelianization of B and the abelianization of A.

(iii) Suppose that $L = A$ is finite, with A acting on itself by left multiplication. For any group G for which there exists an extension

$$1 \longrightarrow B \longrightarrow G \longrightarrow A \longrightarrow 1,$$

there exists a subgroup of $B \wr A$ isomorphic to G.

Proof. For the first assertion, consider the natural mapping from the centre of B to the subgroup of $B \wr_L A$ consisting of pairs $(f, 1)$, with $f : L \longrightarrow Z(B)$ a constant function from L to the centre of B, and with $1 \in A$. It is easy to check that this mapping is an isomorphism onto the centre of $B \wr_L A$.

For a group G, let $g \longmapsto \bar{g}$ denote the quotient homomorphism from G onto its abelianization $G^{ab} = G/[G,G]$. If L is finite, there is a natural homomorphism

$$\begin{cases} (B \wr_L A)^{ab} \longrightarrow B^{ab} \times A^{ab} \\ \\ (f,a) \quad\quad \longmapsto \left(\prod_{l \in L} \overline{f(l)}, \bar{a} \right) \end{cases}$$

and we leave it to the reader to check that this is an isomorphism onto. (See also [Bass4–96], page 137.)

The third assertion is a standard theorem of Kaloujnine and Krasner; see e.g. Th. 6.2.8 in [KarMe–85]. $\square$

5. Proposition. *With the notation of Items 2 and 3, we have for each $k \geq 1$ an isomorphism*

$$G^{(d)}/St_{G^{(d)}}(k) \approx \left(G^{(d)}/St_{G^{(d)}}(k-1) \right) \wr_{L(1)} Sym(d)$$

and the quotient $St_{G^{(d)}}(k-1)/St_{G^{(d)}}(k)$ is a group of order $(d!)^{d^{k-1}}$.

In particular, if $d = 2$, the order $|G^{(2)}/St_{G^{(2)}}(k)|$ of the automorphism group of $T^{(2)}(k)$ is

$$|G^{(2)}/St_{G^{(2)}}(k)| = 2^{2^k - 1}$$

for each $k \geq 0$. (The finite tree $T^{(2)}(k)$ has been defined in Item 2.)

Proof. This is straightforward. Observe that, for each $j \in \{0, 1, \ldots, d-1\}$, the group $G^{(d)}/St_{G^{(d)}}(k-1) \approx Aut\left(T^{(d)}(k-1)\right)$ can be seen as acting on $T^{(d)}(k)$, fixing any vertex $(j_1, j_2, \ldots, j_l)$ with $l \in \{1, \ldots, k\}$ and $j_1 \neq j$.

Observe also that the quotient $St_{G^{(d)}}(k-1)/St_{G^{(d)}}(k)$ is naturally in bijection with the set of all mappings from $L(k-1)$ to $Sym(d)$. $\square$

In case $d = 2$, we have[1]

$G^{(2)}/St_{G^{(2)}}(1) \approx \mathbb{Z}/2\mathbb{Z}$,

$G^{(2)}/St_{G^{(2)}}(2) \approx (\mathbb{Z}/2\mathbb{Z}) \wr (\mathbb{Z}/2\mathbb{Z})$ is a dihedral group of order 8,

$G^{(2)}/St_{G^{(2)}}(3) \approx \left(G^{(2)}/St_{G^{(2)}}(2) \right) \wr (\mathbb{Z}/2\mathbb{Z})$ is of order 2^7.

Indeed, for each $k \geq 1$, the group $G/St_{G^{(2)}}(k)$ is an iterated wreath product of groups of order two, and thus also a Sylow 2-subgroup of the symmetric group $Sym\left(2^k\right)$. (The last statement is the particular case for the prime $p = 2$ of a theorem of Kaloujnine on Sylow p-subgroups of $Sym\left(p^k\right)$; see Theorem 7.27 in [Rotma–95].)

---

[1]We denote by $\mathbb{Z}/2\mathbb{Z}$ the group of order 2, even if it is appropriate to consider it as the multiplicative group $\{1, -1\}$ rather than as the additive group $\{0, 1\}$.

6. Proposition. *We have*

$$\bigcap_{k=1}^{\infty} St_{G^{(d)}}(k) = \{1\}.$$

In particular, the group $G^{(d)}$ is residually finite, and any finitely-generated subgroup of $G^{(d)}$ is Hopfian.

Proof. The first claim is a straightforward consequence of the definitions. For the notion of a Hopfian group and for the result of Mal'cev from which the last claim follows, see Complement III.19. $\square$

7. Automorphisms of infinite order, and the adding machine. Consider again an integer $d \geq 2$, the infinite d-ary tree $T^{(d)}$, and its vertex set $T^{(d)}$, as well as the symmetric group $Sym(d)$ of $\{0, 1, \ldots, d-1\}$. To each mapping

$$\alpha : T^{(d)} \longrightarrow Sym(d)$$

one associates a permutation g_α of $T^{(d)}$ defined inductively by

$$g_\alpha(x_\emptyset) = x_\emptyset$$
$$g_\alpha(j) = (\alpha(x_\emptyset))(j)$$
$$g_\alpha(j_1, \ldots, j_k) = \left(g_\alpha(j_1, \ldots, j_{k-1}), (\alpha(j_1, \ldots, j_{k-1}))(j_k) \right)$$

with the second line for all $j \in L^{(d)}(1) = \{0, \ldots, d-1\}$ and the third one for all $(j_1, \ldots, j_k) \in L^{(d)}(k)$, for all $k \geq 2$.

One checks that the permutation g_α of the set $T^{(d)}$ so defined is an automorphism of the tree $T^{(d)}$. (This shows again that the automorphism group $G^{(d)}$ of this tree is uncountable; a first argument was given above, at the beginning of Item 2.)

Now let ϵ denote the circular permutation $j \mapsto j+1 \pmod{d}$ of $\{0, \ldots, d-1\}$. Let $\beta : T^{(d)} \longrightarrow Sym(d)$ be defined by $\beta(x_\emptyset) = \epsilon$ and

$$\beta(j_1, \ldots, j_k) = \begin{cases} \epsilon & \text{if } j_1 = \ldots = j_k = d-1 \\ 1 & \text{otherwise} \end{cases}$$

for $k \geq 1$. We have for example

$$g_\beta(j_1) = j_1 + 1$$

at the first level,

$$g_\beta(j_1, j_2) = \begin{cases} (j_1 + 1, j_2) & \text{if } j_1 \neq d-1 \\ (0, j_2 + 1) & \text{if } j_1 = d-1 \end{cases}$$

at the second level, and

$$g_\beta(j_1, j_2, j_3) = \begin{cases} (j_1 + 1, j_2, j_3) & \text{if} \quad j_1 \neq d - 1 \\ (0, j_2 + 1, j_3) & \text{if} \quad j_1 = d - 1 \quad \text{and} \quad j_2 \neq d - 1 \\ (0, 0, j_3 + 1) & \text{if} \quad j_1 = j_2 = d - 1 \end{cases}$$

at the third level.

It is clear that the permutation of $L^{(d)}(k)$ induced by g_β is a cycle of order d^k for all $k \geq 1$; it follows that g_β is an automorphism of infinite order of the tree $\mathcal{T}^{(d)}$.

The automorphism g_β appears in many places and, at least in some form, goes back to von Neumann and Kakutani; see Chapter 6 in [Fried–70] and Chapter 3 in [Nadka–95]. It is known under various names, including *"odometer"* and *"adding machine"*; see also [Bass4–96], pages 5, 25, 116, and [Weiss–81].

The group of automorphisms of $\mathcal{T}^{(d)}$ contains non-abelian free subgroups; see in particular [BruSi] and [Olijn].

Each proper subgroup P of $Sym(d)$ gives rise to an interesting group of automorphisms of $\mathcal{T}^{(d)}$, consisting of automorphisms of the form g_α with α a mapping from $T^{(d)}$ to P.

8. Remarks. (i) The space Ω of infinite rays of $\mathcal{T}^{(2)}$ can be viewed as the space of binary expansions of numbers in $[0, 1]$. Let Ω' denote the space obtained from Ω by erasing eventually constant sequences of 0 's and 1 's. Then Ω' is in natural bijection with the complement in $[0, 1]$ of the set

$$\mathbb{D} = \left\{ t \in [0, 1] \mid \text{there exist integers } a, b \geq 0 \text{ with } t = a2^{-b} \right\}$$

of dyadic numbers. The group $G^{(2)}$ *does not* leave Ω' invariant; but its subgroup Γ of the next section does, and may therefore be seen as a group of transformations of $[0, 1] \setminus \mathbb{D}$. This is the context in which the group Γ of the next section was defined in [Grigo–80].

(ii) One may also view $Aut(\mathcal{T}^{(2)})$ as the Galois group of an extension of $\mathbb{Q}$ associated to explicit sequences of polynomials. See [Stoll–92], and papers by R.W.K. Odoni quoted there.

VIII.B. The group Γ as an answer to one of Burnside's problems

9. Grigorchuk's group $\Gamma = \langle a, b, c, d \rangle$. Let us define four automorphisms of the binary tree $\mathcal{T} = \mathcal{T}^{(2)}$ of Item 1. If $j \in \{0, 1\}$, we define $\bar{j}$ by $\bar{0} = 1$ and $\bar{1} = 0$.

The *automorphism a* maps a non-empty $(0, 1)$-sequence to the sequence obtained by changing its first element:

$$a(j_1, j_2, \ldots, j_k) = (\bar{j}_1, j_2, \ldots, j_k).$$

It is clear that $a^2 = 1$, where $1 \in \Gamma$ indicates the identity transformation of $\mathcal{T}$, and a has a unique fixed point which is the root $x_\emptyset$ of $\mathcal{T}$.

The *three automorphisms* b, c, d of $\mathcal{T}$ are defined simultaneously and recursively (on the level k) as follows: they coincide with 1 on the root $x_\emptyset$ and on the two vertices x_0, x_1 of level 1, and

$$ b = (a, c) \qquad c = (a, d) \qquad d = (1, b) $$

with the notation of Item 2. To spell out these equalities, we have

$$ \begin{cases} b(0, j_2, j_3, \ldots, j_k) = (0, \bar{j}_2, j_3, \ldots, j_k) \\ b(1, j_2, j_3, \ldots, j_k) = (1, c(j_2, j_3, \ldots, j_k)) \end{cases} $$

$$ \begin{cases} c(0, j_2, j_3, \ldots, j_k) = (0, \bar{j}_2, j_3, \ldots, j_k) \\ c(1, j_2, j_3, \ldots, j_k) = (1, d(j_2, j_3, \ldots, j_k)) \end{cases} $$

$$ \begin{cases} d(0, j_2, j_3, \ldots, j_k) = (0, j_2, j_3, \ldots, j_k) \\ d(1, j_2, j_3, \ldots, j_k) = (1, b(j_2, j_3, \ldots, j_k)). \end{cases} $$

Example:

$$ b(1, 1, 1, 0, 1) = (1, c(1, 1, 0, 1)) = (1, 1, d(1, 0, 1)) = (1, 1, 1, b(0, 1)) $$
$$ = (1, 1, 1, 0, 0). $$

By definition, the "first" *Grigorchuk group* of the present chapter is the group Γ of automorphisms of the tree $\mathcal{T}$ generated by a, b, c, d, and we will write

$$ \Gamma = \langle a, b, c, d \rangle. $$

10. Observations. (i) Γ **is a quotient of** $(\mathbb{Z}/2\mathbb{Z}) * \mathbb{V}$. *We have* $a^2 = b^2 = c^2 = d^2 = 1$ *and* $bc = cb = d$, *so that* $\{1, b, c, d\}$ *constitutes a subgroup of* Γ *which is a "Vierergruppe"* $\mathbb{V}$. (From now on, we denote Klein's Vierergruppe, namely the group $(\mathbb{Z}/2\mathbb{Z}) \times (\mathbb{Z}/2\mathbb{Z})$ of order 4, by $\mathbb{V}$.)

Indeed, on the one hand, the equality $a^2 = 1$, already noted, is obvious. On the other hand, one shows that

$$ b^2(x) = c^2(x) = d^2(x) = x $$
$$ bc(x) = cb(x) = d(x) $$

for each vertex $x \in \mathcal{T}$ by induction on the level of x. [Hint: show together $bc = cb = d$, $cd = dc = b$, and $db = bd = c$.] It follows that Γ is a quotient of the free product of $(\mathbb{Z}/2\mathbb{Z}) = \{1, a\}$ with $\mathbb{V} = \{1, b, c, d\}$. This free product may also be seen as the group of isometries of the hyperbolic plane generated by the three reflections a, b, c which fix the sides of a triangle with angles $\frac{\pi}{2}, 0, 0$.

(ii) **The definition of** $\{b, c, d\}$ **revisited.** Let 1 and ϵ denote the two elements of the symmetric group $Sym(2)$ of $\{0, 1\}$. Define three maps β, γ, δ from the vertex set T to $Sym(2)$ by

$$\beta(j_1, \ldots, j_k) = \begin{cases} \epsilon & \text{if } (j_1, \ldots, j_k) = (1, \ldots, 1, 0) \text{ and } k - 1 \equiv 0 \text{ or } 1 \pmod{3} \\ 1 & \text{otherwise} \end{cases}$$

$$\gamma(j_1, \ldots, j_k) = \begin{cases} \epsilon & \text{if } (j_1, \ldots, j_k) = (1, \ldots, 1, 0) \text{ and } k - 1 \equiv 0 \text{ or } 2 \pmod{3} \\ 1 & \text{otherwise} \end{cases}$$

$$\delta(j_1, \ldots, j_k) = \begin{cases} \epsilon & \text{if } (j_1, \ldots, j_k) = (1, \ldots, 1, 0) \text{ and } k - 1 \equiv 1 \text{ or } 2 \pmod{3} \\ 1 & \text{otherwise} \end{cases}$$

(so that $k - 1$ is in each case the number of "1"), and let $g_\beta, g_\gamma, g_\delta$ be the corresponding automorphisms of T, as defined in Item 7. We leave it to the reader to check that

$$g_\beta = (a, g_\gamma) \qquad g_\gamma = (a, g_\delta) \qquad g_\delta = (1, g_\beta)$$

so that also

$$g_\beta = b \qquad g_\gamma = c \qquad g_\delta = d.$$

[Proceed again by induction on the level of $x \in T$ to check that $b_\beta(x) = (a, g_\gamma)(x), \ldots, b_\delta(x) = (a, g_\beta)(x)$ for all $x \in T$.]

(iii) **Automata.** There are *finite state automata* which describe the automorphisms a, b, c, d. See [HopUl–79] for a basic book on automata, [MulSc–81] and [MulSc–83] for early connections with group theory, § 23 in [KarMe–85] for an introduction to this connection, and Section 2.3 in [BarGr–c]. Here, short of any systematic exposition of automata, let us briefly describe one example.

Let $\mathcal{A}$ be the automaton represented in Figure 2: it is a directed graph, each directed edge is labelled either 0 or 1, each of the five vertices is labelled with an element of the symmetric group $\{1, \epsilon\}$ acting on $\{0, 1\}$, and each vertex has precisely two outgoing edges, one labelled 0 and one labelled 1 (a label $0, 1$ indicates *two* edges, one with each label).

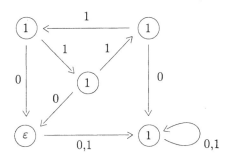

FIGURE 2.   An automaton for Γ

Consider first $\mathcal{A}$ together with its top left vertex as "initial state". To each finite 0-1-sequence $(j_1, j_2, \ldots, j_k)$, namely to each vertex of the tree $\mathcal{T}^{(2)}$, the automaton marked at this initial state associates two things: first, a directed path in the graph, say with vertex-sequence $(\xi_1, \xi_2, \ldots, \xi_{k+1})$, second, another finite 0-1-sequence $(j'_1, j'_2, \ldots, j'_k)$, defined as follows: ξ_1 is the initial vertex, the list of labels on the edges of the path is precisely $(j_1, j_2, \ldots, j_k)$, and j'_i is the image of j_i by the label of the vertex ξ_i $(1 \le i \le k)$. Then the transformation

$$(j_1, j_2, \ldots, j_k) \longmapsto (j'_1, j'_2, \ldots, j'_k)$$

is the automorphism b of $\mathcal{T}^{(2)}$.

We leave it as an exercise to check that $\mathcal{A}$ together with the central [respectively top right, bottom left, bottom right] vertex as initial state defines the automorphism c [respectively d, a, the identity] of $\mathcal{T}^{(2)}$.

11. Remark. The group Γ is obviously generated by 3 elements, more precisely by a and by any two of $\{b, c, d\}$. We will show below that 2 generators are not enough (Proposition 22), though nearly enough (Proposition 23), and that any two elements of $\{a, b, c, d\}$ generate a finite subgroup of Γ (Proposition 16).

12. Reduced words. As the canonical homomorphism

$$(\mathbb{Z}/2\mathbb{Z}) * \mathbb{V} \longrightarrow \Gamma$$

is onto (Observation 10.i), any positive word w in the letters a, b, c, d defines both an element in $\mathbb{Z}/2\mathbb{Z} * \mathbb{V}$ and an element in Γ. (We only need positive words, namely words without $a^{-1}, b^{-1}, c^{-1}, d^{-1}$, because all four generators are of order 2.)

Recall from the theory of free products in Chapter II that any element in $(\mathbb{Z}/2\mathbb{Z}) * \mathbb{V}$ can be represented uniquely by a *reduced word*, namely by a word of the form

$$u_0 a u_1 a u_2 a \ldots u_{l-1} a u_l$$

with $l \ge 0$, with $u_1, \ldots, u_{l-1} \in \{b, c, d\}$ and with $u_0, u_l \in \{\emptyset, b, c, d\}$. Moreover, for an arbitrary word w in $\{a, b, c, d\}$, the *reduction* $r(w)$ of w is the unique word representing the same element as w in $(\mathbb{Z}/2\mathbb{Z}) * \mathbb{V}$. For example

$$r(ab^3 cda^2 cdab) = abab.$$

Of course several distinct reduced words may represent the same element of Γ; this is for example the case for $adada$ and dad, distinct in $(\mathbb{Z}/2\mathbb{Z}) * \mathbb{V}$, but equal in Γ (see Proposition 16 below).

13. The subgroup $St_\Gamma(1)$ and the homomorphism $\psi = (\phi_0, \phi_1)$. For each $k \ge 0$, let $St_\Gamma(k)$ denote the subgroup of Γ of those elements which fix vertices of level $\le k$; observe that $St_\Gamma(k) = \Gamma \cap St_{G^{(2)}}(k)$ is normal in Γ. One has of course $St_\Gamma(0) = \Gamma$, and $St_\Gamma(1)$ is of index 2 in Γ. We will show below that

$\Gamma/St_\Gamma(2) \approx G^{(2)}/St_{G^{(2)}}(2)$ is a dihedral group of order 8 (Exercise 74),

$\Gamma/St_\Gamma(3) \approx G^{(2)}/St_{G^{(2)}}(3)$ is of order 2^7 (proof of Proposition 22),

$\Gamma/St_\Gamma(4)$ is a subgroup of order 2^{12} in the group

$G^{(2)}/St_{G^{(2)}}(4)$ of order 2^{15} (Proposition 41).

In the present Item, we concentrate on the subgroup $St_\Gamma(1)$ of index 2.

A word in $\{a, b, c, d\}$ represents an element in $St_\Gamma(1)$ if and only if it has an even number of occurrences of a. It follows that $St_\Gamma(1)$ is generated by the 6 elements

$$b = (a, c) \qquad\qquad aba = (c, a)$$
$$c = (a, d) \qquad\qquad aca = (d, a)$$
$$d = (1, b) \qquad\qquad ada = (b, 1).$$

Observe that $St_\Gamma(1)$ can be generated by four elements, e.g. b, c, aba, aca.

The restriction to $St_\Gamma(1)$ of the homomorphism $\psi = (\phi_0, \phi_1)$ of Item 2 provides a homomorphism $St_\Gamma(1) \longrightarrow \Gamma \times \Gamma$ for which we use the same notation.

14. Proposition. *The homomorphism*

$$\psi = (\phi_0, \phi_1) : St_\Gamma(1) \longrightarrow \Gamma \times \Gamma$$

is injective and is given by

$$\phi_0(b) = a \quad \phi_1(b) = c \qquad\qquad \phi_0(aba) = c \quad \phi_1(aba) = a$$
$$\phi_0(c) = a \quad \phi_1(c) = d \qquad\qquad \phi_0(aca) = d \quad \phi_1(aca) = a$$
$$\phi_0(d) = 1 \quad \phi_1(d) = b \qquad\qquad \phi_0(ada) = b \quad \phi_1(ada) = 1$$

In particular, $\phi_j : St_\Gamma(1) \longrightarrow \Gamma$ *is onto for* $j \in \{0, 1\}$.

Proof: straightforward from the definitions; observe that both the image of ϕ_0 and the image of ϕ_1 contain $\{a, b, c, d\}$. $\square$
(See Theorem 28 for the image of ψ, which is of index 8 in $\Gamma \times \Gamma$.)

15. Corollary. *The group* Γ *is infinite.*

Proof: the group Γ has a *proper* subgroup $St_\Gamma(1)$ which is mapped *onto* Γ by ϕ_0. $\square$

16. Proposition. *(i) The subgroup* $\langle a, d \rangle$ *of* Γ *generated by* a *and* d *is a dihedral group of order 8.*

(ii) $\langle a, c \rangle$ *is a dihedral group of order 16.*

(iii) $\langle a, b \rangle$ *is a dihedral group of order 32.*

Proof. *(i)* As $d = (1, b)$ and $ada = (b, 1)$, we have $(ad)^2 = (da)^2 = (b, b)$, which is of order 2. This shows *(i)*.

(*ii*) Similarly we have $(ac)^2 = (da, ad)$ and $(ca)^2 = (ad, da)$; by (*i*), these are both of order 4 and $(ac)^4 = (ca)^4$. This shows (*ii*).

(*iii*) Similarly we have $(ab)^2 = (ca, ac)$ and $(ba)^2 = (ac, ca)$; by (*ii*), these are both of order 8 and $(ab)^8 = (ba)^8$. This shows (*iii*) and concludes the proof. $\square$

The following result and its proof will be sharpened in Theorem 70 below. As Γ is given together with the generating set $\{a, b, c, d\}$, recall for the proof that each element $\gamma \in \Gamma$ has a well-defined *length* $\ell(\gamma) \in \mathbb{N}$ (see Section IV.A).

17. Theorem. *The group Γ is a 2-group. In other words, for any $\gamma \in \Gamma$, there exists an integer $N \geq 0$ such that*

$$\gamma^{2^N} = 1.$$

Proof. Let k denote the length of γ and let w be a word of length k representing γ. We clearly have $\gamma = 1$ if $k = 0$ and $\gamma^2 = 1$ if $k = 1$. Moreover $\gamma^{16} = 1$ if $k \leq 2$ by Proposition 16. We then proceed by induction on k, so that we assume that $k \geq 3$ and that the theorem holds for words not longer than $k - 1$. We may also assume that the word w is reduced (see Item 12).

Suppose first that the length k of w is odd. Then either the first and last letters of w are both a, in which case γ is conjugate to the element $a\gamma a$ represented by a word of length $k - 2$; or the first letter u and the last letter v of w are both in $\{b, c, d\}$, in which case γ is conjugate to the element $u\gamma u$ represented by a word of length $k - 1$ or $k - 2$. Thus in all cases with k odd, the order of γ is the order of some element of Γ represented by a strictly shorter word, and the induction hypothesis applies.

We assume now that the length k of w is even. Upon replacing γ by one of $b\gamma b, c\gamma c, d\gamma d$ if necessary, we may also assume that w begins with a and is thus of the form

$$w = au_1 au_2 \dots au_l$$

with $l = \frac{k}{2}$ and $u_1, \dots, u_l \in \{b, c, d\}$.

If $l = 2m$ is even, we have $\gamma \in St_\Gamma(1)$ and

$$\psi(\gamma) = \psi(au_1 a)\psi(u_2) \dots \psi(au_{2m-1}a)\psi(u_{2m}) = (\gamma_0, \gamma_1)$$

where the last equality *defines* γ_0 and γ_1. The formulas of Proposition 14 for $\phi_j(aua)$ and $\phi_j(u)$, with $j \in \{0, 1\}$ and $u \in \{b, c, d\}$, show that γ_0 and γ_1 are each represented by a word of length at most $2m = \frac{k}{2}$. The induction hypothesis implies that there exist $L, M \geq 0$ such that $\gamma_0^{2^L} = 1 = \gamma_1^{2^M}$; as ψ is injective, the order of γ is the least common multiple of the orders of γ_0 and γ_1, and is in particular a power of 2.

If finally l is odd, so that $k = 4m - 2$ for some integer $m \geq 2$, then γ^2 is represented by the word

$$ww = (au_1 a)u_2 \dots u_{2m-2}(au_{2m-1}a)\, u_1(au_2 a) \dots (au_{2m-2}a)u_{2m-1}$$

of length $8m - 4$. We have, as in the previous step,

$$\psi(\gamma^2) = \psi(au_1a)\psi(u_2) \ldots \psi(u_{2m-2})\psi(au_{2m-1}a)$$
$$\psi(u_1)\psi(au_2a) \ldots \psi(au_{2m-2}a)\psi(u_{2m-1})$$
$$= (\alpha, \beta)$$

for some $\alpha, \beta \in \Gamma$, both of length $\leq 4m - 2$. We now distinguish 3 subcases.

(i) There exists $j \in \{1, \ldots, 2m - 1\}$ such that $u_j = d$. As $\psi(au_ja) = (b, 1)$ and $\psi(u_j) = (1, b)$, each of α, β is in fact represented by a word of length at most $4m - 3 = k - 1$, and the induction hypothesis applies to γ^2, showing that the order of γ^2 is a power of 2.

(ii) There exists $j \in \{1, \ldots, 2m - 1\}$ such that $u_j = c$. Then $\psi(au_ja) = (d, a)$ and $\psi(u_j) = (a, d)$, so that each of α, β is represented either by a word of length $4m - 2$ involving d, in which case (i) applies, or by a shorter word, in which case the induction hypothesis applies.

(iii) There exists $j \in \{1, \ldots, 2m - 1\}$ such that $u_j = b$. Then $\psi(au_ja) = (c, a)$ and $\psi(u_j) = (a, c)$, and one ends the arguments using (ii), as we did for (ii) using (i).

[Alternatively, if neither (i) nor (ii) applies, then w is an even word in a and b, so that the order of γ is at most 16 by the previous proposition.]

In all cases (i), (ii) and (iii), we have shown that the order of γ^8 is a power of 2, and the proof is complete. $\square$

Observe that, for $\gamma \in St_\Gamma(1), \gamma \neq 1$ and $\psi(\gamma) = (\gamma_0, \gamma_1)$, we have $\gamma_0 \neq \gamma$ and $\gamma_1 \neq \gamma$. This follows from the inequality

$$\ell(\gamma_j) \leq \frac{\ell(\gamma) + 1}{2} \qquad (j = 0, 1)$$

used in the previous proof (and repeated in Lemma 46), and from a direct inspection for the three elements b, c, d of $St_\Gamma(1)$ of length 1.

18. Remark. Under closer inspection, the proof of Theorem 17 provides for the order of an element $\gamma \in \Gamma$ an exponential bound of the form

$$\ell(\gamma) \leq k \qquad \Longrightarrow \qquad order(\gamma) \leq 2^{\frac{3}{4}k + \frac{5}{2}}$$

but this is far from optimal. By Theorem 4.1 of [Grigo–85], there exists a polynomial P such that

$$\ell(\gamma) \leq k \qquad \Longrightarrow \qquad order(\gamma) \leq P(k),$$

and we will show in Theorem 70 below (following Leonov and Lysionok[2]) that one may take $P(k) = 2k^3$.

---

[2]Other spelling: Lysenok.

19.   On Burnside problems.   In 1902, Burnside [Burns–02] asked the
following question, later known as the *general Burnside problem:*

> Let Γ be a finitely-generated group such that each of its elements has
> finite order; is Γ necessarily a finite group ?

The answer is "yes" under the *additional assumption* that Γ is a subgroup of
$GL(n, \mathbb{K})$ for some integer n and for some field $\mathbb{K}$; after some partial results by
Burnside himself, this was first proved by Schur in 1911 when $\mathbb{K}$ is of characteristic zero, and the proof was adapted to the general case by Kaplansky; see
e.g. § 9 in [Lam–91]. But *the answer is "no" in general,* as was first shown by
Golod and Shafarevich [Golod–64]. There is a nice exposition of this due to
Herstein; see Theorem 8.1.4 in [Herst–68], as well as [Ol's–95a].

The group of the present chapter again provides a negative answer to the
general Burnside problem; see also [GupS–83b] and [Sergi–91]. There is a third
class of examples illustrating this negative answer, due to Gromov and (hence)
quite geometric, involving quotients of hyperbolic groups; see Items 0.2(A) and
4.5.C in [Gromo–87], and [GhyHf–90]. It is moreover known that, given any
non-elementary hyperbolic group Γ, there exists an integer $n = n(\Gamma)$ such that
the quotient Γ/Γ^n is infinite, where Γ^n denotes the subgroup of Γ generated by
n^{th} powers (see [IvaOl–96], as well as [Ol's–91a]).

The next corollary is a consequence of Corollary 15 and Theorem 17, together
with the Burnside – Schur – Kaplansky result recalled above.

Corollary.   *For any integer $n \geq 1$ and any field $\mathbb{K}$, the image of any homomorphism*

$$\phi : \Gamma \longrightarrow GL(n, \mathbb{K})$$

is finite. In particular, the group Γ of Grigorchuk is not a "linear group".

A variant of the previous problem is the so-called *Burnside problem:*

> Let Γ be a finitely-generated group of finite exponent, say N;
> is Γ necessarily a finite group ?

(Recall that a group Γ is of *exponent* N if $\gamma^N = 1$ for all $\gamma \in \Gamma$.) The answer
is "yes" for $N \in \{2, 3, 4, 6\}$: this is a very easy exercise if $N = 2$, a rather easy
exercise and a result of Burnside if $N = 3$, a theorem of Sanov (1940) if $N = 4$,
and a theorem of M. Hall (1958) if $N = 6$; the answer is unknown for other small
values of N. (For $N \leq 4$, see Theorems 5.24 and 5.25 in [MagKS–66].) And *the
answer is "no" in general;* this was shown by Novikov and Adian in 1968 for N a
sufficiently large odd integer. (Of course, "large" is a function of time; it means
$N \geq 665$ in [Adian–75]; clearly it is enough to assume that N is a multiple
of such a "large" odd integer.) In 1992, Ivanov and Lysionok announced that
the answer to the Burnside problem is also "no" for all sufficiently large N;
detailed proofs appear in [Ivano–94] and [Lysen–96]; in case N is a power of 2,
"sufficiently large" means $N \geq 2^{48}$ in Ivanov's paper and $N \geq 2^{13}$ in Lysionok's
paper. See also [Ivano–98], as well as Math. Reviews 95h:20051 on [Ivano–94]
(by A.Yu. Ol'shanskii) and Math. Reviews 97j:20037 on [Lysen–96] (by V.S.
Guba). As far as I understand, experts agree at the time of writing that any
$N \geq 2000$ is "sufficiently large".

A third problem has been considered by group theorists since the late 1930s, and is known as the *restricted Burnside problem* (a terminology due to W. Magnus):

> *For given natural numbers N and k, are there only finitely many finite groups of exponent N generated by k elements ?*

The answer is known to be "yes" for N a power of a prime, due to work of Kostrikin and Zel'manov; assuming the classification of finite simple groups, this implies that the answer is "yes" for any N. It follows that,

> *a finitely-generated residually finite group of finite exponent is finite.*

See [Zelma–91], [Zelma–92], [Feit–95], and [VauZe–99].

Both the negative answer to the Burnside problem and the positive answer to the restricted Burnside problem are, at present, much beyond the level of this book. Related open problems are discussed in [Zelma–95] and [Zelma–99].

VIII.C. On some subgroups of Γ

The following result is a particular case of Proposition 8.1 in [Grigo–84] (which applies to other groups as well).

20. Proposition. *The centre of Γ is trivial.*

Proof. Let $\gamma \in \Gamma$, $\gamma \neq 1$.

Assume first that $\gamma \notin St_\Gamma(1)$, so that $\gamma = ha$ for some $h = (h_0, h_1) \in St_\Gamma(1)$. Then

$$\psi(\gamma d\gamma^{-1}) = \psi(h)\psi(ada)\psi(h^{-1}) = (h_0, h_1)(b, 1)(h_0, h_1)^{-1} = (h_0 b h_0^{-1}, 1)$$
$$\neq (1, b) = \psi(d)$$

so that γ is not central.

Assume now that $\gamma \in St_\Gamma(1)$, and let k be the word length of γ. If $\gamma \in \{b, c, d\}$ then γ is not central by Proposition 16. We may thus assume that $k \geq 2$, and we proceed by induction on k. We have $\psi(\gamma) = (\gamma_0, \gamma_1) \in \Gamma \times \Gamma$. We assume furthermore that $\gamma_0 \neq 1$ (the argument for the case $\gamma_1 \neq 1$ is similar). Thus

$$1 \leq \ell(\gamma_0) \leq \frac{\ell(\gamma) + 1}{2} < \ell(\gamma) = k$$

(see the observation which follows Theorem 17). By the induction hypothesis, there exists $h_0 \in \Gamma$ such that $h_0 \gamma_0 \neq \gamma_0 h_0$. By Proposition 14 there exist $h \in St_\Gamma(1)$ and $h_1 \in \Gamma$ such that $\psi(h) = (h_0, h_1)$. Thus

$$\psi(h\gamma) = (h_0 \gamma_0, h_1 \gamma_1) \neq (\gamma_0 h_0, \gamma_1 h_1) = \psi(\gamma h)$$

and γ is not central in Γ. $\square$

21. Remark. More generally, it is known that any conjugacy class of Γ distinct from $\{1\}$ is infinite [Rozhk–94].

22. Proposition. *The group Γ has a quotient which is an elementary abelian 2-group of order 8. In particular, Γ cannot be generated by two elements.*

Proof. As Γ is generated by 3 elements of order 2 (for example a, b and c), any abelian quotient of Γ is of order at most 8.

For the present proof, let us denote by $1, 2, \ldots, 8$ the vertices

$$(0,0,0), \, (0,0,1), \, \ldots, \, (1,1,1)$$

of level 3 and by $\pi : \Gamma \longrightarrow Sym(8)$ the natural homomorphism from Γ to the symmetric group of these 8 vertices. One has

$$\begin{aligned}
\pi(a) &= (1,5)\,(2,6)\,(3,7)\,(4,8) \\
\pi(b) &= (1,3)\,(2,4)\,(5,6) \\
\pi(c) &= (1,3)\,(2,4) \\
\pi(d) &= (5,6).
\end{aligned}$$

We now leave it to the reader to check that the image of π is a wreath product

$$\Big((\mathbb{Z}/2\mathbb{Z}) \wr (\mathbb{Z}/2\mathbb{Z}) \Big) \wr (\mathbb{Z}/2\mathbb{Z})$$

of order 2^7. (This is also the full automorphism group of the finite tree $\mathcal{T}^{(2)}(3)$, and also a 2-Sylow subgroup of $Sym(8)$.) As the abelianization of this wreath product is

$$(\mathbb{Z}/2\mathbb{Z}) \times (\mathbb{Z}/2\mathbb{Z}) \times (\mathbb{Z}/2\mathbb{Z})$$

by Proposition 4, this concludes the proof. $\square$

More precisely, *the commutator subgroup $[\Gamma, \Gamma]$ is of index 8 in Γ*, and the abelianization $\Gamma/[\Gamma, \Gamma]$ can be identified with the abelian 2-group

$$\left\{ (a', b', c', d') \in \mathbb{F}_2^4 \mid b' + c' + d' = 0 \right\};$$

see Remark 31.d.

23. Proposition. *The subgroup $\langle b, ac \rangle$ of Γ generated by b and ac is of index 2 in Γ. Similarly, each of $\langle c, ad \rangle$ and $\langle d, ab \rangle$ is of index 2 in Γ.*

Proof. The subgroup $\langle b, ac \rangle$ of Γ is normal, because it contains each of

$$\begin{array}{ll}
aba = (ac)b(ac)^{-1} & a(ac)a = (ac)^{-1} \\
bbb & b(ac)b \\
cbc = b & c(ac)c = (ac)^{-1}.
\end{array}$$

Now let p denote the canonical projection of Γ onto $\Gamma/\langle b, ac \rangle$. We have $p(b) = 1$, hence $p(c) = p(d)$, and $p(ac) = 1$, hence $p(a) = p(c)$. Thus this quotient group is of order at most 2. The previous proposition implies that $\langle b, ac \rangle$ is a proper subgroup of Γ, hence $[\Gamma : \langle b, ac \rangle] = 2$.

One shows similarly that conjugates of c and ad by the generators a, c, d of Γ are in $\langle c, ad \rangle$ [respectively that conjugates of d and ab by the generators a, b, d are in $\langle d, ab \rangle$] so that $\langle c, ac \rangle$ and $\langle d, ab \rangle$ are also of index 2 in Γ. □

24. Remark. It is known (§ 7 in [Rozhk–94]) that there are precisely 7 subgroups of index 2 in Γ, which are the three subgroups of the previous proposition, the three subgroups

$$\langle a, b, cac \rangle \qquad \langle a, c, dad \rangle \qquad \langle a, d, bab \rangle$$

and the stabilizer

$$St_\Gamma(1) = \langle b, c, aba, aca \rangle$$

of level 1 (see Item 13).

25. The normal subgroup B of Γ. Let B denote the *normal* subgroup of Γ generated by b.

Proposition. *The group B is of index 8 in Γ, and is generated by the four elements*

$$b, \quad u = aba, \quad v = (bada)^2, \quad w = (abad)^2.$$

Proof. We follow [Rozhk–94]. Set $B_1 = \langle b, u, v, w \rangle$. In the quotient group Γ/B, the four generators of B_1 are clearly mapped to the unit element. Hence $b \in B_1 < B$. We claim that B_1 is normal in Γ, so that $B_1 = B$.

To prove the claim, one has to show that the conjugates of b, u, v, w by any of a, b, c are in B_1. Conjugates of b and conjugates by a or b are obvious, so that the claim is proved by the following 3 computations:

$$cuc = bd\,aba\,db = bw^{-1}abab = bw^{-1}ub$$
$$cvc \stackrel{*}{=} c\,b\,dadad\,b\,dadad\,c = adabadab = v^{-1}$$
$$cwc = bd\,abad\,abad\,c = bw^{-1}b$$

where $\stackrel{*}{=}$ indicates that the relation $(ad)^4 = 1$ has been used (see Proposition 16). For later use (proof of Proposition 30.i), let us also compute

$$dvd \stackrel{*}{=} d\,b\,dadad\,b\,dadad\,d = badabada = v.$$

We now compute the index of $B = B_1$ in Γ.

In the quotient Γ/B, the generator b is mapped onto 1, so c and d are mapped onto the same element. Thus Γ/B is generated by the images of a and d. It follows from Proposition 16.i that the order of Γ/B divides 8.

Now let $\pi : \Gamma \longrightarrow Sym(8)$ be as in the proof of Proposition 22, with $Sym(8)$ being viewed as the symmetric group of the 8 vertices in $L(3)$. The image $\pi(B)$ is generated by

$$\pi(b) = (1,3)\,(2,4)\,(5,6)$$
$$\pi(u) = (1,2)\,(5,7)\,(6,8)$$
$$\pi(v) = (1,2)\,(3,4)$$
$$\pi(w) = (5,6)\,(7,8).$$

We observe that $\pi(v)$ and $\pi(w)$ are central in $\pi(B)$ and generate a subgroup of order 4. As

$$(\pi(b)\pi(u))^2 = \pi(v)\pi(w), \quad \pi(b)^2 = 1, \quad \pi(u)^2 = 1,$$

it follows that $\pi(B)$ is of order at most 16.

We know from the proof of Proposition 22 that the image $\pi(\Gamma)$ is of order 2^7, and this implies that

$$|\Gamma/B| \geq |\pi(\Gamma)/\pi(B)| \geq \frac{2^7}{2^4} = 8.$$

As $|\Gamma/B|$ also divides 8, we have $|\Gamma/B| = 8$, and thus $|\pi(B)| = 16$. $\square$

26. Corollary. *The group* Γ *is a semi-direct product*

$$\Gamma = B \rtimes D$$

with B *as in the previous item and with* $D = \langle a, d \rangle$ *the dihedral group of order 8 of Proposition 16.*

Proof. The subgroup of Γ generated by B and D is Γ itself, because it contains b, a, and d. As $|B\backslash\Gamma| = 8$ and $|D| = 8$, we have $\Gamma = BD$ and $B \cap D = \{1\}$. As B is normal in Γ, we have $\Gamma = B \rtimes D$. $\square$

27. Notation. Set

$$D^{diag} = \langle (a,d)\,,\, (d,a) \rangle < \Gamma \times \Gamma$$
$$D^{right} = \langle (1,a)\,,\, (1,d) \rangle < \Gamma \times \Gamma.$$

Both D^{diag} and D^{right} are isomorphic to the subgroup $D = \langle a, d \rangle$ of Γ, and in particular are of order 8. Observe that $D^{diag} \cap D^{right} = \{1\}$. Set also

$$D^{\sharp\sharp} = \langle D^{diag}, D^{right} \rangle < \Gamma \times \Gamma.$$

As D^{right} is clearly normal in $D^{\sharp\sharp}$, the latter is a semi-direct product

$$D^{\sharp\sharp} = D^{diag} \ltimes D^{right}$$

and is of order 64. It now follows from Corollary 26 that

$$(B \times B) \cap (D^{diag} \ltimes D^{right}) = \{1\}$$

and that $B \times B$ is a normal subgroup of index 64 in $\Gamma \times \Gamma$. Thus, we have shown that

$$\Gamma \times \Gamma = (B \times B) \rtimes (D^{diag} \ltimes D^{right}).$$

Note however that $\Gamma \times \Gamma$ is *not* a semi-direct product of $(B \times B) \rtimes D^{diag}$ and D^{right}, because neither $(B \times B) \rtimes D^{diag}$ nor D^{right} is normal in $\Gamma \times \Gamma$ (see Exercise 77).

28. Theorem. *The image of the injective homomorphism*

$$\psi = (\phi_0, \phi_1) : St_\Gamma(1) \longrightarrow \Gamma \times \Gamma$$

of Proposition 14 is the subgroup $(B \times B) \rtimes D^{diag}$ *of index 8 of* $\Gamma \times \Gamma$.

Proof. The formulas of Proposition 14 show that the image of ψ is a subgroup of $\Gamma \times \Gamma$ which contains on the one hand $(b, 1)$ and $(1, b)$, and on the other hand (a, d) and (d, a).

For any $\gamma_1 \in \Gamma$, we know from Proposition 14 that there exist $\gamma \in St_\Gamma(1)$ and $\gamma_2 \in \Gamma$ such that $\psi(\gamma) = (\gamma_1, \gamma_2)$. Hence $Im(\psi)$ contains $(\gamma_1 b \gamma_1^{-1}, 1)$. It follows from the definition of B that $Im(\psi)$ also contains $B \times \{1\}$. Similarly $Im(\psi)$ contains $\{1\} \times B$. As (a, d) and (d, a) generate D^{diag}, this shows that $Im(\psi)$ contains $(B \times B) \rtimes D^{diag}$.

The same formulas of Proposition 14 show that the 6 generators b, c, d, aba, aca, ada of $St_\Gamma(1)$ are mapped into $(B \times B) \rtimes D^{diag}$, so that the proof is complete. $\square$

Recall from Item 2 that for each $k \geq 1$ there is an injective homomorphism

$$\psi_k = (\phi_{0,0,\ldots,0,0}, \ldots, \phi_{1,1,\ldots,1,1}) : \begin{cases} St_\Gamma(k) \longrightarrow & \Gamma^{2^k} \\ \gamma \longmapsto (\gamma_{0,0,\ldots,0,0}, \ldots, \gamma_{1,1,\ldots,1,1})_k \end{cases}$$

In the proof of Proposition 30, we will use the homomorphism ψ_2, about which we need a lemma.

29. Lemma. *The composition of* $\psi_2 : St_\Gamma(2) \longrightarrow \Gamma \times \Gamma \times \Gamma \times \Gamma$ *with any projection* $\Gamma^4 \longrightarrow \Gamma$ *is onto.*

Proof. This follows from the formulas

$$\psi_2(cadab) = (\psi(aba), \psi(dc)) = (c, a, a, c)_2$$
$$\psi_2(acadaba) = (\psi(dc), \psi(aba)) = (a, c, c, a)_2$$
$$\psi_2((ac)^4) = (\psi(dada), \psi(adad)) = (b, b, b, b)_2$$

and from the fact that $\{a, b, c\}$ generates Γ. $\square$

For ψ_3, see the digression at the end of the proof of Proposition 30.iv.

30. The normal subgroup K of Γ. Let K denote the *normal* subgroup of Γ generated by $(ab)^2$.

Proposition. *(i) The group K is generated by the three elements*

$$t = (ab)^2, \quad v = (bada)^2, \quad w = (abad)^2.$$

(ii) The group K is a subgroup of index 2 in the group B of Proposition 25, and thus a subgroup of Γ of index 16.

(iii) We have $St_\Gamma(3) < K < St_\Gamma(1)$ and $[K : St_\Gamma(3)] = 8$.

(iv) We have $St_\Gamma(5) < [K, K] < St_\Gamma(3)$. (See Remark 31.c for the index.)

(v) We have $K \times K < \psi(K)$. (See also Exercise 81.)

(vi) For each $k \geq 1$, the image of $\psi_k : St_\Gamma(k) \longrightarrow (\Gamma)^{2^k}$ contains $(K)^{2^k}$.

(vii) The element $(ab)^8 \in K$, of order 2, is in $St_\Gamma(4)$ and

$$\psi_3\left((ab)^8\right) = (b, b, b, b, b, b, b, b)_3 \in \Gamma^8,$$
$$\psi_4\left((ab)^8\right) = (a, c, a, c, \ldots, a, c)_4 \in \Gamma^{16}.$$

Proof. (i) Set $K_1 = \langle t, v, w \rangle$. As K is the normal subgroup of Γ generated by t, we have

$$td^{-1}t^{-1}d = abab\, d\, baba\, d = w \in K$$
$$awa = v \in K$$

and thus $K_1 < K$. We claim that K_1 is a normal subgroup of Γ, so that $K_1 = K$. The claim follows from the computations below, where $\overset{*}{=}$ denotes an equality from the proof of Proposition 25, thus an equality which is a consequence of the relation $(ad)^4 = 1$, and where each equality may depend on the previous ones.

$$\begin{cases} ata & = t^{-1} \in K_1 \\ btb & = t^{-1} \in K_1 \\ dtd & = dababd = dabad(abaaba)b = w^{-1}t \in K_1. \end{cases}$$

$$\begin{cases} ava & = w \in K_1 \\ bvb & = (adab)^2 = v^{-1} \in K_1 \\ dvd & \overset{*}{=} v \in K_1. \end{cases}$$

$$\begin{cases} awa & = v \in K_1 \\ bwb & = bavab = babv^{-1}bab = baba\, w^{-1}\, abab = t^{-1}w^{-1}t \in K_1 \\ dwd & = dabadaba = abvba = av^{-1}a = w^{-1} \in K_1. \end{cases}$$

(ii) From Proposition 25, we know that B/K is generated by the images of b and aba, and that these images are identical (because $abab \in K$). Hence $|B/K| \leq 2$, and we have to show that $K \neq B$. We offer two different arguments.

First argument. Consider the subgroup $\langle d, ab \rangle$ of Proposition 23, which is of index 2 in Γ, and in particular normal in Γ. As the quotient $\Gamma/\langle d, ab \rangle$ is generated by the image of b (which is identical to those of c and of a), we have $B \not< \langle d, ab \rangle$. As K, the normal subgroup generated by $(ab)^2$, is inside the normal subgroup $\langle d, ab \rangle$, we have $K \neq B$.

Second argument. Consider the canonical projection $\pi : \Gamma \longrightarrow \Gamma/St_\Gamma(3)$. Recall from the proofs of Propositions 22 and 25 that $\Gamma/St_\Gamma(3)$, of order 2^7, and $\pi(B)$, of order 2^4, are both subgroups of $Sym(8)$. One easily checks that, as a subgroup of $Sym(8)$, the image $\pi(K)$ is of order 2^3, so that $K \neq B$.

(iii) The inclusion $K < St_\Gamma(1)$ is obvious from the definition of K. Now, from *(ii)*, we have on the one hand

$$[\pi(\Gamma) : \pi(K)] = [\Gamma : K St_\Gamma(3)] = 2^4$$

and on the other

$$[\Gamma : K] = 2^4.$$

It follows that the obvious inclusion $K < K St_\Gamma(3)$ is in fact an equality, so that $St_\Gamma(3) < K$. Observe that

$$[K : St_\Gamma(3)] = \frac{[\Gamma : St_\Gamma(3)]}{[\Gamma : K]} = \frac{2^7}{2^4} = 8.$$

(iv) The commutator subgroup $[K, K]$ is generated *as a normal subgroup* of Γ by the commutators of the generators of K, namely by $[t, v]$, $[v, w]$, and $[w, t]$. One computes first

$$\psi(t) = (ca, ac) , \quad \psi(v) = (abab, 1) , \quad \psi(w) = (1, abab)$$

and then

$$\psi([t, v]) = ([\phi_0(t), \phi_0(v)] , [\phi_1(t), \phi_1(v)])$$
$$= ([ca, (ab)^2] , [ac, 1])$$
$$= ((adab)^2, 1)$$
$$\psi([v, w]) = ([\phi_0(v), \phi_0(w)] , [\phi_1(v), \phi_1(w)])$$
$$= ([(ab)^2, 1] , [1, (ab)^2])$$
$$= (1, 1)$$
$$\psi([w, t]) = ([\phi_0(w), \phi_0(t)] , [\phi_1(w), \phi_1(t)])$$
$$= ([1, ca] , [(ab)^2, ac])$$
$$= (1 , babadabac).$$

In particular

$$\psi([K, K]) < St_\Gamma(1) \times St_\Gamma(1)$$

and we may apply ψ again. As $\psi(adab) = (ba, c)$, the image $\psi_2([K, K])$ contains[3]

$$\psi_2([t, v]) = \left(\psi((adab)^2) , \psi(1) \right) = ((ba)^2 , 1 , 1 , 1)_2 .$$

---

[3] Observe that $[t, v] \in St_\Gamma(3)$; by a similar computation, $[w, t] \in St_\Gamma(3)$; hence $[K, K] < St_\Gamma(3)$.

Now $[K, K]$ is a normal subgroup of $St_\Gamma(2)$ (indeed a normal subgroup of Γ). It follows from Lemma 29 that $\psi_2([K, K])$ also contains the group

$$\langle (ab)^2 \rangle^\Gamma \times \{1\} \times \{1\} \times \{1\}$$

where $\langle (ab)^2 \rangle^\Gamma$ denotes the *normal* subgroup of Γ generated by $(ab)^2$; but this is precisely the group $K \times \{1\} \times \{1\} \times \{1\}$.

One shows similarly that $\psi_2([K, K])$ contains

$$\psi_2(b[t, v]b) = \Big(\psi((daba)^2), \psi(1) \Big) = (1, baba, 1, 1)_2$$

$$\psi_2(a[t, v]a) = \Big(\psi(1), \psi((adab)^2) \Big) = (1, 1, baba, 1)_2$$

$$\psi_2(ab[t, v]ba) = \Big(\psi(1), \psi((daba)^2) \Big) = (1, 1, 1, baba)_2$$

so that $\psi_2([K, K])$ also contains

$$\{1\} \times K \times \{1\} \times \{1\} \qquad \{1\} \times \{1\} \times K \times \{1\} \qquad \{1\} \times \{1\} \times \{1\} \times K$$

by the same argument as above. Therefore $\psi_2([K, K])$ contains $K \times K \times K \times K$. Thus $\psi_2(St_\Gamma(5)) = St_\Gamma(3) \times St_\Gamma(3) \times St_\Gamma(3) \times St_\Gamma(3) < \psi_2([K, K])$ by Claim (iii), and Claim (iv) follows.

(Digression : from the computation of the images by ψ_3 of $(ab)^8$, $(ac)^4$, $[t, v]$, $b[t, v]b$, $a[t, v]a$ and $ab[t, v]ba$, it follows that the composition of ψ_3 with any of the 8 projections $\Gamma^8 \longrightarrow \Gamma$ is onto.)

(v) From the formulas of Proposition 14, we have

$$\psi\left((abad)^2\right) = (1, abab).$$

For any $\gamma_2 \in \Gamma$, we know from the same proposition that there exist $\gamma \in St_\Gamma(1)$ and $\gamma_1 \in \Gamma$ such that $\psi(\gamma) = (\gamma_1, \gamma_2)$; hence $\psi(K)$ also contains

$$\psi\left(\gamma(abad)^2\gamma^{-1}\right) = \left(1, \, \gamma_2(ab)^2\gamma_2^{-1}\right).$$

It follows that $\psi(K)$ contains $\{1\} \times K$. Also $\psi(K) = \psi(aKa)$ contains $K \times \{1\}$, and Claim (v) follows.

(Exercise: show that $[\psi(K) : K \times K] = 4$. Hint: $[\Gamma \times \Gamma : K \times K] = 2^8$ and $[\Gamma \times \Gamma : \psi(K)] = [\Gamma \times \Gamma : \psi(St_\Gamma(1))][St_\Gamma(1) : K] = 2^6$.)

(vi) This follows from (v) by induction on k.

(vii) This is a straightforward consequence of the formulas of Proposition 14. (We have $(ab)^{16} = 1$ by Proposition 16.) $\square$

31. Remarks. (a) In the proof of (iv) above, we have shown that $\psi_2([K, K])$ contains $K \times K \times K \times K$. On the other hand, the group $[K, K]$ is generated as a normal subgroup of Γ by $[t, v]$ and $[w, t]$ (recall that $[v, w] = 1$); as $\psi_2([t, v]) =$

$(baba, 1, 1, 1)_2$ and $\psi_2([w, t]) = (1, 1, 1, cabad)_2$ are both in $K \times K \times K \times K$, it follows that $\psi_2([K, K])$ is contained in $K \times K \times K \times K$. Thus we have

$$\psi_2([K, K]) = K \times K \times K \times K.$$

(b) The index of $Im(\psi_2)$ is given by

$$\left[\Gamma^4 : \psi_2\left(St_\Gamma(2)\right)\right] = 2^9.$$

Indeed, as the index of $St_\Gamma(2)$ in $St_\Gamma(1)$ is 4, the index of $\psi\left(St_\Gamma(2)\right)$ in $\psi\left(St_\Gamma(1)\right) = (B \times B) \rtimes D^{diag}$ is also 4, so that

$$\left[\Gamma \times \Gamma : \psi\left(St_\Gamma(2)\right)\right] = 4 \cdot 8 = 32 \quad \text{and} \quad \left[St_\Gamma(1) \times St_\Gamma(1) : \psi\left(St_\Gamma(2)\right)\right] = 8.$$

Consequently and as the index of $(\psi \times \psi)\left(St_\Gamma(1) \times St_\Gamma(1)\right)$ in Γ^4 is $8 \cdot 8 = 2^6$, the index of $\psi_2\left(St_\Gamma(2)\right)$ in Γ^4 is $8 \cdot 2^6 = 2^9$.

(c) The index of $[K, K]$ is given by

$$\left[St_\Gamma(2) : [K, K]\right] = 2^7 \quad \text{or} \quad \left[K : [K, K]\right] = 2^6.$$

Indeed, as $\psi_2([K, K])$ is of index 2^{16} in Γ^4 by Remark (a) and Proposition 30.ii, the commutator subgroup $[K, K]$ is of index $2^{16}/2^9 = 2^7$ in $St_\Gamma(2)$ by Remark (b); the formula for $[K : [K, K]]$ follows. Using Proposition 41 below, this also shows that $[[K, K] : St_\Gamma(5)] = 2^{12}$.

(d) Denote by $(\gamma_n(G))_{n \geq 1}$ the lower central series of a group G, with $\gamma_1(G) = G$ and $\gamma_{n+1}(G) = [\gamma_n(G), G]$ for $n \geq 1$.

Proposition 22 shows that $[\Gamma : \gamma_2(\Gamma)] \geq 8$. Now $K < \gamma_2(\Gamma)$ by definition of K and, as $[a, d] \notin B$ by Proposition 16, we have $[a, d] \notin K$ and $[\Gamma : \gamma_2(\Gamma)] < [\Gamma : K] = 16$. Thus $[\Gamma : \gamma_2(\Gamma)] = 8$; also $\gamma_2(\Gamma)$ is generated by K and $[a, d]$.

It is known that $\gamma_3(\Gamma)$ is of index 4 in $\gamma_2(\Gamma)$. More generally, the index of $\gamma_{n+1}(\Gamma)$ in $\gamma_n(\Gamma)$ has been computed for each $n \geq 1$; see [Rozhk–96] and Theorem 6.6 in [BarGr–a]; in particular, this index is always 2 or 4 for $n \geq 2$. See also [Grigo], for both the derived series and the lower central series of Γ.

(e) A group G is said to be of *finite width* if

$$\sup_{n \geq 1} \operatorname{rank}\left(\gamma_n(G)/\gamma_{n+1}(G)\right) < \infty,$$

where "rank" means "smallest number of generators". Thus Γ has finite width. The definition has interesting variants. See [KlaLP–97] and [BarGr–a].

Estimates on the width of a group provide lower estimates on its growth, like those of Remark 66.ii. Width estimates for free groups go back to Magnus: see [Bourb–72], Theorem 3 in § II.5 and Theorem II in § II.3, as well as [Jacob–92]; they can be used to give another proof of our Proposition VII.13 on the uniformly exponential growth of non-abelian free groups.

(f) Let $\mathcal{T}$ be an infinite rooted tree which is *spherically symmetric:* this means that there exists a sequence $(m_k)_{k \geq 1}$ such that, for a vertex x of $\mathcal{T}$ at distance $k - 1 \geq 0$ from the root, the number of neighbours of x at distance k from the root is m_k; we will assume that $m_k \geq 2$ for all $k \geq 1$. Let H be a group of automorphisms of $\mathcal{T}$. For each vertex x of $\mathcal{T}$, define the *rigid stabilizer of x* to be the subgroup $Rist_H(x)$ of H consisting of those $h \in H$ which restrict to the identity outside the subtree $\mathcal{T}_x$ of vertices "below x". For each integer $k \geq 1$, define the *rigid stabilizer of level k* to be the subgroup $Rist_H(k)$ of H generated by the $Rist_H(x)$ for all x at distance k from the root. A group H is a *branch group* if it can act faithfully on a spherically symmetric tree $\mathcal{T}$ as above in such a manner that

- H is transitive on each level of $\mathcal{T}$,
- $Rist_H(k)$ is of finite index in H for all $k \geq 1$.

In case of the Grigorchuk group Γ acting on the binary tree $\mathcal{T}^{(2)}$, corresponding to the constant sequence $m_k = 2$ for all $k \geq 1$, we have

Γ *is a branch group*

by Proposition 30.vi.

Here is one recent result which, among others, shows that the notion of branch group is important. Recall that an infinite group Γ is *just infinite* if any normal subgroup of Γ distinct from 1 is of finite index, and that Γ is *hereditary just infinite* if it is residually finite and if any subgroup of finite index in Γ is just infinite; for these, see [Wilso].

Let Γ be an infinite finitely-generated group which is just infinite. Then one of the following holds:

(i) Γ is a branch group,

(ii) Γ contains an abelian subgroup of finite index
    (see [McCar-68] and [McCar-70]),

(iii) Γ contains a subgroup of finite index which is a direct product of finitely many copies of a simple group,

(iv) Γ contains a subgroup of finite index which is a direct product of finitely many copies of a hereditary just infinite group.

We refer to [Grigo] for further comments and for the proof of this theorem. For branch groups, see also [BarGr-c].

32. Theorem. *For any $n \in \mathbb{N}$, there exists $\gamma \in \Gamma$ such that*

$$\gamma^{2^n} \neq 1.$$

Proof. We will prove more precisely the following result, which yields the theorem by induction on n.

Claim. *Consider an integer $n \geq 1$ and an element $\gamma_n \in K$ such that*

$$\gamma_n^{2^n} \neq 1.$$

Then there exists an element $h_n \in St_\Gamma(5) < K$ *such that*

$$\psi_5(h_n) = (\gamma_n, 1, \ldots, 1)_5 \in K^{32},$$

and the element $\gamma_{n+1} = (ab)^8 h_n \in K$ *satisfies*

$$\gamma_{n+1}^{2^{n+1}} \neq 1.$$

For any $(\gamma_{0,0,0,0,0}, \ldots, \gamma_{1,1,1,1,1}) \in K^{32}$, there exists by Proposition 30.vi an element $h \in St_\Gamma(5) < K$ such that

$$\psi_5(h) = (\gamma_{0,0,0,0,0}, \ldots, \gamma_{1,1,1,1,1})_5 \in K^{32}.$$

In particular, there exists $h_n \in K$ such that

$$\psi_5(h_n) = (\gamma_n, 1, 1, \ldots, 1)_5 \in K^{32},$$

with $31 = 2^5 - 1$ coordinates equal to 1; for the argument below, it is convenient to write this as

$$\psi_4(h_n) = \Big((\gamma_n, 1), (1,1), \ldots, (1,1)\Big)_4 \in K^{16},$$

with $15 = 2^4 - 1$ coordinates equal to $(1,1)$.

Set $\gamma_{n+1} = (ab)^8 h_n$, which is in K because $(ab)^2 \in K$ and $h_n \in K$. We have

$$\psi_4\left(\gamma_{n+1}^2\right) =$$
$$\Big(a\,(\gamma_n,1)\,a\,(\gamma_n,1)\,,\, c\,(1,1)\,c\,(1,1)\,,\, a\,(1,1)\,a\,(1,1)\,,\, \ldots\,,\, c\,(1,1)\,c\,(1,1)\Big)_4 =$$
$$\Big(\gamma_n\,,\, \gamma_n\,,\, 1 \ldots\,,\, 1\Big)_5$$

(with 30 coordinates equal to 1). Consequently

$$\psi_4\left(\gamma_{n+1}^{2^{n+1}}\right) = \left(\gamma_n^{2^n}, \gamma_n^{2^n}, 1, \ldots, 1\right)_5 \neq 1 \in \Gamma^{32}$$

so that the proof of the claim is complete. $\square$

33. Remark. Let G be an infinite finitely-generated p-group. A theorem by Zel'manov already mentioned in Item 19 shows that G cannot be both residually finite and of finite exponent; see [Zelma–91], [Zelma–92]. As Γ is residually finite (Proposition 6), this provides another proof of Theorem 32.

VIII.D. Congruence subgroups

Recall from Proposition 6 that

$$\bigcap_{k=1}^{\infty} St_{\Gamma}(k) = \{1\}.$$

34. Definition. A *congruence subgroup* of Γ is a subgroup which contains $St_{\Gamma}(k)$ for k large enough.

Observe that congruence subgroups are of finite index. The main result of the present section is that any non-trivial *normal* subgroup of Γ is of finite index, and indeed is a congruence subgroup (compare with the beginning of Section III.D). We also prove results on the finite quotients corresponding to the $St_{\Gamma}(k)$'s.

35. Notation. Let k be an integer, $k \geq 1$. Recall from Item 2 that $T(k)$ denotes the finite subtree of T spanned by the vertices of level $\leq k$. We denote by Γ_k the finite quotient $\Gamma/St_{\Gamma}(k)$, and we view it as a subgroup

$$\Gamma_k < Aut(T(k))$$

of the group of automorphisms of $T(k)$. As we have already observed in Item 13 and in the proof of Proposition 22,

  $\Gamma_1 = Aut(T(1))$ is a group of order 2,
  $\Gamma_2 = Aut(T(2))$ is a dihedral group of order 8,
  $\Gamma_3 = Aut(T(3))$ is a group of order 2^7.

We will show below that Γ_k is a proper subgroup of $Aut(T(k))$ for any $k \geq 4$.

36. Proposition. *For each $k \geq 1$, the action of Γ on the vertices of the kth level $L(k)$ is transitive.*

In other words, the action of Γ_k on the leaves of $T(k)$ is transitive.

Proof. The claim is obvious for $k = 1$ because $a \in \Gamma$ exchanges the two vertices of $L(1)$. We proceed by induction on k, assuming that $k \geq 2$ and that Γ acts transitively on $L(k-1)$. Consider two vertices $x = x_{j_1,\ldots,j_k}$ and $x' = x_{j'_1,\ldots,j'_k}$ in $L(k)$.

If $j_1 = j'_1$, there exists by the induction hypothesis $\gamma_{j_1} \in \Gamma$ such that $\gamma_{j_1}(x_{j_2,\ldots,j_k}) = x_{j'_2,\ldots,j'_k}$, and by Proposition 14 there exists $\gamma \in St_{\Gamma}(1)$ such that $\phi_{j_1}(\gamma) = \gamma_{j_1}$. Thus $\gamma(x_{j_1,\ldots,j_k}) = x_{j'_1,\ldots,j'_k}$.

If $j_1 \neq j'_1$, set $y = a(x')$, and proceed for x, y as above for x, x'. $\square$

37. Ends. We denote by Ω the space of *ends* of the tree T, and we identify Ω with the set of one-way infinite $(0,1)$-sequences, as in Remark 8.i.

Two ends $(i_k)_{k\geq 1}$, $(j_k)_{k\geq 1}$ are said to be *equivalent* if there exists $k_0 \geq 1$ such that $i_k = j_k$ for all $k \geq k_0$. Below, we denote by $[\omega]$ the equivalence class of an end $\omega \in \Omega$.

For a vertex $x \in T$, set

$$St_\Gamma(x) = \{\gamma \in T \mid \gamma(x) = x\}.$$

For $\gamma \in St_\Gamma(x)$, we denote by $\gamma|T_x$ the automorphism obtained by restriction of γ to the subtree T_x spanned by the vertices "below x". Recall from Item 1 that there is a natural isomorphism δ_x from the tree T onto the subtree T_x.

38. Lemma. *For any $x \in T$ and $\gamma \in \Gamma$, there exists $\gamma_x \in St_\Gamma(x)$ such that*

$$\delta_x^{-1}(\gamma_x|T_x)\delta_x = \gamma.$$

Proof. Let k denote the level of x. If $k = 0$, there is nothing to prove; if $k = 1$, the lemma is contained in Proposition 14. We assume from now on that $k \geq 2$, and we assume inductively that the lemma holds for every vertex of level $\leq k - 1$.

Let $x = x_{j_1,\ldots,j_k}$. Assume first that $j_1 = 0$. As x is of level $k - 1$ inside T_0, there exists by the induction hypothesis an automorphism $\gamma_{x,0}$ of T_0 such that $\delta_x^{-1}\gamma_{x,0}\delta_x = \gamma$. By Proposition 14, there exist $\gamma_x \in St_\Gamma(1)$ and $\gamma_{x,1} \in \Gamma$ such that $\psi(\gamma_x) = (\gamma_{x,0}, \gamma_{x,1})$. Then γ_x has the desired properties.

The argument is similar in case $j_1 = 1$. $\square$

39. Corollary. *The orbits of the natural action of the group Γ on the space of ends Ω coincide with the equivalence classes of ends (in the sense of Item 37).*

Moreover, the action of Γ on each orbit is faithful.

Proof. Observe first that Γ preserves equivalence classes, because its generators do so. Consider then two ends $\omega_0 = (i_k)_{k\geq 1}$ and $\omega = (j_k)_{k\geq 1}$ in the same equivalence class; we have to show that there exists $\gamma \in \Gamma$ such that $\gamma\omega = \omega_0$.

If $\omega = \omega_0$, there is nothing to show. Otherwise, let l denote the largest k such that $j_k \neq i_k$. By Lemma 38, there exists $\gamma \in St_\Gamma(j_1,\ldots,j_{l-1})$ which acts as a on the subtree $T_{j_1,\ldots,j_{l-1}}$ of T; in particular

$$\gamma(j_1,\ldots,j_{l-1},j_l,j_{l+1},\ldots) = (j_1,\ldots,j_{l-1},\overline{j_l},j_{l+1},\ldots).$$

In other words, if $\gamma\omega = (j_k')_{k\geq 1}$ and if l' is the largest k such that $j_k' \neq i_k$, one has $l' \leq l - 1$.

The same argument repeated as often as necessary (at most l times) shows that ω and ω_0 are in the same Γ-orbit.

For the second claim, consider an element $\gamma \in \Gamma, \gamma \neq 1$, and an equivalence class $[\omega_0] \in \Omega$. There exist $k \geq 1$ and $x = (j_1,\ldots,j_k) \in L(k)$ such that $\gamma(x) \neq x$, and there exists $\omega \in [\omega_0]$ such that the first k terms of the infinite sequence ω are precisely $(j_1, \ldots, j_k)$; thus $\gamma\omega \neq \omega$. $\square$

It is known that the stabilizer of an end in Γ is a *weakly maximal subgroup* of Γ [BarGr–b]. This means that any subgroup of Γ properly containing such a stabilizer is a subgroup of finite index in Γ (see Item III.24).

40. On the finite quotients Γ_k. For each $k \geq 1$, denote by a_k, b_k, c_k, d_k the natural images of a, b, c, d in the finite group Γ_k, constituting a generating set (Γ_k is defined in Item 35). Denote by Δ_k the subgroup of Γ_k consisting of those automorphisms fixing the vertices x_0, x_1 of the first level; it is also the subgroup of Γ_k of elements which can be written as words in a_k, b_k, c_k, d_k with an even number of a_k's. Moreover:

for each $k \geq 3$, the normal subgroup B_k of Γ_k generated by b_k
is of index 8 in Γ_k (see Proposition 25),

for each $k \geq 3$, the group Γ_k is a semi-direct product of B_k
and a dihedral group D_k of order 8 (see Corollary 26),

for each $k \geq 4$, there is an injective homomorphism $\psi_k : \Delta_k \longrightarrow \Gamma_{k-1} \times \Gamma_{k-1}$
whose image is of index 8 and is the semi-direct product of $B_{k-1} \times B_{k-1}$
and of a dihedral group D_{k-1}^{diag} of order 8 (see Theorem 28).

These facts imply the following result.

41. Proposition. *We have*

$$|\Gamma_k| = \frac{1}{4}|\Gamma_{k-1}|^2 \qquad \text{for all} \quad k \geq 4$$

and consequently

$$|\Gamma_k| = 2^{5 \cdot 2^{k-3}+2} \qquad \text{for all} \quad k \geq 3.$$

In particular, Γ_4 is a subgroup of $Aut(\mathcal{T}(4)) \approx Aut(\mathcal{T}(3)) \wr (\mathbb{Z}/2\mathbb{Z})$ of order 2^{12} and of index 2^3.

If $G^{(2)} = Aut(\mathcal{T})$ is the group of Item 2, it follows from Propositions 5 and 41 that

$$\lim_{k \to \infty} \frac{\log [\Gamma : St_\Gamma(k)]}{\log [G^{(2)} : St_{G^{(2)}}(k)]} = \lim_{k \to \infty} \frac{5 \cdot 2^{k-3} + 2}{2^k - 1} = \frac{5}{8}.$$

This is the *Hausdorff dimension* of the closure $\overline{\Gamma}$ of Γ in the profinite group G, in the sense of [BarnS–97]. It is known that this closure $\overline{\Gamma}$ is a profinite group isomorphic to the 2-completion of Γ (Theorem 10 in [Grigo]).

42. Theorem. *Any normal subgroup of* Γ *distinct from* $\{1\}$ *is a congruence subgroup (see Definition 34).*

Proof. Consider $\gamma \in \Gamma, \gamma \neq 1$, and the *normal* subgroup $N = \langle\gamma\rangle^\Gamma$ of Γ generated by γ. Let $k \geq 0$ be the integer defined by

$$\gamma \in St_\Gamma(k), \quad \gamma \notin St_\Gamma(k+1),$$

and set

$$\psi_k(\gamma) = (\gamma_{0,0,\ldots,0,0}, \gamma_{0,0,\ldots,0,1}, \cdots, \gamma_{1,1,\ldots,1,1})_k \in \Gamma^{2^k}.$$

By definition of k, there exists a component of $\psi_k(\gamma)$ which is not in $St_\Gamma(1)$; we assume from now on that $\gamma_{0,0,\ldots,0,0} \notin St_\Gamma(1)$, and write

$$\gamma_{0,0,\ldots,0,0} = ha$$

with $h \in St_\Gamma(1)$, and $\psi(h) = (h_0, h_1)$.

Choose first $x \in K$. By Proposition 30.vi, there exists $u \in St_\Gamma(k+1)$ such that

$$\psi_{k+1}(u) = (x, 1, \ldots, 1)_{k+1} = ((x, 1), (1, 1), \ldots, (1, 1))_k.$$

Then

$$\psi_k([\gamma, u]) = (w, 1, \ldots, 1)_k$$

with

$$\begin{aligned}
w &= \gamma_{0,0,\ldots,0,0}^{-1}(x,1)^{-1}\gamma_{0,0,\ldots,0,0}(x,1) = ah^{-1}(x^{-1},1)ha(x,1)\\
&= a(h_1^{-1}x^{-1}h_1,1)a(x,1) = (x, h_1^{-1}x^{-1}h_1).
\end{aligned}$$

Then choose $y \in K$ and let $v \in St_\Gamma(k+1)$ such that

$$\psi_{k+1}(v) = (y, 1, \ldots, 1)_{k+1}.$$

Then

$$\psi_k([[\gamma, u], v]) = (z_1, z_2, 1, \ldots, 1)_{k+1}$$

with

$$\begin{aligned}
(z_1, z_2) &= w^{-1}(y^{-1},1)w(y,1)\\
&= (x^{-1}, h_1^{-1}xh_1)(y^{-1},1)(x, h_1^{-1}x^{-1}h_1)(y,1)\\
&= ([x,y],1).
\end{aligned}$$

Observe that $[[\gamma, u], v] \in N$. As x and y are arbitrary in K it follows that, for any element $z \in [K, K]$, there exists $t \in N$ such that

$$\psi_{k+1}(t) = (z, 1, \ldots, 1)_{k+1}.$$

Hence $\psi_{k+1}(N)$ contains $[K, K] \times \{1\} \times \ldots \times \{1\}$, with $2^{k+1} - 1$ factors $\{1\}$.

As Γ is transitive on the level $L(k+1)$ by Proposition 36, the image $\psi_{k+1}(N)$ also contains the product of one copy of $[K, K]$ and $2^{k+1} - 1$ copies of $\{1\}$ in any order, so that

$$[K, K] \times [K, K] \times \ldots \times [K, K] < \psi_{k+1}(N)$$

(with 2^{k+1} factors $[K, K]$). Hence

$$St_\Gamma(5) \times \ldots \times St_\Gamma(5) < \psi_{k+1}(N)$$

(with 2^{k+1} factors $St_\Gamma(5)$) by Proposition 30.iv. As

$$\psi_{k+1}(St_\Gamma(k+6)) = St_\Gamma(5) \times \ldots \times St_\Gamma(5),$$

we have

$$St_\Gamma(k+6) < N,$$

and the proof is complete. $\square$

VIII.E. Word problem and non-existence of finite presentations

Let S denote the generating set $\{a, b, c, d\}$ of Γ, let S^* be the set of all finite words in letters of S (including the empty word $\emptyset$), and let $\begin{cases} S^* \longrightarrow \Gamma \\ w \longmapsto \underline{w} \end{cases}$ denote the natural mapping. Our first goal in this section is to describe an algorithm for deciding when a word $w \in S^*$ represents $\underline{w} = 1 \in \Gamma$, and our second goal is to show that Γ is *not* finitely presentable.

43. Notation for sets of words. For a word $w \in S^*$, we denote by $|w|_a$ the number of occurrences of a in w, and similarly for $|w|_b$, $|w|_c$, $|w|_d$, $|w|_{c,d} = |w|_c + |w|_d$, and so on. We denote by the same letter ℓ the word length

$$\ell : \begin{cases} S^* \longrightarrow \mathbb{N} \\ w \longmapsto |w|_a + |w|_b + |w|_c + |w|_d \end{cases}$$

on words and the length function

$$\ell : \begin{cases} \Gamma \longrightarrow \mathbb{N} \\ \gamma \longmapsto \inf \{\ell(w) \mid w \in S^*, \underline{w} = \gamma\} \end{cases}$$

on the group.

Let S_{red}^* denote the subset of S^* of all reduced words and let $r : S^* \longrightarrow S_{red}^*$ denote the reduction mapping. (See Item 12. Though S_{red}^* can be viewed as the group $(\mathbb{Z}/2\mathbb{Z}) * \mathbb{V}$, we will not use this before Item 50.) Set

$$S_{red}^{*, a \equiv 0} = \{w \in S_{red}^* \mid |w|_a \text{ is even }\}$$

and observe that any $w \in S_{red}^{*,a\equiv0}$ may be considered as a word in

$$b\ ,\quad c\ ,\quad d\ ,\quad aba\ ,\quad aca\ ,\quad ada$$

in a unique way. Observe also that, for $w \in S_{red}^{*}$, we have

$$\underline{w} \in St_\Gamma(1) \quad \Longleftrightarrow \quad w \in S_{red}^{*,a\equiv0}.$$

44. The mappings $\tilde{\phi}_0, \tilde{\phi}_1 : S_{red}^{*,a\equiv0} \longrightarrow S^*$. The formulas of Proposition 14 suggest that one defines $\tilde{\phi}_0, \tilde{\phi}_1$ by

$$
\begin{array}{ll}
\tilde{\phi}_0(b) \ = a & \tilde{\phi}_1(b) = c \\
\tilde{\phi}_0(c) \ = a & \tilde{\phi}_1(c) = d \\
\tilde{\phi}_0(d) \ = \emptyset & \tilde{\phi}_1(d) = b \\
\tilde{\phi}_0(aba) = c & \tilde{\phi}_1(aba) = a \\
\tilde{\phi}_0(aca) = d & \tilde{\phi}_1(aca) = a \\
\tilde{\phi}_0(ada) = b & \tilde{\phi}_1(ada) = \emptyset
\end{array}
$$

and for longer words recursively by

$$\tilde{\phi}_j(uw') \ = \ \tilde{\phi}_j(u)\tilde{\phi}_j(w')$$

for $u \in \{b, c, d, aba, aca, ada\}$, w' a word in $\{b, c, d, aba, aca, ada\}$ shorter than uw'.

45. Observation. *(i) For $w \in S_{red}^{*,a\equiv0}$, we have*

$$\tilde{\phi}_0(w) \ = \ \phi_0(\underline{w}) \qquad and \qquad \tilde{\phi}_1(w) \ = \ \phi_1(\underline{w}).$$

(ii) For $w \in S_{red}^{*}$, we have

$$|w|_a \quad is\ even\ and \quad \tilde{\phi}_j(w) = 1 \in \Gamma \quad for \quad j = 0, 1 \quad \Longleftrightarrow \quad \underline{w} = 1 \in \Gamma.$$

Proof. Claim (*i*) is a straightforward consequence of the definitions of ϕ_0, ϕ_1 (Proposition 14) and of $\tilde{\phi}_0, \tilde{\phi}_1$ (just above). Claim (*ii*) follows from the injectivity of the map $\psi = (\phi_0, \phi_1) : St_\Gamma(1) \longrightarrow \Gamma \times \Gamma$. $\square$

46. Lemma. Let $w \in S_{red}^{*,a\equiv 0}$, $w \neq \emptyset$, and let $w_j = r(\tilde{\phi}_j(w))$ for $j = 0, 1$.

(i) We have

$$\ell(w_j) \leq \frac{\ell(w) + 1}{2} \qquad (j = 0, 1).$$

(ii) If, moreover, w begins with a and ends with some $u \in \{b, c, d\}$, or vice-versa, then

$$\ell(w_j) \leq \frac{\ell(w)}{2} \qquad (j = 0, 1).$$

Set $\gamma = \underline{w}$ and $\gamma_j = \phi_j(\gamma)$, or equivalently $\gamma_j = \underline{w_j}$, for $j = 0, 1$.

(iii) We have

$$\ell(\gamma_j) \leq \frac{\ell(\gamma) + 1}{2} \qquad (j = 0, 1).$$

(iv) If, moreover, γ is not conjugate to an element of shorter length in Γ, then

$$\ell(\gamma_j) \leq \frac{\ell(\gamma)}{2} \qquad (j = 0, 1).$$

Proof: Clear. $\square$

47. Algorithm. Let $w \in S^*$. To decide whether or not $\underline{w} = 1 \in \Gamma$, proceed as follows.

(i) Compute $|w|_a$.

     If $|w|_a$ is odd then $\underline{w} \neq 1 \in \Gamma$.

     If $|w|_a$ is even, compute $r(w)$ as defined in Item 12;

         if $r(w)$ is empty then $\underline{w} = 1 \in \Gamma$,

         if $\ell(r(w)) \geq 1$ go to (ii).

(ii) Compute $w_j = r(\tilde{\phi}_j(w))$ for $j = 0, 1$ and return to (i), which should be performed for both w_0 and w_1.

It is an easy consequence of Lemma 46 that the algorithm terminates.

For $w \in S_{red}^*$, define inductively reduced words $w_{\underline{j}}$, where $\underline{j}$ is a finite $(0, 1)$-sequence, as follows:

$$w_{\emptyset} = w$$

$$w_{j_1} \begin{cases} = r\left(\tilde{\phi}_{j_1}(w)\right) & \text{if } w \in S_{red}^{*,a\equiv 0} \\ \text{is not defined otherwise} \end{cases}$$

$\cdots$

$$w_{j_1,\ldots,j_{k-1},j_k} \begin{cases} = r\left(\tilde{\phi}_{j_k}(w_{j_1,\ldots,j_{k-1}})\right) & \text{if } w_{j_1,\ldots,j_{k-1}} \text{ is defined and is in } S_{red}^{*,a\equiv 0} \\ \text{is not defined otherwise.} \end{cases}$$

The algorithm can be visualized as a branching process represented by a binary tree; $\underline{w} = 1 \in \Gamma$ if and only if there exists $n \geq 0$ such that $w_{j_1,\ldots,j_k}$ is defined whenever $k \leq n$ and such that $w_{j_1,\ldots,j_n}$ is the empty word for any sequence $(j_1,\ldots,j_n)$ of length n (it follows that $w_{\underline{j}}$ is defined and is the empty word for any sequence $\underline{j}$ of length $\geq n$).

48. Example. Figure 3 shows two examples of the above algorithm.

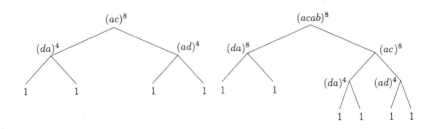

FIGURE 3.   Binary trees of Algorithm 47 for $(ac)^8$ and $(acab)^8$

49. Corollary. *The word problem is soluble in* Γ.

For generalities on the word problem, see for example [Still–82] and Chapter 12 in [Rotma–95]. Recall from Item V.26 the conjecture according to which a group with soluble word problem can be embedded in a finitely-presented simple group; with this in mind, C. Röver has shown how to embedd Γ (and many other "similar" groups) in finitely-presented simple groups [Röver–99].

It is also known that the conjugacy problem is soluble in Γ (Leonov [Leon–98a], Rozhkov).

50. The groups K_n. Recall from Item 12 that the set S^*_{red} is in natural bijection with the group

$$F \doteq (\mathbb{Z}/2\mathbb{Z}) * \mathbb{V},$$

where we write (improperly[4]) $\mathbb{Z}/2\mathbb{Z} = \{1, a\}$ and $\mathbb{V} = \{1, b, c, d\}$. Thus the restriction to S^*_{red} of the mapping $S^* \longrightarrow \Gamma$ introduced just before Item 43 may now be viewed as a homomorphism

$$\pi : \begin{cases} F \longrightarrow \Gamma \\ w \longmapsto \underline{w} \end{cases}$$

which is onto. To any $w \in Ker(\pi)$ there corresponds a family $(w_{\underline{j}})$, where the index $\underline{j}$ describes finite $(0,1)$-sequences, defined as in Item 47 (observe that $|w_{j_1,\ldots,j_{k-1}}|_a$ cannot be odd, because otherwise $\underline{w} \neq 1$).

---

[4]See the footnote to the proof of Proposition 5.

Definition. For each $n \geq 0$, let K_n denote the subset of $\{w \in S_{red}^* \mid \underline{w} = 1\}$, identified with a subset of $Ker(\pi)$, consisting of the words w such that $w_{j_1, \ldots, j_n} = \emptyset$ for all $j_1, \ldots, j_n \in \{0, 1\}$.

Examples. We have $(ad)^4$ and $(da)^4$ in K_1; also, $(ac)^8 \in K_2$, $(ac)^8 \notin K_1$ and $(acab)^8 \in K_3$, $(acab)^8 \notin K_2$, as the algorithm shows.

51. Lemma. *For each $n \geq 0$, the set K_n is a normal subgroup of F and*

$$K_0 = \{1\} < K_1 < \ldots < K_n < K_{n+1} < \ldots < \bigcup_{n=0}^{\infty} K_n = Ker(\pi).$$

(About our notation, recall that $K_n < K_{n+1}$ means that K_n is a subgroup of K_{n+1}, which could coincide with K_{n+1}. However, it will be shown in Item 54 that K_n is a *proper* subgroup of K_{n+1}.)

Proof. The only non-obvious fact to check is that the subgroup K_n is normal for each $n \geq 0$. As this is obvious for $n = 0$, we proceed by induction on n, assuming that $n \geq 1$ and that K_{n-1} is normal in F. Observe that

$$K_n = \{ w \in F \mid |w|_a \text{ is even and } w_0, w_1 \in K_{n-1} \}$$

with $w_j = r\left(\tilde{\phi}_j(w)\right)$ for $j = 0, 1$ (or $w_j = \overline{\phi}_j(w)$ with the notation of Item 52 below).

We claim that, for $w \in K_n$, we have $awa, bwb, cwc, dwd \in K_n$.

The claim for awa follows from the fact that $|awa|_a \equiv |w|_a \pmod{2}$, and from the fact that, for $|w|_a$ even, one has $(awa)_0 = w_1$ and $(awa)_1 = w_0$.

The claim for bwb, cwc, dwd follows from

$$|bwb|_a = |cwc|_a = |dwd|_a = |w|_a,$$

from the fact that, when $|w|_a$ is even, one has

$$
\begin{aligned}
(bwb)_0 &= aw_0a & (bwb)_1 &= cw_1c \\
(cwc)_0 &= aw_0a & (cwc)_1 &= dw_1d \\
(dwd)_0 &= w_0 & (dwd)_1 &= bw_1b
\end{aligned}
$$

and from the induction hypothesis.

We have already observed that Lemma 46 implies that Algorithm 47 terminates, so that one has the equality $\cup_{n=0}^{\infty} K_n = Ker(\pi)$. $\square$

52. The homomorphisms $\overline{\phi}_0, \overline{\phi}_1 : F^{a \equiv 0} \longrightarrow F$. Let $F^{a \equiv 0}$ be the kernel of the homomorphism $F \longrightarrow \mathbb{Z}/2\mathbb{Z}$ which maps a to 1 and b, c, d to 0. If F is identified as above with the set S_{red}^* of reduced words in $\{a, b, c, d\}$, then $F^{a \equiv 0}$ corresponds to the subset $S_{red}^{*, a \equiv 0}$ of words with an even number of a. On the other hand $F^{a\equiv 0}$ has a set $\{b, c, d, aba, aca, ada\}$ of 6 generators and is a free product

$$F^{a\equiv 0} = \{1, b, c, d\} * \{1, aba, aca, ada\} \approx \mathbb{V} * \mathbb{V}$$

of two Vierergruppen. One defines two homomorphisms $\overline{\phi}_0, \overline{\phi}_1$ from $F^{a\equiv 0}$ to F by

$$\begin{array}{ll}
\overline{\phi}_0(b) = a & \overline{\phi}_1(b) = c \\
\overline{\phi}_0(c) = a & \overline{\phi}_1(c) = d \\
\overline{\phi}_0(d) = 1 & \overline{\phi}_1(d) = b \\
\overline{\phi}_0(aba) = c & \overline{\phi}_1(aba) = a \\
\overline{\phi}_0(aca) = d & \overline{\phi}_1(aca) = a \\
\overline{\phi}_0(ada) = b & \overline{\phi}_1(ada) = 1.
\end{array}$$

Thus, for a reduced word $w \in S_{red}^{*,a\equiv 0}$, corresponding to an element in the group $F^{a\equiv 0}$, and for $j \in \{0, 1\}$, the reduction of the word $\tilde{\phi}_j(w)$ defined in Item 44 corresponds to the image in the group F by $\overline{\phi}_j$ of this group element; we will write (with some misuse of notation) $\overline{\phi}_j(w) = r\left(\tilde{\phi}_j(w)\right)$.

53. The endomorphism $\overline{\sigma}$ of F. This is defined by

$$\overline{\sigma}(a) = aca, \quad \overline{\sigma}(b) = d, \quad \overline{\sigma}(c) = b, \quad \overline{\sigma}(d) = c.$$

From the presentation $F = \langle a, b, c, d \mid a^2 = b^2 = c^2 = d^2 = bcd = 1 \rangle$, it is clear that the mapping from $\{a, b, c, d\}$ to F defined by the formulas above extends to an endomorphism of the group F itself.

Proposition. *The image of $\overline{\phi}_0 \overline{\sigma}$ is the infinite dihedral group generated in F by a and d. One has $\overline{\phi}_1 \overline{\sigma} = id_F$; in particular $\overline{\sigma}$ is injective and $\overline{\phi}_1$ is onto.*

Proof. The first claim follows from the formulas

$$\overline{\phi}_0 \overline{\sigma}(a) = d, \quad \overline{\phi}_0 \overline{\sigma}(b) = 1, \quad \overline{\phi}_0 \overline{\sigma}(c) = a, \quad \overline{\phi}_0 \overline{\sigma}(d) = a$$

and the second one from

$$\overline{\phi}_1 \overline{\sigma}(a) = a, \quad \overline{\phi}_1 \overline{\sigma}(b) = b, \quad \overline{\phi}_1 \overline{\sigma}(c) = c, \quad \overline{\phi}_1 \overline{\sigma}(d) = d.$$

$\square$

54. Lemma. *One has $K_n \neq K_{n+1}$ for all $n \geq 0$. In other words, the inclusions of Lemma 51 are strict.*

Proof. Let us write $\tilde{\sigma} : S_{red}^* \longrightarrow S_{red}^*$ for $\overline{\sigma}$ viewed as a transformation of reduced words rather than as an endomorphism of the group F. We claim that the word $\tilde{\sigma}^n\left((ad)^4\right)$ defines an element in K_{n+1} and not in K_n for all $n \geq 0$, where $\tilde{\sigma}^n$ denotes the nth iterate of $\tilde{\sigma}$. For $n \leq 2$, the claim can be checked by easy applications of Algorithm 47, as was already observed in the examples of Item 50. We proceed now by induction on n, assuming that $n \geq 3$ and that the claim holds up to $n - 1$.

For $\tilde{\sigma}^n((ad)^4)$, Step (i) of the algorithm provides the number $|\tilde{\sigma}^n((ad)^4)|_a$ which is a multiple of 8 and which is in particular even. As $\tilde{\sigma}$ maps reduced words to reduced words, we have also

$$\ell\left(\tilde{\sigma}^n\left((ad)^4\right)\right) \geq 1.$$

Step (ii) consists in computing

$$w_j = r\left(\tilde{\phi}_j\left(\tilde{\sigma}^n\left((ad)^4\right)\right)\right)$$

for $j = 0, 1$. Consider first w_0. One the one hand, by the first part of Proposition 53, w_0 is a reduced word in the two letters a, d, so that $\underline{w}_0$ is in the subgroup $\langle a, d \rangle$ of Γ. On the other hand, because of the exponent 4 in $(ad)^4$, the element $\underline{w}_0$ is a fourth power. As $\langle a, d \rangle$ is of exponent 4 by Proposition 16, we have $\underline{w}_0 = 1$. Consider now w_1. As the composition $\overline{\phi}_1\overline{\sigma}$ is the identity by the second part of Proposition 53, we have

$$w_1 = r\left(\tilde{\sigma}^{n-1}\left((ad)^4\right)\right).$$

By the induction hypothesis, we have $w_1 \in K_n$, $w_1 \notin K_{n-1}$.

Hence $w \in K_{n+1}$, $w \notin K_n$. $\square$

55. Theorem. *The group Γ is not finitely presentable.*

Proof. If Γ were finitely presentable, it would have a presentation of the form

$$\Gamma = \left\langle a, b, c, d \; \middle| \; \begin{array}{c} a^2 = b^2 = c^2 = d^2 = bcd = 1 \\ r_1 = r_2 = \ldots = r_k = 1 \end{array} \right\rangle.$$

In other words Γ would be a quotient of the free product $F = \{1, a\} * \{1, b, c, d\}$ by a normal subgroup generated (as a normal subgroup) by a finite number of elements $r_1, \ldots, r_k$. But this would be in contradiction with Lemmas 51 and 54. $\square$

56. Lysionok presentation. I. Lysionok [Lysen–85] has shown that Γ has the presentation

$$\Gamma = \left\langle a, b, c, d \; \middle| \; \begin{array}{c} a^2 = b^2 = c^2 = d^2 = bcd = 1 \\ w_n^4 = (w_n w_{n+1})^4 = 1 \quad (n \geq 0) \end{array} \right\rangle$$

where $(w_n)_{n \geq 0}$ is the sequence of words defined inductively by $w_0 = ad$ and $w_{n+1} = \tilde{\sigma}(w_n)$. Here as in the proof of Lemma 54, the mapping $\tilde{\sigma} : S_{red}^* \longrightarrow S_{red}^*$ is defined by $\tilde{\sigma}(a) = aca$, $\tilde{\sigma}(b) = d$, $\tilde{\sigma}(c) = b$, $\tilde{\sigma}(d) = c$ and $\tilde{\sigma}(ww') = \tilde{\sigma}(w)\tilde{\sigma}(w')$ for a reduced word ww' obtained by concatenation from two shorter words w and w'.

R. Grigorchuk has shown that Lysionok's relations are independent [Grigo–99].

57. Corollary. *The endomorphism* $\overline{\sigma} : F \longrightarrow F$ *of Item 53 induces an injective homomorphism* $\sigma : \Gamma \longrightarrow \Gamma$ *with image of infinite index. In particular, the group* Γ *is not co-Hopfian.*

Proof. Recall that $\pi : F \longrightarrow \Gamma$ denotes the natural projection. Let $w \in F$. By Proposition 53, we have

$$(\pi, \pi) \, (\overline{\phi}_0, \overline{\phi}_1) \, \overline{\sigma}(w) = \left(\pi\overline{\phi}_0\overline{\sigma}(w), \pi(w)\right) \in D \times \Gamma$$

where $D = \langle a, d \rangle$ is the finite dihedral subgroup of Γ introduced in Proposition 16.

Suppose moreover that $w \in Ker(\pi)$, so that

$$(\pi, \pi) \, (\overline{\phi}_0, \overline{\phi}_1) \, \overline{\sigma}(w) \in D \times \{1\}.$$

As $(\pi, \pi) \, (\overline{\phi}_0, \overline{\phi}_1) = (\phi_0, \phi_1) \, (\pi | F^{a \equiv 0})$, we have

$$(\pi, \pi) \, (\overline{\phi}_0, \overline{\phi}_1) \, \overline{\sigma}(w) \in (D \times \{1\}) \cap \left(Im\,(\phi_0, \phi_1)\right).$$

Now $(\phi_0, \phi_1) = \psi$ and $(D \times \{1\}) \cap Im(\psi) = \{1\}$ by Theorem 28. It follows that

$$(\pi, \pi) \, (\overline{\phi}_0, \overline{\phi}_1) \, \overline{\sigma}(w) = 1 \quad \text{for all} \quad w \in Ker(\pi).$$

As ψ is injective, we also have $\left(\pi | F^{a \equiv 0}\right) \overline{\sigma}(w) = 1$. This shows that $\overline{\sigma}(Ker(\pi)) \subset Ker(\pi)$, so that $\sigma : \Gamma \longrightarrow \Gamma$ is well defined; observe also that $\sigma(\Gamma)$ is a subgroup of $St_\Gamma(1)$. The following diagram shows the various homomorphisms involved in the previous argument.

$$
\begin{array}{ccccc}
F & \xrightarrow{\ \overline{\sigma}\ } & F^{a \equiv 0} & \xrightarrow{\ (\overline{\phi}_0, \overline{\phi}_1)\ } & F \times F \\
\downarrow{\scriptstyle \pi} & & \downarrow{\scriptstyle \pi | F^{a \equiv 0}} & & \downarrow{\scriptstyle \pi \times \pi} \\
\Gamma & \xrightarrow{\ \sigma\ } & St_\Gamma(1) & \xrightarrow{\ \psi = (\phi_0, \phi_1)\ } & \Gamma \times \Gamma
\end{array}
$$

Now, as

$$\psi(Im(\sigma)) = (\pi, \pi) \, (\overline{\phi}_0, \overline{\phi}_1) \, (Im(\overline{\sigma})) < D \times \Gamma,$$

as $\psi(\Gamma)$ is of finite index in $\Gamma \times \Gamma$ by Theorem 28, and as D is of infinite index in Γ, the image of σ is of infinite index in Γ.

And finally σ is injective because $\psi(\sigma(\gamma)) \in D \times \{\gamma\}$ for any $\gamma \in \Gamma$. $\square$

VIII.F. Growth

In this section, we consider the length function $\ell : \Gamma \longrightarrow \mathbb{N}$ corresponding to the generating set $\{a, b, c, d\}$ of Γ defined in Item 9. Recall that

$$\psi_3 : \begin{cases} St_\Gamma(3) \longrightarrow & \Gamma^8 \\ \gamma \longmapsto (\gamma_{i_1,i_2,i_3})_{i_1,i_2,i_3 \in \{0,1\}} \end{cases}$$

denotes the third iterate of the homomorphism ψ of Proposition 14.

58. Reduction of words. Let w be a positive word in a, b, c, d with $|w|_a$ even. We use the notation $|w|_a, \ldots, |w|_{d,b}$ of Item 43.

The formulas of Item 44 provide two words $\tilde{\phi}_0(w), \tilde{\phi}_1(w)$; as these are in general not reduced, even if w is reduced, it is important to distinguish $\tilde{\phi}_j(w)$ from its reduction $r(\tilde{\phi}_j(w))$ as defined in Item 12 ($j = 0, 1$).

The process of reduction of some word w to the corresponding reduced word involves a finite number of elementary reductions, each of one of the two following types

(i) $uu' \rightsquigarrow u''$ for some permutation (u, u', u'') of (b, c, d),
    reducing length by 1,
(ii) $uu \rightsquigarrow$ empty word, for some $u \in \{a, b, c, d\}$,
    reducing length by 2.

We will denote by $\rho(w)$ the *weighted number of elementary reductions* of w, namely the sum of the number of reductions of type (i) and twice the number of reductions of type (ii). As an element of the free product $(\mathbb{Z}/2\mathbb{Z}) * \mathbb{V}$ has a unique normal form, the number $\rho(w)$ does not depend on the order of the elementary reductions involved in the reduction (see Item 12). Thus, for example, each of the reductions

$$bbcc \longmapsto cc \longmapsto \emptyset$$
$$bbcc \longmapsto bdc \longmapsto cc \longmapsto \emptyset$$

gives $\rho(bbcc) = 4$.

In general, each of the quantities $|w|_a, |w|_b, |w|_c, |w|_d$ does change during the process of reduction. However, one has the following result.

Proposition. *For any word w in $\{a, b, c, d\}$:*

$$|r(w)|_a \equiv |w|_a \pmod{2}$$
$$|r(w)|_{b,c} \equiv |w|_{b,c} \pmod{2}$$
$$|r(w)|_{c,d} \equiv |w|_{c,d} \pmod{2}$$
$$|r(w)|_{d,b} \equiv |w|_{d,b} \pmod{2}.$$

Proof: straightforward. $\square$

It follows that $w \longmapsto |w|_a$, $w \longmapsto |w|_{b,c}$, $w \longmapsto |w|_{c,d}$ and $w \longmapsto |w|_{d,b}$ are homomorphisms. For example, the last one can be viewed as the composition of the abelianization from Γ onto $\{(a',b',c',d') \in \mathbb{F}_2^4 \mid b' + c' + d' = 0\}$ with the projection $(a',b',c',d') \longmapsto b' + c'$ from $\mathbb{F}_2^4$ onto $\mathbb{F}_2$ (see the remark which follows Proposition 22).

The following is Lemma 3.5 in [Grigo–85].

59. Main Lemma. *With the previous notation, we have*

$$\sum_{i_1,i_2,i_3=0,1} \ell(\gamma_{i_1,i_2,i_2}) \leq \frac{3}{4}\ell(\gamma) + 8$$

for all $\gamma \in St_\Gamma(3)$.

Proof. As the lemma is obvious for $\gamma = 1$, we assume that $\gamma \neq 1$. Let $w \in S_{red}^*$ be a reduced word of length $\ell(w) = \ell(\gamma)$ representing γ. Using the notation of Item 43, we have

$$(1) \qquad \frac{\ell(\gamma) - 1}{2} \leq |w|_{b,c,d} \leq \frac{\ell(\gamma) + 1}{2}$$

(as already observed in Item 12), and the number $|w|_a$ is even. It is straightforward to check that $\ell(w) \geq 4$, and that w has one of the following four forms:

$$
\begin{array}{lll}
w = & au_1a \, u_2 \, au_3a \, u_4 \, \ldots \, au_{2m-1}a \, u_{2m} & \text{of length} \quad 4m \\
w = & au_1a \, u_2 \, au_3a \, u_4 \, \ldots \, au_{2m-1}a & \text{of length} \quad 4m-1 \\
w = u_0 \, au_1a \, u_2 \, au_3a \, u_4 \, \ldots \, au_{2m-1}a & \text{of length} \quad 4m \\
w = u_0 \, au_1a \, u_2 \, au_3a \, u_4 \, \ldots \, au_{2m-1}a \, u_{2m} & \text{of length} \quad 4m+1
\end{array}
$$

where the u_j's are in $\{b,c,d\}$. The formulas of Item 44 yield

$$\ell(\tilde{\phi}_0(w)) + \ell(\tilde{\phi}_1(w)) \leq \ell(\gamma) + 1 - |w|_d$$
$$|\tilde{\phi}_0(w)|_{c,d} + |\tilde{\phi}_1(w)|_{c,d} = |w|_{b,c}.$$

For $i_1, i_2, i_3 \in \{0,1\}$, we set

$$w_{i_1} = r\left(\tilde{\phi}_{i_1}(w)\right) , \quad w_{i_1,i_2} = r\left(\tilde{\phi}_{i_2}(w_{i_1})\right) , \quad w_{i_1,i_2,i_3} = r\left(\tilde{\phi}_{i_3}(w_{i_2,i_3})\right)$$

and we analyse these words in three steps, as follows.

Step one. Denote by

$$\rho_1 = \rho(\tilde{\phi}_0(w)) + \rho(\tilde{\phi}_1(w))$$

the weighted number of elementary reductions involved in reducing $\tilde{\phi}_0(w)$ to w_0 and $\tilde{\phi}_1(w)$ to w_1. Then

$$(2) \qquad \ell(w_0) + \ell(w_1) \leq \ell(\gamma) + 1 - |w|_d - \rho_1$$

and

$$|w_0|_{c,d} + |w_1|_{c,d} \geq |w|_{b,c} - 2\rho_1.$$

Using (1) and the equality $|w|_{b,c,d} = |w|_{b,c} + |w|_d$, we deduce from the last inequality that

(3) $$|w_0|_{c,d} + |w_1|_{c,d} \geq \frac{\ell(\gamma) - 1}{2} - |w|_d - 2\rho_1.$$

Step two. Denote by

$$\rho_2 = \rho(\tilde{\phi}_0(w_0)) + \rho(\tilde{\phi}_1(w_0)) + \rho(\tilde{\phi}_0(w_1)) + \rho(\tilde{\phi}_1(w_1))$$

the weighted number of elementary reductions involved in reducing

$$\tilde{\phi}_0(w_0), \quad \tilde{\phi}_0(w_1), \quad \tilde{\phi}_1(w_0), \quad \tilde{\phi}_1(w_1)$$

to $w_{0,0}, w_{0,1}, w_{1,0}, w_{1,1}$. On the one hand, in the same way as for (2) above, we now have

(4) $$\sum_{i_1,i_2=0,1} \ell(w_{i_1,i_2}) \leq \ell(\gamma) + 3 - |w|_d - \rho_1 - |w_0|_d - |w_1|_d - \rho_2$$

by using (2). On the other hand, in the same way as for (3) above, we now have

$$\sum_{i_1,i_2=0,1} |w_{i_1,i_2}|_d \geq |w_0|_c + |w_1|_c - 2\rho_2$$

$$= |w_0|_{c,d} + |w_1|_{c,d} - |w_0|_d - |w_1|_d - 2\rho_2$$

so that

(5) $$\sum_{i_1,i_2=0,1} |w_{i_1,i_2}|_d \geq \frac{\ell(\gamma) - 1}{2} - |w|_d - 2\rho_1 - |w_0|_d - |w_1|_d - 2\rho_2$$

by using (3).

Step three. Arguing a third time as we did for (2) above, we have

$$\sum_{i_1,i_2,i_3=0,1} \ell(w_{i_1,i_2,i_3}) \leq \sum_{i_1,i_2=0,1} (\ell(w_{i_1,i_2}) + 1 - |w_{i_1,i_2}|_d)$$

so that

(6) $$\sum_{i_1,i_2,i_3=0,1} \ell(w_{i_1,i_2,i_3}) \leq \ell(\gamma) + 7 - |w|_d - \rho_1 - |w_0|_d - |w_1|_d - \rho_2$$
$$- \frac{\ell(\gamma) - 1}{2} + |w|_d + 2\rho_1 + |w_0|_d + |w_1|_d + 2\rho_2$$
$$\leq \frac{\ell(\gamma)}{2} + 8 + \rho_1 + \rho_2$$

by using (4) and (5).

We now distinguish two cases.

Let us first assume that $\rho_1 + \rho_2 \leq \frac{\ell(\gamma)}{4}$. Then the inequality of the lemma follows from (6).

Let us now assume that $\rho_1 + \rho_2 > \frac{\ell(\gamma)}{4}$. Then

$$\sum_{i_1,i_2=0,1} \ell(w_{i_1,i_2}) \leq \frac{3\ell(\gamma)}{4} + 3$$

by (4) and the inequality of the lemma follows again because

$$\ell(w_{i_1,i_2,0}) + \ell(w_{i_1,i_2,1}) \leq \ell(w_{i_1,i_2}) + 1$$

for each $(i_1, i_2) \in \{0,1\}^2$. $\square$

60. Lemma. *Let Δ be a group generated by a finite set S and let Δ_0 be a subgroup of finite index, say $n = [\Delta : \Delta_0]$. Let $\ell : \Delta \longrightarrow \mathbb{N}$ denote the word length defined by S. For each $k \in \mathbb{N}$, denote by*

$$\beta(k) = |\{\delta \in \Delta \mid \ell(\delta) \leq k\}|$$

the value at k of the growth function of (Δ, S) and by

$$\beta_0(k) = |\{\delta \in \Delta_0 \mid \ell(\delta) \leq k\}|$$

the value at k of the relative growth function of Δ_0.

Then one has

$$\beta(k) \leq n\beta_0(k + n - 1)$$

for all $k \geq 0$.

Proof. The coset space $\Delta_0 \setminus \Delta$ is naturally a metric space for the distance defined by

$$d(\Delta_0 x, \Delta_0 y) = \inf \left\{ \ell\left(x^{-1}\delta y\right) \mid \delta \in \Delta_0 \right\}.$$

As S generates Δ, two points x, y of $\Delta_0 \setminus \Delta$ can always be connected by a chain $x_0 = x, x_1, \ldots, x_m = y$ with $d(x_{j-1}, x_j) = 1$ for $j \in \{1, \ldots, m\}$. As $|\Delta_0 \setminus \Delta| = n$, the diameter of $\Delta_0 \setminus \Delta$ is at most $n - 1$. Thus one can choose representatives $\delta_1, \delta_2, \ldots, \delta_n$ in Δ of the n classes modulo Δ_0 such that $\ell(\delta_j) \leq j - 1$ for $j \in \{1, \ldots, n\}$.

Now any $x \in \Delta$ can be written as $x = \delta_j^{-1} y$ for some $j \in \{1, \ldots, n\}$ and $y \in \Delta_0$, and we have $\ell(y) \leq \ell(\delta_j) + \ell(x) \leq n - 1 + \ell(x)$. It follows that

$$|\{x \in \Delta \mid \ell(x) \leq k\}| \leq n |\{y \in \Delta_0 \mid \ell(y) \leq k + n - 1\}|$$

for all $k \geq 0$. $\square$

[One can write a shorter proof if one assumes basic facts about Schreier systems; for these, see e.g. Section 2.3 in [MagKS–66].]

Concerning this lemma, see also the result from [VdDW–84b] quoted in Item VII.30.

61. Theorem. *The group Γ has subexponential growth.*

Proof. For all $k \geq 0$, set

$$\beta(k) = |\{\gamma \in \Gamma \mid \ell(\gamma) \leq k\}|$$
$$\beta_0(k) = |\{\gamma \in St_\Gamma(3) \mid \ell(\gamma) \leq k\}|$$

where $\ell : \Gamma \longrightarrow \mathbb{N}$ is the word length with respect to $\{a, b, c, d\}$. Set also

$$\omega = \lim_{k \to \infty} \sqrt[k]{\beta(k)}$$

as in Definition VI.51. Choose $\epsilon > 0$. It follows from Proposition VI.56 that there exists an integer k_0 such that

$$(\omega - \epsilon)^k \leq \beta(k) \leq (\omega + \epsilon)^k$$

as soon as $k \geq k_0$. Writing C for $\beta(k_0)$, we have a fortiori

(1) $$\beta(k) \leq C(\omega + \epsilon)^k$$

for all $k \geq 0$.

As $\psi_3 : St_\Gamma(3) \longrightarrow \Gamma^8$ is injective, Lemma 59 implies that

(2) $$\beta_0(k) \leq \sum \beta(k_1)\beta(k_2)\ldots\beta(k_8)$$

for all $k \geq 0$, where the summation is over all $(k_1, \ldots, k_8) \in \mathbb{N}^8$ with $k_1 + \ldots + k_8 \leq \frac{3}{4}k + 8$. Recall that $St_\Gamma(3)$ is of index 2^7 in Γ (proof of Proposition 22). By Lemma 60, Inequalities (1) and (2) imply that

$$\beta(k) \leq 2^7 \beta_0(k + 2^7 - 1) \leq D \sum (\omega + \epsilon)^{k_1 + \ldots + k_8}$$

for all $k \geq 0$, where D denotes the constant $2^7 C^8$ and where the summation is now over all $(k_1, \ldots, k_8) \in \mathbb{N}^8$ with $k_1 + \ldots + k_8 \leq \frac{3}{4}\left(k + 2^7 - 1\right) + 8$. Let $P(k)$ denote the number of 8-uples $(k_1, \ldots, k_8)$ of positive integers involved in this summation; it is well known that P is a polynomial (see e.g. Chap. II.5 of [Felle–50]). Consequently

$$\beta(k) \leq D\,P(k)(\omega + \epsilon)^{\frac{3}{4}(k + 2^7 - 1) + 8}$$

for all $k \geq 0$. As $\lim_{k \to \infty} \sqrt[k]{D P(k)(\omega + \epsilon)^{\frac{3}{4}(2^7 - 1) + 8}} = 1$, this implies that

$$\omega \leq \lim_{k \to \infty} \sqrt[k]{(\omega + \epsilon)^{\frac{3}{4}k}} = (\omega + \epsilon)^{\frac{3}{4}}.$$

As this holds for any $\epsilon > 0$, we have $\omega \leq \omega^{\frac{3}{4}}$. Thus $\omega = 1$ and the proof is complete. $\square$

62. Lemma. *Let* $\alpha : \mathbb{N} \longrightarrow \mathbb{R}_+$ *be a growth function such that* $\alpha(0) = 1$, $\alpha(1) \geq 2$ *and*

$$\alpha^2(k) \overset{s}{\prec} \alpha(k).$$

Then there exists a constant $\sigma > 0$ *such that*

$$e^{k^\sigma} \overset{s}{\prec} \alpha(k).$$

Proof. Recall that the hypothesis on α means that $\alpha^2(k) \leq \alpha(\lambda k)$ for some constant $\lambda > 1$ and for all integers $k \geq 1$ (see Definition VI.29). We set

$$\sigma = \frac{1}{\log_2(\lambda)}.$$

Define a function $\beta : \mathbb{R}_+ \longrightarrow \mathbb{R}_+$ by $\beta(t) = \alpha(\lceil t \rceil)$, where $\lceil t \rceil$ denotes the smallest integer such that $t \leq \lceil t \rceil$ (see VI.21). We have $\beta^2(t) \leq \beta(\lambda t)$ for all $t \geq 0$.

Let $t \geq 1$, and set $k = [\log_\lambda(t)]$. We have

$$\beta(t) \geq \beta^2\left(\frac{t}{\lambda}\right) \geq \beta^4\left(\frac{t}{\lambda^2}\right) \geq \ldots \geq \beta^{2^k}\left(\frac{t}{\lambda^k}\right).$$

Now on the one hand $\beta(t\lambda^{-k}) \geq \beta(1) = \alpha(1) \geq 2$; and on the other hand

$$\log_\lambda(t)log_2(\lambda) = \log_2(t), \qquad \text{hence} \qquad 2^k \geq \frac{1}{2}2^{\log_\lambda t} = \frac{1}{2}t^\sigma.$$

It follows that

$$\beta(t) \geq 2^{\left(\frac{1}{2}t^\sigma\right)} \qquad \text{for all} \qquad t \in \mathbb{R}, \ t \geq 1$$

and a fortiori that

$$\alpha(k) \geq \sqrt{2}^{(k^\sigma)} \qquad \text{for all} \qquad k \in \mathbb{N}, \ k \geq 1.$$

$\square$

63. Proposition. *Let* Γ *be a finitely-generated group which is quasi-isometric to* $\Gamma \times \Gamma$, *and let* $\beta(k)$ *denote a growth function of* Γ. *Then there exists a constant* $\sigma > 0$ *such that*

$$e^{k^\sigma} \overset{s}{\prec} \beta(k).$$

Proof. This is a straightforward consequence of Lemmas 60 and 62. $\square$

64. Remark. For examples of finitely-generated groups Δ isomorphic to $\Delta \times \Delta$, see [Meier–82].

65. Theorem. *The group Γ is of intermediate growth.*

Proof. This follows from the following three facts:

Γ has subexponential growth (Theorem 61),
Γ and $\Gamma \times \Gamma$ are commensurable (Theorem 28),
Γ is not of polynomial growth (Proposition 63). $\square$

66. Remarks. (i) There are other ways of seeing that Γ cannot be of polynomial growth.

The elementary theory of finitely-generated nilpotent groups shows that such a group is finite if and only if all its elements have finite order (see for example Theorem 9.17 in [MacDo–88]). Thus, it follows from Corollary 15 and Theorem 17 that Γ has no nilpotent subgroup of finite index, and from Gromov's theorem VII.29 that Γ is not of polynomial growth. Proposition 63 above of course shows more.

(ii) For the Grigorchuk group Γ, a more detailed study shows that

$$e^{\sqrt{k}} \overset{s}{\prec} \beta_{\Gamma}(k) \overset{s}{\prec} e^{k^{\theta}}$$

for $\theta = \log_{32}(31)$. See Theorem 3.2 in [Grigo–85].

Recently, Laurent Bartholdi has shown that one may take $\theta = \log_{2/\eta}(2) \cong 0.7674$ and where $\eta \cong 0.811$ is the real root of the polynomial $T^3 + T^2 + T - 2$ [Barth–98]. Both Yu.G. Leonov and Laurent Bartholdi have shown that $e^{k^{\sigma}} \prec \beta_{\Gamma}(k)$, respectively with $\sigma = \log_{(87/22)} 2 \cong 0.504$ and $\sigma = 0.5157$; see [Leon–98b], [Leon], and [Barth].

(iii) The following more general result has already been quoted in Complement VII.29: for any finitely-generated infinite group Γ which is residually-p for some p, or which is residually nilpotent and which is not almost nilpotent, the growth function β satisfies

$$e^{\sqrt{k}} \overset{s}{\prec} \beta(k)$$

(subradical growth).

(iv) As already indicated in Problem VII.29, it is conjectured that the growth function β of any finitely-generated group which is not of polynomial growth satisfies

$$e^{\sqrt{k}} \overset{s}{\prec} \beta(k).$$

(v) Let us quote the following results, even though the analogy could well have no deep meaning.

Let G be a subgroup of the full symmetric group of a set Ω. For each integer $n \geq 1$, set

$$\Omega^{n} = \{\ n\text{-uples of elements of } \Omega \ \}$$

$$\Omega^{(n)} = \{\ n\text{-uples of } distinct \text{ elements of } \Omega \ \}$$

$$\Omega^{\{n\}} = \{\ \text{subsets of } \Omega \text{ containing } n \text{ elements} \ \}$$

and observe that G acts on each of Ω^n, $\Omega^{(n)}$, and $\Omega^{\{n\}}$. For a given n, the following properties are equivalent:

the number F_n^* of G-orbits on Ω^n is finite,

the number F_n of G-orbits on $\Omega^{(n)}$ is finite,

the number f_n of G-orbits on $\Omega^{\{n\}}$ is finite.

(For example, $f_n \leq F_n \leq n! f_n$.) Define the permutation group G to be *oligomorphic* if these numbers are finite for all $n \geq 1$.

Theorem (M. Pouzet, 1981; H.D. Macpherson, 1985). *Let G be an oligomorphic permutation group. Then*
either there exists an integer d and constants $c_1, c_2 > 0$ such that

$$c_1 n^d \leq f_n \leq c_2 n^d$$

for all $n \geq 1$,
or, for any $\epsilon > 0$, one has

$$f_n \geq exp\left(n^{\frac{1}{2}-\epsilon}\right)$$

for n large enough.

See Theorem 5.23 in [Camer–99].

67. Corollary. *The growth series of Γ with respect to $\{a, b, c, d\}$ is transcendental, and moreover has the unit circle as natural boundary.*

Proof. If a power series with integral coefficients is algebraic and its radius of convergence is equal to 1, this series is necessarily rational, by a result of Fatou. It follows from Theorem 65 that the growth series of Γ is not algebraic. (See page 368 of [Fatou–06], or N° 167 in Part VIII of [PólSz–76]. A simpler result is that a rational power series has coefficients which are either of polynomial growth or of exponential growth; see Section 7.3 in [GraKP–91]; this and Theorem 65 imply that the growth series of Γ is not rational.)

Morevoer, a theorem of Carlson [Carls–21] shows that a power series with integral coefficients that converges inside the unit disc is either a rational function, or has the unit circle as natural boundary (in particular cannot be algebraic irrational). □

68. Corollary. *The group Γ is not in the class EG of Day.*

Proof. In the terminology of [Day–57] (clarified by [Chou–80]), finitely-generated groups which are in the class EG of "elementary amenable" groups are either of polynomial growth or of exponential growth. □

It is known that the HNN-extension of Γ by the endomorphism σ of Corollary 57 is an amenable group which is not in the class EG of Day and which is moreover *finitely presented* ([Grigo–96], [Grigo–98]).

69. Cyclic reduction. Given a generating subset S of some group, with $1 \notin S$, a reduced word in the letters of $S \cup S^{-1}$ is said to be *cyclically reduced* if either it is of length ≤ 1 or if the product of its first letter with its last letter is not an element of $\{1\} \cup S \cup S^{-1}$. In the particular example of a reduced word $w \in S^*_{red}$, one can obtain a cyclic reduction $s(w)$ as follows.

Let $w \in S^*_{red}$ be a reduced word in $\{a, b, c, d\}$. Set

$$s_a(w) = \begin{cases} w' & \text{if } w \text{ is of the form } aw'a, \\ w & \text{otherwise,} \end{cases}$$

and

$$s_u(w) = \begin{cases} w' & \text{if } w \text{ is of the form } uw'u \text{ for some } u \in \{b, c, d\}, \\ w'u'' & \text{if } w \text{ is of the form } uw'u' \\ & \text{where } (u, u', u'') \text{ is a permutation of } (b, c, d), \\ w & \text{otherwise.} \end{cases}$$

Then

$$s(w) = (s_u s_a)^N (w) \quad \text{for } N \text{ large enough}$$

(for example for $N \geq \ell(w)$); in particular $s_a(w) = s_u(w) = s(w) = w$ if $\ell(w) \leq 2$. Observe that, for $w \in S^*_{red}$, the words $s_a(w), s_u(w)$ and $s(w)$ are again in S^*_{red}.

For example, $s(bac) = ad$ and $s(acada) = ab$; the first of these shows that one may have $s(w) \neq w$ even if w is cyclically reduced.

In the next proposition, $k \equiv l$ means $k \equiv l \pmod 2$. Recall that $\underline{w}$ denotes the element of Γ defined by a word w; here $o(\gamma) \in \mathbb{N}$ denotes the order of an element $\gamma \in \Gamma$.

Proposition. *For $w \in S^*_{red}$ and $s(w)$ as above, we have*

(i) $|s(w)|_a \equiv |w|_a$, $|s(w)|_{b,c} \equiv |w|_{b,c}$,
     $|s(w)|_{b,d} \equiv |w|_{b,d}$, $|s(w)|_{c,d} \equiv |w|_{c,d}$.

(ii) $o\left(\underline{s(w)}\right) = o\left(\underline{w}\right)$.

(iii) *Assume that $s(w) = w$; if $|w|_a$ is even, then*

$$\ell\left(\tilde{\phi}_j(w)\right) \leq \frac{1}{2}\ell(w) \quad \text{and} \quad o(\underline{w}) = \max\left\{o\left(\underline{\tilde{\phi}_0(w)}\right), o\left(\underline{\tilde{\phi}_1(w)}\right)\right\};$$

if $|w|_a$ is odd, then

$$\ell\left(\tilde{\phi}_j(ww)\right) \leq \ell(w) \quad \text{and} \quad o(\underline{ww}) = \max\left\{o\left(\underline{\tilde{\phi}_0(ww)}\right), o\left(\underline{\tilde{\phi}_1(ww)}\right)\right\}.$$

Proof. For (i) and (ii), see the formulas of Item 44. For the length inequalities in (iii), see Lemma 46; the equality of the orders in (iii) follows from the injectivity of the homomorphism $\psi = (\phi_0, \phi_1) : St_\Gamma(1) \longrightarrow \Gamma$.

Observe that $\tilde{\phi}_0(w)$ and $\tilde{\phi}_1(w)$ need not have the same order; for example, if $w = dabadaba$, then $\phi_0(w) = 1$ has order 1 and $\phi_1(w) = (ba)^2$ has order 8. $\square$

The following theorem is contained in results due to Yurij Leonov [Leon–98b] and Igor Lysionok [Lysen]. One method of proof is similar to that used for Lemma 59.

70. Theorem. *For any $\gamma \in \Gamma$, the order $o(\gamma)$ and the length $\ell(\gamma)$ of γ with respect to $\{a, b, c, d\}$ satisfy*

$$o(\gamma) \text{ is a power of } 2,$$
$$o(\gamma) \leq 2\ell(\gamma)^3.$$

Proof. Let $\gamma \in \Gamma$. Choose a reduced word $w \in S^*_{red}$ such that $\underline{w} = \gamma$ and $\ell(w) = \ell(\gamma)$. There is no loss of generality if we assume γ cyclically reduced, indeed if we assume that $s(w) = w$. Define a family $(w_j)_{j \in T}$ of reduced words indexed by the set T of finite $(0, 1)$-sequences, inductively on the length of j, as follows. First set

$$w_\emptyset = w.$$

Then consider an integer $k \geq 0$, a sequence $(j_1, \ldots, j_{k+1}) \in T$, and assume that $w_{j_1,\ldots,j_k}$ has already been defined. If $(j_1, \ldots, j_k)$ is of *even type*, by which we mean that $|w_{j_1,\ldots,j_k}|_a$ is even, set

$$w_{j_1,\ldots,j_{k+1}} = s\left(r\left(\tilde{\phi}_{j_{k+1}}(w_{j_1,\ldots,j_k})\right)\right).$$

If $(j_1, \ldots, j_k)$ is of *odd type*, by which we mean that $|w_{j_1,\ldots,j_k}|_a$ is odd, set

$$w_{j_1,\ldots,j_{k+1}} = s\left(r\left(\tilde{\phi}_{j_{k+1}}(w_{j_1,\ldots,j_k} w_{j_1,\ldots,j_k})\right)\right)$$

where $w_{j_1,\ldots,j_k} w_{j_1,\ldots,j_k}$ is the word obtained by concatenation of two copies of $w_{j_1,\ldots,j_k}$.

Claim A. Let $(j_1, \ldots, j_{k+3}) \in T$ be such that the three sequences

$$(j_1, \ldots, j_k) \quad , \quad (j_1, \ldots, j_{k+1}) \quad , \quad (j_1, \ldots, j_{k+2})$$

are of odd type. Then $(j_1, \ldots, j_{k+3})$ is of even type.

Indeed, because of the formulas of Item 44 and the rules of reduction, the hypothesis of the claim implies that

$$|w_{j_1,\ldots,j_k}|_{b,c} \equiv |w_{j_1,\ldots,j_{k+1}}|_a \pmod{2},$$
$$|w_{j_1,\ldots,j_k}|_{b,d} \equiv |w_{j_1,\ldots,j_{k+1}}|_{b,c} \equiv |w_{j_1,\ldots,j_{k+2}}|_a \pmod{2}$$

so that both $|w_{j_1,\ldots,j_k}|_{b,c}$ and $|w_{j_1,\ldots,j_k}|_{b,d}$ are odd. If $(j_1,\ldots,j_{k+3})$ were of odd type, one would have

$$|w_{j_1,\ldots,j_k}|_{c,d} \equiv |w_{j_1,\ldots,j_{k+1}}|_{b,d} \equiv |w_{j_1,\ldots,j_{k+2}}|_{b,c} \equiv |w_{j_1,\ldots,j_{k+3}}|_a \pmod{2},$$

so that $|w_{j_1,\ldots,j_k}|_{c,d}$ and $|w_{j_1,\ldots,j_k}|_{b,c} + |w_{j_1,\ldots,j_k}|_{b,d} + |w_{j_1,\ldots,j_k}|_{c,d}$ would also be odd. But this is absurd because

$$|w_{j_1,\ldots,j_k}|_{b,c} + |w_{j_1,\ldots,j_k}|_{b,d} + |w_{j_1,\ldots,j_k}|_{c,d} \equiv$$
$$2\left(|w_{j_1,\ldots,j_k}|_b + |w_{j_1,\ldots,j_k}|_c + |w_{j_1,\ldots,j_k}|_d\right)$$

is even, and this proves Claim A.

Proposition 69 implies that, if $(j_1,\ldots,j_k)$ is of even type, then

$$\ell\left(w_{j_1,\ldots,j_{k+1}}\right) \leq \frac{1}{2}\ell\left(w_{j_1,\ldots,j_k}\right)$$
$$o\left(\underline{w}_{j_1,\ldots,j_k}\right) = \max\left\{o\left(\underline{w}_{j_1,\ldots,j_k,0}\right), o\left(\underline{w}_{j_1,\ldots,j_k,1}\right)\right\}$$

and, if $(j_1,\ldots,j_k)$ is of odd type, then

$$\ell\left(w_{j_1,\ldots,j_{k+1}}\right) \leq \ell\left(w_{j_1,\ldots,j_k}\right)$$
$$o\left(\underline{w}_{j_1,\ldots,j_k}\right) = 2\max\left\{o\left(\underline{w}_{j_1,\ldots,j_k,0}\right), o\left(\underline{w}_{j_1,\ldots,j_k,1}\right)\right\}.$$

It follows from Claim A that

$$\ell\left(w_{j_1,\ldots,j_{k+4}}\right) \leq \frac{1}{2}\ell\left(w_{j_1,\ldots,j_k}\right)$$
$$o\left(\underline{w}_{j_1,\ldots,j_k}\right) \text{ divides } 8\,o\left(\underline{w}_{j_1,\ldots,j_k,j,j',j'',j'''}\right)$$

for any $(k+4)$-uple $(j_1,\ldots,j_{k+4})$ and for any $j,j',j'',j''' \in \{0,1\}$.

Claim B. Let $w \in S^*_{red}$ be a reduced and cyclically reduced word representing an element $\gamma \in \Gamma, \gamma \neq 1$ of length $\ell(\gamma) = \ell(w)$ and let n be an integer such that

$$2^{n-1} \leq \ell(w) < 2^n.$$

Then $o(\gamma) \leq 2\ell(\gamma)^3$.

Indeed, for $n = 1$, namely for $\ell(w) = 1$, it is straightforward to check that Claim B holds. We proceed by induction on n, assuming that $n \geq 2$ and that the claim holds up to $n-1$.

Let $(j_1,j_2,j_3,j_4) \in \{0,1\}^4$, and let γ_{j_1,j_2,j_3,j_4} denote the element of Γ represented by the word w_{j_1,j_2,j_3,j_4} associated to w. By Claim A we have

$$\ell\left(w_{j_1,j_2,j_3,j_4}\right) \leq \frac{1}{2}\ell(w) < 2^{n-1}$$
$$o(\gamma) \leq 8\,o\left(\gamma_{j_1,j_2,j_3,j_4}\right).$$

By the induction hypothesis, this implies that

$$o(\gamma) \leq 8\,o\,(\gamma_{j_1,j_2,j_3,j_4}) \leq 16\ell\,(w_{j_1,j_2,j_3,j_4})^3 \leq 2\ell(w)^3$$

and the proof is complete. $\square$

L. Bartholdi and Z. Šuník recently announced that $o(\gamma) \leq 4\ell(\gamma)^2$.

71. Lower bounds for the maximum order functions. For an integer n, define $o(n)$ to be the maximum of the orders $o(\gamma)$ for γ with length $\ell(\gamma) \leq n$. Thus Theorem 70 shows that $o(n) \leq 2n^3$. It has been shown by Lysionok [Lysen] that

$$o(n) \geq 3\sqrt{n}$$

for all $n \geq 0$. (The method of proof of Theorem 32 above shows that $o(n) \geq Cn^{\frac{1}{5}}$ for an appropriate constant C.)

VIII.G. Exercises and complements

72. Exercise. Check that, with the notation of Items 2 and 7 and for all $k \geq 0$, the subquotient $St_{G^{(d)}}(k)/St_{G^{(d)}}(k+1)$ of $G^{(d)}$ can be identified with the group of all mappings from $L(k)$ to the symmetric group $Sym(d)$.

73. Exercise. For an element $\gamma \in \Gamma$ and an integer $k \geq 0$, let $Fix_k(\gamma)$ denote the number of vertices in the level $L(k)$ which are fixed by γ. For example, for the generator a of Γ, we have

$$Fix_k(a) = \begin{cases} 1 & \text{if } k = 0 \\ 0 & \text{if } k \geq 1. \end{cases}$$

Let $u \in \{b, c, d\}$. Clearly, $Fix_0(u) = 1$ and $Fix_1(u) = 2$. Show that

$$Fix_{k+1}(b) = Fix_k(c)$$
$$Fix_{k+1}(c) = Fix_k(d)$$
$$Fix_{k+1}(d) = 2^k + Fix_k(b)$$

for all $k \geq 1$. Check that this implies that $Fix_k(u)$ grows exponentially with k if u is one of b, c, d, and in particular that

$$\sum_{k=0}^{\infty} Fix_k(b)z^k = \frac{1 - 2z^2 - 3z^3}{(1 - z^3)(1 - 2z)}.$$

There are similar rational functions, with the same denominator, for the series $\sum_{k=0}^{\infty} Fix_k(c)z^k$ and $\sum_{k=0}^{\infty} Fix_k(d)z^k$.

More generally one can show that, for any $\gamma \in \Gamma$, the series $\sum_{k=0}^{\infty} Fix_k(\gamma) z^k$
is of the form $\frac{P(z)}{(1-z^3)(1-2z)}$, for some polynomial $P(z)$.

[If $\gamma \in \Gamma, \gamma \notin St_\Gamma(1)$, then $Fix_k(\gamma) = Fix_k(a)$ for all $k \geq 0$. If $\gamma \in St_\Gamma(1)$,
$\psi(\gamma) = (\gamma_0, \gamma_1)$, use induction.]

74. Exercise. Let $\pi_2 : \Gamma \longrightarrow Sym(4)$ denote the natural homomorphism
from Γ to the symmetric group of the 4 vertices of level 2. Show that the image
of π_2 is a dihedral group of order 8.

75. Exercise. Consider the homomorphism $\pi_3 : \Gamma \longrightarrow Sym(8)$ defined in
the proof of Proposition 22, as well as the homomorphisms $\pi_2 : \Gamma \longrightarrow Sym(4)$
and $\pi_1 : \Gamma \longrightarrow Sym(2)$ corresponding to the actions of Γ on the levels $L(2)$ and
$L(1)$ of the tree T.

(i) Show that the natural homomorphism from $Im(\pi_3)$ to $Im(\pi_2)$ is *onto*,
that its kernel contains the images of $d, ada, cadac,$ and $acadaca,$ and consequently that its kernel is isomorphic to $(\mathbb{Z}/2\mathbb{Z})^4$.

(ii) Show that the image of π_3 is isomorphic to $((\mathbb{Z}/2\mathbb{Z}) \wr (\mathbb{Z}/2\mathbb{Z})) \wr (\mathbb{Z}/2\mathbb{Z})$.

(iii) Denote by $\chi_j : Sym(2^j) \longrightarrow \{1, -1\}$ the signature homomorphism ($j = 1, 2, 3$). Observe that π_2 and π_1 factor via the image of π_3, so that χ_1 and χ_2
can be viewed as defined on $\pi_3(\Gamma)$. Show that the homomorphism

$$\begin{cases} \pi_3(\Gamma) \longrightarrow (\mathbb{Z}/2\mathbb{Z}) \times (\mathbb{Z}/2\mathbb{Z}) \times (\mathbb{Z}/2\mathbb{Z}) \\ \sigma \longmapsto (\chi_1(\sigma), \chi_2(\sigma), \chi_3(\sigma)) \end{cases}$$

induces an isomorphism from the abelianization $\pi_3(\Gamma)/[\pi_3(\Gamma), \pi_3(\Gamma)]$ onto the
elementary 2-group $(\mathbb{Z}/2\mathbb{Z})^3$.

(iv) With the notation of Proposition 14, show that the homomorphism

$$\begin{cases} St_\Gamma(1) \longrightarrow (\mathbb{Z}/2\mathbb{Z}) \times (\mathbb{Z}/2\mathbb{Z}) \times (\mathbb{Z}/2\mathbb{Z}) \times (\mathbb{Z}/2\mathbb{Z}) \\ \gamma \longmapsto (\chi_2\phi_0(\gamma), \chi_3\phi_0(\gamma), \chi_2\phi_1(\gamma), \chi_3\phi_1(\gamma)) \end{cases}$$

is onto [hint: compute the images of b, c, aba, aca], and induces an isomorphism
from the abelianization $St_\Gamma(1)/[St_\Gamma(1), St_\Gamma(1)]$ onto $(\mathbb{Z}/2\mathbb{Z})^4$.

It follows that the minimal number of generators of $St_\Gamma(1)$ is four.

(v) Using the result quoted in Remark 24, deduce from (iii) that $St_\Gamma(1)$ is a
characteristic subgroup of Γ.

(vi — Complement) In unpublished work, V. Nekrashevych, and R. Grigorchuk and S. Sidki show the following result: the natural map from the
normalizer of Γ in $Aut(T)$ to the group of automorphisms of Γ is in fact onto;
this implies that $St_\Gamma(k)$ is a characteristic subgroup of Γ for any $k \geq 1$.

76. Exercise. Check that

$$K < B < \langle b, ac \rangle \cap St_\Gamma(1) < \langle b, ac \rangle < \Gamma$$

constitute a matriochka where each group is of index 2 in the next one.

[Indication: see Propositions 23, 25, 30, as well as Section 4.4 of [KarMe–85] for the term "matriochka".]

77. Exercise. Check that, with the notation of Item 26, the subgroup $(B \times B) \rtimes D^{diag}$ is *not* normal in $\Gamma \times \Gamma$.

[Hint. If it were, as $(a, d) = \psi(c) \in Im(\psi) = (B \times B) \rtimes D^{diag}$, one would have $(1, a)(a, d)(1, a)(a, d) = (1, (ad)^2) \in (B \times B) \rtimes D^{diag}$, which is impossible by Theorem 28.]

78. Exercise. (i) Show that the smallest normal subgroup D of Γ containing d is isomorphic to $B \times B$, and of index 16 in Γ.

(ii) Show that the smallest normal subgroup C of Γ containing c is of index 8 in Γ.

[Hint. (i) As $d = (1, b)$ and $ada = (b, 1)$, we have $D \approx B \times B$. For the index, observe that $[St_\Gamma(1) : D] = [Im(\psi) : B \times B]$, and use the result of Theorem 28. For (ii), the argument is analogous.]

79. Complement. Z. Šuni'k has checked that the smallest normal subgroup Γ containing a is the kernel of the composition

$$\Gamma \longrightarrow \{(a', b', c', d') \in \mathbb{F}_2^4 \mid b' + c' + d' = 0\} \longrightarrow \{(b', c', d') \in \mathbb{F}_2^3 \mid b' + c' + d' = 0\}$$

(see the remark which follows Proposition 22) and is the subgroup of Γ generated by a, bab, cac, dad.

80. Exercise. The purpose of this exercise is to identify $\psi(B)$ and $\psi(K)$; see [BarGr–b].

(i) Show that $\phi_j(B)$ is normal in Γ for $j \in \{0, 1\}$.

(ii) Show that $\psi(B)$ contains $K \times K$. [Hint: compute $\psi(v)$ and $\psi(w)$, where v and w are as in Item 25.]

(iii) Show that the index $[\psi(B) : (K \times K)]$ is 8. [Hint: use the value $[\Gamma : K] = 16$.]

(iv) Let C^{diag} denote the dihedral subgroup of $\Gamma \times \Gamma$ of order 16 generated by $\psi(b) = (a, c)$ and $\psi(aba) = (c, a)$. Show that $(K \times K) \cap C^{diag}$ is a group of order 2 which contains $((ac)^8, (ac)^8)$ and that $\psi(B)$ is generated by $K \times K$ and C^{diag}.

(v) Similarly it can be shown that $\psi(K)$ is generated by $K \times K$, which is of index 4 in $\psi(K)$, and the element $\psi(abab) = (ac, ca)$, which is of order 8 and which has its fourth power in $K \times K$.

81. Exercise. The purpose of this exercise, also due to L. Bartholdi, is to identify the inverse image of $K \times K$ by ψ^{-1}; notation is as in Proposition 30.

(i) Check that

$$
\begin{aligned}
\psi(v) &= (t, 1) & \psi(w) &= (1, t) \\
\psi\left(v^{-1}t^{-1}vt\right) &= (v, 1) & \psi\left(w^{-1}twt^{-1}\right) &= (1, v) \\
\psi\left(vtv^{-1}t^{-1}\right) &= (w, t) & \psi\left(wt^{-1}w^{-1}t\right) &= (1, w)
\end{aligned}
$$

so that $\psi^{-1}(K \times K)$ is generated by

$$
v, \quad tvt^{-1}, \quad t^{-1}vt, \quad w, \quad twt^{-1}, \quad t^{-1}wt.
$$

Thus $\psi^{-1}(K \times K)$ coincides with the normal subgroup of K generated by v and w, denoted below by $\langle v, w \rangle^K$.

(ii) Show that $\langle v, w \rangle^K$ is of index 4 in K, by computing

$$
\begin{aligned}
\left[K : \langle v, w \rangle^K\right] &= \frac{1}{8}\left[St_\Gamma(1) : \langle v, w \rangle^K\right] = \frac{1}{8}\left[Im(\psi) : (K \times K)\right] \\
&= \frac{1}{64}\left[(\Gamma \times \Gamma) : (K \times K)\right].
\end{aligned}
$$

82. Exercise. Consider an integer $k \geq 4$. The purpose of this exercise is to describe the kernel Ker_k of the natural homomorphism

$$
\Gamma_k = \Gamma/St_\Gamma(k) \longrightarrow \Gamma_{k-1} = \Gamma/St_\Gamma(k-1).
$$

See Items 40 and 41.

(i) Recall that $(ab)^2 \in K < St_\Gamma(1)$. As $\psi((ab)^2) = (ca, ac)$, the element $(ab)^2$ acts on $T(2)$ as (a, a). Using the fact that the image of $\psi_{k-2} : St_\Gamma(k-2) \longrightarrow \Gamma^{2^{k-2}}$ contains $K^{2^{k-2}}$, show that for any $(i_1, \ldots, i_{k-2}) \in L(k-2)$, the automorphism $\gamma_{(i_1, \ldots, i_{k-2})}$ of $T(k)$ defined by

$$
\gamma_{(i_1, \ldots, i_{k-2})}(j_1, \ldots, j_k) = \begin{cases} (j_1, \ldots, j_{k-2}, j_{k-1}, \bar{j}_k) & \text{if } (j_1, \ldots, j_{k-2}) = \\ & \qquad (i_1, \ldots, i_{k-2}) \\ (j_1, \ldots, j_{k-2}, j_{k-1}, j_k) & \text{otherwise} \end{cases}
$$

is in Γ_k. Check that $\left(\gamma_{(i_1, \ldots, i_{k-2})}\right)_{(i_1, \ldots, i_{k-2}) \in L(k-2)}$ generate a subgroup of Ker_k, say A_k, isomorphic to $(\mathbb{Z}/2\mathbb{Z})^{2^{k-2}}$.

(ii) Recall also that $(ac)^4 \in K \cap St_\Gamma(3)$. As $\psi_3((ac)^4) = (a, c, a, c, a, c, a, c)$, the element $(ac)^4$ acts on $T(4)$ as $(a, 1, a, 1, a, 1, a, 1)$. Show that this provides as in (i) a subgroup of Ker_k, say B_k, isomorphic to $(\mathbb{Z}/2\mathbb{Z})^{2^{k-4}}$.

(iii) Check that any element of A_k commutes with any element of B_k, that $A_k \oplus B_k$ is a subgroup of Ker_k of order $2^{5 \cdot 2^{k-4}}$, and deduce from Proposition 41 that it coincides with Ker_k itself.

More economically, we know from Proposition 41 that $|Ker_k| = 2^{5 \cdot 2^{k-4}}$. As $x^2 = 1$ for any $x \in Ker_k$, this shows that Ker_k is an elementary abelian 2-group of the indicated order.

83. Exercise. For $k \geq 0$, set $a_{(k)} = (1, \ldots, 1, a)_k \in \Gamma^{2^k}$. Let G_k be the subgroup of $Aut\left(T^{(2)}\right)$ generated by $a_{(0)}, \ldots, a_{(k-1)}$. Check that G_k is an iterated wreath product

$$(\mathbb{Z}/2\mathbb{Z})^{\wr^k} = \left(\ldots \left((\mathbb{Z}/2\mathbb{Z}) \wr (\mathbb{Z}/2\mathbb{Z}) \right) \wr \ldots \right) \wr (\mathbb{Z}/2\mathbb{Z})$$

of k copies of $\mathbb{Z}/2\mathbb{Z}$, and also the full automorphism group of the finite tree $T^{(2)}(k)$ spanned in $T^{(2)}$ by the vertices of level at most k.

84. Exercise. The purpose of this exercise is to show that, for any $k \geq 0$, the iterated wreath product $(\mathbb{Z}/2\mathbb{Z})^{\wr^k}$ is also a subgroup of Γ. It then follows from Proposition 4.iii that

any finite 2-group is isomorphic to a subgroup of Γ.

The strategy is to find elements $\kappa_{(0)}, \ldots, \kappa_{(k-1)}$ sharing some properties with the $a_{(j)}$'s of the previous exercise. The group K of Proposition 30 plays an essential role.

(i) Set $\kappa_{(0)} = (ab)^8 \in K$ and $u = (0,0,0,0) \in L(4)$. Check that $\kappa_{(0)}$ is of order 2, that it fixes the vertex u, and that $\delta_u^{-1} \kappa_{(0)} \delta_u = a$ if δ_u is, as in Item 1, the isomorphism from the tree T to the subtree T_u.

(ii) Using Proposition 30, define inductively $\kappa_{(k)} \in K$ for $k \geq 1$ by

$$\psi_4 \left(\kappa_{(k)} \right) = \left(\kappa_{(k-1)}, 1, \ldots, 1 \right)_4 \in K^{16}.$$

If F_k denotes the subgroup of Γ generated by $\kappa_{(0)}, \ldots, \kappa_{(k-1)}$, check that

$$F_k = F_{k-1} \wr (\mathbb{Z}/2\mathbb{Z})$$

so that F_k is indeed isomorphic to $(\mathbb{Z}/2\mathbb{Z})^{\wr^k}$.

(iii) Let us quote a result of D. Held [Held–66]: an infinite 2-group contains infinite abelian groups.

85. Problem. As Γ is a quotient of the free group of rank 3, it is a fortiori a quotient of the fundamental group of a surface of genus $g \geq 3$.

Is Γ a quotient of the fundamental group of a surface of genus 2 ?

86. Problem. I believe that the automorphism group of the Grigorchuk group is *not* finitely generated. Prove this.

87. Project. For p an *odd* prime, Passman and Temple have obtained results about finite-dimensional representations of p-groups analogous in a sense to the Grigorchuk group [PasTe–96]. Work out the theory for $p = 2$, and in particular for the Grigorchuk group.

> The gods have imposed upon my writing the yoke
> of a foreign tongue that was not sung at my cradle.
> *"Was dies heissen will, weiss jeder,*
> *Der im Traum pferdlos geritten."*
> (Preface of [Weyl–39], and Gottfried Keller)

REFERENCES

Numbers in brackets {} indicate the item(s) in the book where the reference occurs

ABCKT-96. J. Amorós, M. Burger, K. Corlette, D. Kotschick, and D. Toledo, *Fundamental groups of compact Kähler manifolds*, Mathematical Surveys and Monographs **44**, Amer. Math. Soc., 1996 {V.32}.

Abels-86. H. Abels, *Finite presentability of S-arithmetic groups*, in "Proceedings of Groups — St Andrews 1985", E.F. Robertson and C.M. Campbell, Editors, Cambridge Univ. Press (1986), 128–134 {V.20.viii}.

Abels-91. H. Abels, *Finiteness properties of certain arithmetic groups in the function field case*, Israel J. Math. **76** (1991), 113–128 {V.21}.

A'CaB-94. N. A'Campo and M. Burger, *Réseaux arithmétiques et commensurateur d'après G.A. Margulis*, Inventiones Math. **116** (1994), 1–25 {V.20.viii}.

AdeGM-91. S.A. Adeleke, A.M.W. Glass, and L. Morley, *Arithmetic permutations*, J. London Math. Soc. **43** (1991), 255–268 {II.40}.

AdVSr-57. G.M. Adel'son-Vel'skii and Yu. A. Sreider, *The Banach mean on groups*, Uspehi Mat. Nauk. (N.S.) **12**^{6} (1957), 131–136 [Russian original: Uspehi Mat. Nauk (N.S.) 12 1957 no. 6(78) pp. 131–136] {VI.64, VII.34}.

Adian-75. S.I. Adian, *The Burnside problem and identities in groups*, Ergebnisse der Mathematik und ihrer Grenzgebiete, Band 95, Springer, 1979 [Russian original: Nauka, 1975] {VI.20, VII.1, VIII.19}.

Adian-83. S.I. Adian, *Random walks on free periodic groups*, Math. USSR Izvestiya **21**^{3} (1983), 425–434 {VII.35}.

AdiMe-92. S.I. Adian and J. Mennicke, *On bounded generation of $SL(n, \mathbb{Z})$*, Internat. J. Algebra Comput. **2** (1992), 357–365 {III.2}.

Ahlfo-53. L.V. Ahlfors, *Complex analysis*, third edition, McGraw-Hill, 1979 (first published 1953) {I.8}.

AleKi-75. D.V. Alekseevskii and B.N. Kimel'fel'd, *Structure of homogeneous Riemann spaces with zero curvature*, Functional Anal. Appl. **9** (1975), 97–102 {VII.31}.

Alexa-19. J.W. Alexander, *Note on two three-dimensional manifolds with the same group*, Trans. Amer. Math. Soc. **20** (1919), 339–342 {V.27}.

AloBr-95. J.M. Alonso and M.R. Bridson, *Semihyperbolic groups*, Proc. London Math. Soc. **70** (1995), 56–114 {IV.48, V.60}.

Alons-91. J.M. Alonso, *Growth functions of amalgams*, in "Arboreal group theory", R.C. Alperin, Editor, M.S.R.I. Publ. **19**, Springer (1991), 1–34 {VI.8}.

AlpPe-90. R.C. Alperin and B.L. Peterson, *Growth functions for residually torsion-free nilpotent groups*, Proc. Amer. Math. Soc. **109** (1990), 585–587 {VII.39}.

Ancon-90. A. Ancona, *Théorie du potentiel sur les graphes et les variétés*, in "Ecole d'été de probabilités de Saint-Flour XVIII — 1988", P.L. Hennequin, Editor, Lecture Notes in Mathematics **1427**, Springer (1990), 1–112 {I.4}.

AndRV-92. S. Andreadakis, E. Raptis, and D. Varsos, *Hopficity of certain HNN-extensions*, Comm. in Algebra **20** (1992), 1511–1533 {III.21}.

Anton–88.   A. Antonevich, *Linear functional equations. Operator approach*, Birkhäuser, 1996 [Russian original 1988]   {VI.66}.

ArnKr–63.   V. Arnold and A. Krylov, *Uniform distribution of points on a sphere and some ergodic properties of solutions of linear differential equations in a complex region*, Dokl. Akad. Nauk. SSSR **148** (1963), 9–12   {II.41}.

Artin–57.   E. Artin, *Geometric algebra*, Interscience, 1957   {IV.27}.

Arzha–98.   G.N. Arzhantseva, *Generic properties of finitely presented groups and Howson's theorem*, Comm. in Algebra **26** (1998), 3783–3792   {IV.49}.

Arzha.   G.N. Arzhantseva, *On quasiconvex subgroups of word hyperbolic groups*, preprint, University of Geneva, 1999   {IV.49}.

ArzOl–96.   G.N. Arzhantseva and A.Yu. Ol'shanskii, *The class of groups all of whose subgroups with lesser number of generators are free is generic*, Math. Notes **59** (1996), 350–355   {II.43, VII.13}.

AscGu–84.   M. Aschbacher and R. Guralnick, *Some applications of the first cohomology group*, J. of Algebra **90** (1984), 446–460   {III.46}.

Ash–77.   A. Ash, *Deformation retracts with lowest possible dimension of arithmetic quotients of self-adjoint homogeneous cones*, Math. Ann. **225** (1977), 69–76 {V.9}.

Ausla–69.   L. Auslander, *The automorphism group of a polycyclic group*, Annals of Math. **89** (1969), 314–322   {V.22}.

Avez–70.   A. Avez, *Variétés riemanniennes sans points focaux*, C.R. Acad. Sci. Paris, Sér. A **270** (1970), 188–191   {VII.6}.

Avez–72.   A. Avez, *Entropie des groupes de type fini*, C.R. Acad. Sc. Paris, Sér. A, **275** (1972), 1363–1366   {VI.53}.

Avez–76.   A. Avez, *Croissance des groupes de type fini et fonctions harmoniques*, in "Théorie ergodique, Rennes 1973/1974", J.P. Conze and M.S. Keane, Editors, Lecture Notes in Mathematics **532**, Springer (1976), 35–49   {VI.53}.

Bachm–85.   S. Bachmuth, *Automorphisms of solvable groups, Part I*, in "Proceedings of Groups — St Andrews 1985", E.F. Robertson and C.M. Campbell, Editors, Cambridge Univ. Presss (1986), 1–14   {III.4}.

BacHN–97.   R. Bacher, P. de la Harpe, and T. Nagnibeda, *The lattice of integral flows and the lattice of integral cuts on a finite graph*, Bull. Soc. Math. France **125** (1997), 167–198   {VI.10}.

BacHV–97.   R. Bacher, P. de la Harpe, and B. Venkov, *Séries de croissance et séries d'Ehrhart associées aux réseaux de racines*, C.R. Acad. Sci. Paris **325** (1997), 1137–1142   {VI.2}.

BacHV–99.   R. Bacher, P. de la Harpe, and B. Venkov, *Séries de croissance et polynômes d'Ehrhart associées aux réseaux de racines*, Ann. Inst. Fourier **49** (1999), 727–762   {VI.2}.

BacMo–85.   S. Bachmuth and H.Y. Mochizuki, *Automorphisms of solvable groups, Part II*, in "Proceedings of Groups — St Andrews 1985", E.F. Robertson and C.M. Campbell, Editors, Cambridge Univ. Presss (1986), 15–29   {III.4}.

BacVd.   R. Bacher and A.A. Vdovina, *Counting some cellular decompositions of oriented surfaces*, preprint, University of Grenoble, 1998   {V.46}.

BaiPr–97.   Y.G. Baik and S.J. Pride, *On the efficiency of Coxeter groups*, Bull. London Math. Soc. **29** (1997), 32–36   {V.31}.

BalGS–85.   W. Ballmann, M. Gromov, and V. Schroeder, *Manifolds of nonpositive curvature*, Birkhäuser, 1985   {V.20, V.58}.

Ballm–95.   W. Ballmann, *Lectures on spaces of nonpositive curvature (DMV Seminar, Band 25)*, Birkhäuser, 1995   {IV.18}.

Bambe.   J. Bamberg, *Non-free points for groups generated by a pair of 2×2 matrices*, preprint, La Trobe University, 1999   {II.25}.

Barbo–90.   T. Barbot, *Extensions de $\mathbb{Z} \oplus \mathbb{Z}$ par $\mathbb{Z}$*, Mémoire de D.E.A., Ecole Normale Supérieure, Lyon, octobre 1990   {IV.29–30}.

BarCe–a.   L. Bartholdi and T. Ceccherini-Silberstein, *Growth series and random walks on some hyperbolic graphs*, preprint, University of Geneva, 1998   {VI.8, VII.39}.

BarCe–b. L. Bartholdi and T. Ceccherini-Silberstein, *Growth series of some hyperbolic graphs and Salem numbers*, preprint, University of Geneva, 1999   {VII.11}.

BarGh–92. J. Barge and E. Ghys, *Cocycles d'Euler et de Maslov*, Math. Ann. **294** (1992), 235–265   {IV.48}.

BarGr–a. L. Bartholdi and R. Grigorchuk, *Lie methods in growth of groups and groups of finite width*, in "Proceedings of CGAMA Conference at Edinburgh, 1998", N. Gilbert, Editor, London Math. Soc. Lecture Note Ser., Cambridge Univ. Press, to appear   {VII.14, VII.29, VII.39, VIII.31}.

BarGr–b. L. Bartholdi and R. Grigorchuk, *On the parabolic subgroups and growth of certain fractal groups*, preprint, University of Geneva, Spring 1999
{VIII.39, VIII.80}.

BarGr–c. L. Bartholdi and R. Grigorchuk, *On the spectrum of Hecke operators related to some fractal groups*, preprint, University of Geneva, Spring 1999
{VIII.1, VIII.10, VIII.31}.

BarnS–97. Y. Barnea and A. Shalev, *Hausdorff dimension, pro-p-groups, and Kac-Moody algebras*, Trans. Amer. Math. Soc. **349** (1997), 5073–5091   {VIII.41}.

Barth–98. L. Bartholdi, *The growth of Grigorchuk's torsion group*, Internat. Math. Research Notices **20** (1998), 1049–1054   {VI.42, VIII.66}.

Barth–99. L. Bartholdi, *Counting paths in graphs*, l'Enseignement math. **45** (1999), 83–131   {I.11, VII.37}.

Barth. L. Bartholdi, *Lower bounds on the growth of Grigorchuk's torsion group*, Internat. J. Algebra Comput., to appear   {VIII.66}.

BasLu–83. H. Bass and A. Lubotzky, *Automorphisms of groups and of schemes of finite type*, Israel J. Math. **44** (1993), 1–22   {III.18}.

BasLu–94. H. Bass and A. Lubotzky, *Linear - central filtrations on groups*, in "The mathematical legacy of Wilhelm Magnus — Groups, geometry and special functions", W. Abikoff, J.S. Birman, and K. Kuiken, Editors, Contemporary Mathematics **169** (1994), 45–98   {II.42}.

BasMS–67. H. Bass, J. Milnor, and J-P. Serre, *Solution of the congruence subgroup problem for* SL_n *and* Sp_{2n}, Publ. Math. I.H.E.S. **33** (1967), 59–137   {III.2, III.D}.

Bass–72. H. Bass, *The degree of polynomial growth of finitely generated nilpotent groups*, Proc. London Math. Soc. **25** (1972), 603–614   {VII.26}.

Bass–84. H. Bass, *Finitely generated subgroups of* GL_2, in "The Smith conjecture", J.W. Morgan and H. Bass, Editors, Academic Press (1984), 127–136   {III.20}.

Bass–93. H. Bass, *Covering theory for graphs of groups*, J. of Pure and Appl. Algebra **89** (1993), 3–47   {II.18, III.14}.

Bass4–96. H. Bass, M.V. Otero-Espinar, D.N. Rockmore, and C.P.L. Tresser, *Cyclic renormalization and automorphism groups of rooted trees*, Lecture Notes in Mathematics **1621**, Springer, 1996   {VIII.1, VIII.4, VIII.7}.

BauMi–92. G. Baumslag and C.F. Miller III (Editors), *Algorithms and classification in combinatorial group theory*, M.S.R.I. Publ. **23**, Springer, 1992   {V.26}.

Baums–61. G. Baumslag, *Wreath products and finitely presented groups*, Math. Zeitschr. **75** (1961), 22–28   {IV.44}.

Baums–93. G. Baumslag, *Topics in combinatorial group theory*, Birkhäuser, 1993 {VIII.1}.

Baums–99. G. Baumslag, *Some open problems*, in "Summer school in group theory in Banff, 1996", O. Kharlampovich, Editor, CRM Proceedings and Lecture Notes **17** (1999), 1–9   {III.21}.

BauMS–99. G. Baumslag, A.G.Myasnikov and V. Shpilrain, *Open problems in combinatorial group theory*, in "Groups, languages and geometry", R.H. Gilman, Editor, Contemporary Math. **250** (1999), 1–27   {Index of research problems}.

BauPr–78. B. Baumslag and S.J. Pride, *Groups with two more generators than relations*, J. London Math. Soc. **17** (1978), 435–436   {V.31}.

BauRo–84. G. Baumslag and J.E. Roseblade, *Subgroups of direct products of free groups*, J. London Math. Soc. **30** (1984), 44–52   {V.2}.

BauSh–90. G. Baumslag and P.B. Shalen, *Amalgamated products and finitely presented groups*, Comment. Math. Helvetici **65** (1990), 243–254   {Introduction, V.31, VII.9}.

BauSo–62. G. Baumslag and D. Solitar, *Some two-generator one-relator non-Hopfian groups*, Bull. Amer. Math. Soc. **68** (1962), 199–201   {III.21, VI.20}.

Bavar–91. C. Bavard, *Longueur stable des commutateurs*, l'Enseignement math. **37** (1991), 109–150   {IV.3.i}.

Bavar–93. C. Bavard, *La surface de Klein*, Le journal de maths des élèves de l'E.N.S. de Lyon **1** (1993), 13–22   {V.36}.

Bavar. C. Bavard, *Classes minimales de réseaux et rétractions géométriques équivariantes dans les espaces symétriques*, preprint, University of Bordeaux-I, February 1999   {V.9}.

Beard–83. A.F. Beardon, *The geometry of discrete groups*, Graduate Texts in Mathematics **91**, Springer, 1983   {III.25, V.40}.

BehNe–81. G. Behrendt and P.M. Neumann, *On the number of normal subgroups of an infinite group*, J. London Math. Soc. **23** (1981), 429–432   {III.37}.

Behr–79. H. Behr, $SL_3(\mathbb{F}_q[T])$ *is not finitely presentable*, in "Homological group theory, Durham 1977", C.T.C. Wall, Editor, Cambridge Univ. Press (1979), 213–224   {V.21}.

BekCH–94. M. Bekka, M. Cowling, and P. de la Harpe, *Some groups whose reduced C^*-algebra is simple*, Publ. Math. I.H.E.S. **80** (1994), 117–134   {II.43}.

BekMa. B. Bekka and M. Mayer, *Ergodic theory and topological dynamics of group actions on homogeneous spaces*, Lecture notes from a Summer School in Tuczno, 1994, to appear   {I.2}.

Benne–97. C.D. Bennet, *Explicit free subgroups of $Aut(\mathbb{R}, \leq)$*, Proc. Amer. Math. Soc. **125** (1997), 1305–1308   {II.29}.

Benoi–96. Y. Benoist, *Actions propres sur les espaces homogènes réductifs*, Annals of Math. **144** (1996), 315–347   {II.27}.

Benoi–97. Y. Benoist, *Propriétés asymptotiques des groupes linéaires*, GAFA Geom. Funct. Anal. **7** (1997), 1–47   {II.27}.

Benso–87. M. Benson, *On the rational growth of virtually nilpotent groups*, in "Combinatorial group theory and topology", S.M. Gersten and J.R. Stallings, Editors, Ann. Math. Studies **111**, Princeton Univ. Press (1987), 185–196 {VI.2, VI.6}.

BesBr–97. M. Bestvina and N. Brady, *Morse theory and finiteness properties of groups*, Inventiones Math. **129** (1997), 445–470   {II.47, V.23}.

BesF–91a. M. Bestvina and M. Feighn, *Bounding the complexity of simplicial group actions on trees*, Inventiones Math. **103** (1991), 449–469   {III.15}.

BesF–91b. M. Bestvina and M. Feighn, *A counterexample to general accessibility*, in "Arboreal group theory", R.C. Alperin, Editor, M.S.R.I. Publ. **19**, Springer (1991), 133–141   {III.15}.

BesF–95. M. Bestvina and M. Feighn, *Stable actions of groups on real trees*, Inventiones Math. **121** (1995), 287–321   {II.18}.

BesFH–a. M. Bestvina, M. Feighn and M. Handel, *The Tits alternative for $Out(F_n)$ I: Dynamics of exponentially growing automorphisms*, preprint, revised version, January 1999   {II.42}.

BesFH–b. M. Bestvina, M. Feighn and M. Handel, *The Tits alternative for $Out(F_n)$ II: A Kolchin type theorem*, preprint, July 1996   {II.42}.

BeyTa–82. F.R. Beyl and J. Tappe, *Group extensions, representations, and the Schur Multiplicator*, Springer, Lecture Notes in Math. **958**, 1982   {V.31}.

BhaSe–97. R. Bhatia and P. Semrl, *Approximate isometries on Euclidean spaces*, American Math. Monthly **104:6** (1997), 497–504   {IV.45}.

BieNS–87. R. Bieri, W.D. Neumann, and R. Strebel, *A geometric invariant of discrete groups*, Inventiones Math. **90** (1987), 451–477   {Introduction}.

Bigel. S. Bigelow, *Braid groups are linear*, preprint, 1999   {III.20}.

Biggs–74. N.L. Biggs, *Algebraic graph theory second edition*, second edition, Cambridge Univ. Press, 1993 (first published 1974)   {III.39, IV.3.ii}.

Biggs–84. N.L. Biggs, *Presentations for cubic graphs*, in "Computational group theory", M.D. Atkinson, Editor, Academic Press (1984), 57–63 {III.39}.

Biggs–89. N.L. Biggs, *A proof of Serre's theorem*, Discrete Math. **78** (1989), 53–57 {II.18}.

Biggs–99. N.L. Biggs, *The Tutte polynomial as a growth function*, J. Algebraic Combinatorics **10** (1999), 115–133 {VI.10}.

BigLW–76. N.L. Biggs, E.K. Lloyd, and R.J. Wilson, *Graph theory 1736–1936*, Clarendon Press, 1976 {IV.11, IV.15}.

Billi–85. N. Billington, *Growth of groups and graded algebras: erratum*, Comm. in Algebra **13** (1985), 753–755 {VI.2}.

Birma–98. J.S. Birman, *Review of "Braid groups are linear groups"*, Math. Rev., 98h:20061 {III.20}.

Black–98. S. Black, *Asymptotic growth of finite groups*, J. of Algebra **209** (1998), 402–426 {VII.40}.

BloWe–97. J. Block and S. Weinberger, *Large scale homology theories and geometry*, in "Geometric topology, 1993 Georgia International Topology Conference", W.H. Kazez, Editor, Amer. Math. Soc. (1997), 522–569 {V.23}.

BogDy–95. W.A. Bogley and M.N. Dyer, *A group-theoretic reduction of J.H.C. Whitehead's asphericity question*, in "Groups — Korea '94", A.C. Kim and D.L. Johnson, Editors, de Gruyter (1995), 9–14 {II.47}.

Bogle–93. W.A. Bogley, *J.H.C. Whitehead's asphericity question*, in "Two-dimensional homotopy and combinatorial group theory", London Math. Soc. Lecture Note Ser., Cambridge Univ. Press **197** (1993), 309–334 {II.47}.

Bogop. O.V. Bogopolski, *Infinite commensurable hyperbolic groups are bi-Lipschitz equivalent*, Preprint, Bochum and Novosibirsk, August 1996 {IV.14, IV.46}.

Bollo–79. B. Bollobás, *Graph theory, an introductory course*, Graduate Texts in Mathematics **63**, Springer, 1979 {IV.46}.

BonSc. M. Bonk and O. Schramm, *Embeddings of Gromov hyperbolic spaces*, GAFA Geom. Funct. Anal., to appear {IV.45}.

BooHi–74. W.W. Boone and G. Higman, *An algebraic characterization of groups with soluble word problem*, J. Austral Math. Soc. **18** (74), 41–53 {V.26}.

Borel–63. A. Borel, *Compact Clifford-Klein forms of symmetric spaces*, Topology **2** (1963), 111–122 [= Collected Papers II, pages 344–355] {IV.25.ix}.

Borel–69. A. Borel, *Introduction aux groupes arithmétiques*, Hermann, 1969 {III.2, V.20.iii}.

Borel–81. A. Borel, *Commensurability classes and volumes of hyperbolic 3-manifolds*, Ann. Sc. Norm. Super., Cl. Sci. **8** (1981), 1–33 [= Collected Papers III, pages 617–649] {IV.29}.

Bost–92. J.-B. Bost, *Introduction to compact Riemann surfaces, Jacobians and Abelian varieties*, in "From number theory to physics", M. Waldschmidt, P. Moussa, J.-M. Luck, and C. Itzykson, Editors, Springer (1960), 64–211 {V.36}.

Bourb–60. N. Bourbaki, *Topologie générale, chapitre 3, groupes topologiques (théorie élémentaire), chapitre 4, nombres réels*, Hermann, 1960 {IV.22}.

Bourb–68. N. Bourbaki, *Groupes et algèbres de Lie, chapitres 4, 5 et 6*, Hermann, 1968 {V.35, VI.10}.

Bourb–70. N. Bourbaki, *Algèbre, chapitres 1 à 3*, Bourbaki, 1970 {II.16–17, III.53}.

Bourb–72. N. Bourbaki, *Groupes et algèbres de Lie, chapitres 2 et 3*, Hermann, 1972 {III.20, III.37, VII.26, VIII.31}.

Bourb–75. N. Bourbaki, *Groupes et algèbres de Lie, chapitres 7 et 8*, Hermann, 1975 {II.41}.

Bourd–95. M. Bourdon, *Structure conforme au bord et flot géodésique d'un CAT(-1)-espace*, l'Enseignement math. **41** (1995), 63–102 {IV.26}.

Bowdi–91. B.H. Bowditch, *Notes on Gromov's hyperbolicity criterion for path-metric spaces*, in "Group theory from a geometrical viewpoint (26 March — 6 April 1990, Trieste)", E. Ghys, A. Haefliger and A. Verjovsky, Editors, World Scientific (1991), 64–167 {V.60}.

Bowdi–95. B.H. Bowditch, *A short proof that a subquadratic isoperimetric inequality implies a linear one*, Michigan Math. J. **42** (1995), 103–107   {V.57}.

Bowdi–98. B.H. Bowditch, *Continuously many quasiisometry classes of 2-generator groups*, Comment. Math. Helvetici **73** (1998), 232–236   {III.45}.

Bowen–78. R. Bowen, *Entropy and the fundamental group*, in "The structure of attractors in dynamical systems (North Dakota State University, June 1977)", Springer Lecture Notes in Mathematics **668** (1978), 21–29   {VI.53.vi}.

Bozej–80. M. Bozejko, *Uniformly amenable groups*, Math. Ann. **251** (1980), 1–6 {VII.40}.

Brady–99. N. Brady, *Branched coverings of cubical complexes and subgroups of hyperbolic group*, J. London Math. Soc. **60** (1999), 449–460   {V.55}.

Brids–99. M.R. Bridson, *Non-positive curvature in group theory*, in "Groups St Andrews 1997 in Bath, I", C.M. Campbell, E.F. Robertson, N. Ruskuc, and G.C. Smith, Editors, Cambridge Univ. Press (1999), 124–175   {V.55}.

BriGe–96. M. Bridson and S.M. Gersten, *The optimal isoperimetric inequality for torus bundles over the circle*, Quart J. Pure Appl. Math. **47** (1996), 1–23
{IV.29–30, IV.50}.

BriHa–99. M. Bridson and A. Haefliger, *Metric spaces of non-positive curvature*, Die Grundlehren der mathematischen Wissenschaften, Band 319, Springer, 1999
{V.54}.

Brin–96. M.G. Brin, *The chameleon groups of Richard J. Thompson: automorphisms and dynamics*, Publ. Math. I.H.E.S. **84** (1996), 5–33   {V.26}.

Brin–99. M.G. Brin, *The ubiquity of Thompson's group F in groups of piecewise linear homeomorphisms of the unit interval*, J. London Math. Soc. **60** (1999), 449–460
{V.26}.

Brion–96. M. Brion, *Polytopes convexes entiers*, Gazette des mathématiciens **67** (janvier 1996), 21–42   {VI.2}.

BriSq–85. M.G. Brin and C.C. Squier, *Groups of piecewise linear homeomorphisms of the real line*, Inventiones math. **79** (1979), 485–498   {II.48}.

BriWi–99. M.R. Bridson and D.T. Wise, $\mathcal{V}\mathcal{H}$ *complexes, towers and subgroups of* $F \times F$, Math. Proc. Camb. Phil. Soc. **126** (1999), 481–497   {V.2}.

Brown–82. K.S. Brown, *Cohomology of groups*, Graduate Texts in Mathematics **87**, Springer, 1982   {V.23}.

Brown–84. K.S. Brown, *Presentations for groups acting on simply-connected complexes*, J. of Pure and Appl. Algebra **32** (1984), 1–10   {IV.23}.

Brown–92. K.S. Brown, *The geometry of rewriting systems: a proof of the Anick-Groves-Squier theorem*, in "Algorithms and classification in combinatorial group theory", G. Baumslag and C.F. Miller III, Editors, M.S.R.I. Publ. **23**, Springer (1992), 137–163   {Introduction}.

BruBW–79. A.M. Brunner, R.G. Burns, and J. Wiegold, *On the number of quotients, of one kind or another, of the modular group*, Math. Scientist **4** (1979), 93–98
{III.37–38, III.42}.

BruSi. A.M. Brunner and S. Sidki, $GL(n, \mathbb{Z})$ *as a group of finite state automata*, J. of Algebra, to appear   {VIII.7}.

BucHa. M. Bucher and P. de la Harpe, *Free products with amalgamation and HNN-extensions which are of uniformly exponential growth*, Math. Notes, to appear
{VII.18}.

BunNi–51. S. Bundgaard and J. Nielsen, *On normal subgroups with finite index in F-groups*, Mat. Tidsskr. B. **1951** (1951), 56–58 [= J. Nielsen: Collected mathematical papers, volume 2 (1932–1955), pages 320–331]   {V.20}.

Buril–99. J. Burillo, *Quasi-isometrically embedded subgroups of Thompson's group F*, J. of Algebra **212** (1999), 65–78   {V.26}.

BurKl–98. D. Burago and B. Kleiner, *Separated nets in Euclidean space and Jacobians of bilipschitz maps*, GAFA Geom. Funct. Anal. **8** (1998), 273–282   {IV.46}.

BurMa–93. R.G. Burns and O. Macedonska, *Balanced presentations of the trivial group*, Bull. London Math. soc. **25** (1993), 513–526   {II.45, V.6}.

BurMn–99. M. Burger and N. Monod, *Bounded cohomology of lattices in higher rank Lie groups*, J. Eur. Math. Soc. 1 (1999), 199–235 and 338 {IV.50}.

BurMz–a. M. Burger and S. Mozes, *Group acting on trees: from local to global structure*, preprint, ETHZ, 1998 {IV.9, V.26}.

BurMz–b. M. Burger and S. Mozes, *Lattices in product of trees*, preprint, ETHZ, 1999 {IV.9, V.26}.

Burns–02. W. Burnside, *On an unsettled question in the theory of discontinuous groups*, Quart J. Pure Appl. Math. **33** (1902), 230–238 {VIII.19}.

BurZi–85. G. Burde and H. Zieschang, *Knots*, de Gruyter, 1985 {V.16}.

BuxGo. K-U. Bux and C. Gonzalez, *The Bestvina-Brady construction revisited — geometric computation of Σ-invariants for right angled Artin groups*, preprint, Frankfurt, 1997 {V.23}.

Cair5–95. G. Cairns, G. Davies, D. Elton, A. Kolganova, and P. Perversi, *Chaotic group actions*, l'Enseignement math. **41** (1995), 123–133 {III.18}.

Camer–99. P. Cameron, *Permutation groups*, London Math. Soc., Student Texts **45**, Cambridge Univ. Press, 1999 {VIII.66}.

CanCo. J.W. Cannon and G.R. Conner, *The combinatorial structure of the Hawaiian earring group*, preprint, March 1999 {III.6}.

CanFP–96. J.W. Cannon, W.J. Floyd, and W.R. Parry, *Introductory notes on Richard Thompson's groups*, l'Enseignement math. **42** (1996), 215–256 {II.48, V.26, VI.20, VII.1}.

Canno–80. J.W. Cannon, *The growth of the closed surface groups and compact hyperbolic Coxeter groups*, Circulated typescript, University of Wisconsin, 1980 {VI.8–9, VI.19, VII.11}.

Canno–84. J.W. Cannon, *The combinatorial structure of cocompact discrete hyperbolic groups*, Geometriae Dedicata **16** (1984), 123–148 {VI.8}.

Canno–87. J.W. Cannon, *Almost convex groups*, Geometriae Dedicata **22** (1987), 197–210 {IV.49}.

Cann4–89. J.W. Cannon, W.J. Floyd, M.A. Grayson, and W.P. Thurston, *Solvgroups are not almost convex*, Geometriae Dedicata **31** (1989), 291–300 {IV.49}.

CanWa–92. J.W. Cannon and Ph. Wagreich, *Growth functions of surface groups*, Math. Ann. **293** (1992), 239–257 {VI.8, VII.11}.

CarKe–83. D. Carter and G. Keller, *Bounded elementary generation of $SL_n(\mathcal{O})$*, Amer. J. Math. **105** (1983), 673–687 {III.2}.

Carls–21. F. Carlson, *Über Potenzreihen mit ganzzahligen Koeffizienten*, Math. Zeit. **9** (1921), 1–13 {VIII.67}.

Casse–86. J.W.S. Cassels, *Local fields*, London Math. Soc., Student Texts **3**, Cambridge Univ. Press, 1986 {V.20.iii}.

CecGH–99. T. Ceccherini-Silberstein, R. Grigorchuk, and P. de la Harpe, *Amenability and paradoxes for pseudogroups and for discrete metric spaces*, Proc. Steklov Inst. Math. **224** (1999), 57–95 [Russian original: pp. 68–111] {Introduction, IV.3, IV.46, VI.51}.

CecGr–97. T.G. Ceccherini-Silberstein and R.I. Grigorchuk, *Amenability and growth of one-relator groups*, l'Enseignement math. **43** (1997), 337–354 {II.42, VII.8, VII.12}.

CecMS. T. Ceccherini-Silberstein, A. Machì and F. Scarabotti, *Il gruppo di Grigorchuk di crescita intermedia*, preprint, 1999. {VIII}.

ChaMa–82. B. Chandler and W. Magnus, *The history of combinatorial group theory: A case study in the history of ideas*, Springer, 1982 {Introduction, V.8, V.12}.

Champ–95. C. Champetier, *Propriétés statistiques des groupes de présentation finie*, Adv. in Math. **116** (1995), 197–262 {IV.14}.

Champ–94. C. Champetier, *Petite simplification dans les groupes hyperboliques*, Ann. Fac. Sci. Toulouse Math. (6) **3** (1994), 161–221 {V.55}.

Champ. C. Champetier, *L'espace des groupes de type fini*, Topology, to appear; preprint, 1994 {V.10}.

Chand–68. K. Chandrasekharan, *Introduction to analytic number theory*, Die Grundlehren der mathematischen Wissenschaften, Band 148, Springer, 1968   {I.A}.

Chang–97. N. Changey, *Rationalité des séries de croissance complète des groupes de Coxeter*, Mémoire de DEA, Juin 1997   {VI.18}.

Charn–95. R. Charney, *Geodesic automation and growth functions for Artin groups of finite type*, Math. Ann. **301** (1995), 307–324   {VI.20}.

CheEb–75. J. Cheeger and D.G. Ebin, *Comparison theorems in Riemannian geometry*, North Holland, 1975   {V.C, V.51, V.53}.

CheSc–98. P.A. Cherix and G. Schaeffer, *An asymptotic Freiheitssatz for finitely generated groups*, l'Enseignement math. **44** (1998), 9–22   {II.43}.

Chigi–97. N. Chigira, *Generating alternating groups*, Hokkaido Math. J. **26** (1997), 435–438   {III.33}.

Chou–80. C. Chou, *Elementary amenable groups*, Illinois J. Math. **24** (1980), 396–407   {III.35, VII.28, VIII.68}.

CohGl–97. S.D. Cohen and A.M.W. Glass, *Free groups from fields*, J. London Math. Soc. **55** (1997), 309–319   {II.40}.

ColLe–83. D.J. Collins and F. Levin, *Automorphisms and Hopficity of certain Baumslag-Solitar groups*, Arch. Math. **40** (1983), 385–400   {III.21}.

Conde–90. M. Conder, *A surprising isomorphism*, J. of Algebra **129** (1990), 494–501   {III.39}.

Conne–97. G.R. Conner, *A class of finitely generated groups with irrational translation numbers*, Arch. Math. **69** (1997), 265–274   {IV.3.viii}.

CooDP–90. M. Coornaert, T. Delzant, and A. Papadopoulos, *Géométrie et théorie des groupes — les groupes hyperboliques de Gromov*, Lecture Notes in Mathematics **1141**, Springer, 1990   {V.55–57, V.60}.

CooPa–93. M. Coornaert and A. Papadopoulos, *Symbolic dynamics and hyperbolic groups*, Lecture Notes in Mathematics **1539**, Springer, 1993   {V.58, V.60}.

Coorn–93. M. Coornaert, *Mesures de Patterson-Sullivan sur le bord d'un espace hyperbolique au sens de Gromov*, Pacific J. Math. **159** (1993), 241–270   {VII.11}.

CorGr–89. L.J. Corwin and F.P. Greenleaf, *Representations of nilpotent Lie groups and their applications. Part I: Basic theory and examples*, Cambridge Univ. Press, 1989   {VII.26}.

CouSC–93. T. Coulhon and L. Saloff-Coste, *Isopérimétrie pour les groupes et les variétés*, Rev. Mathemática Iberoamericana **9** (1993), 293–314   {VII.32}.

Coxet–48. H.S.M. Coxeter, *Regular polytopes*, 3rd edition, Dover, 1973 (first published 1948)   {V.37}.

Coxet–61. H.S.M. Coxeter, *Introduction to geometry*, second edition, John Wiley & Sons, 1969 (first published 1961)   {V.37}.

CoxMo–57. H.S.M. Coxeter and W.O.J. Moser, *Generators and relations for discrete groups*, 4th edition, Springer, 1980 (first published 1957)   {III.2, IV.3, IV.15}.

CulSh–83. M. Culler and P.B. Shalen, *Varieties of group representations and splitting of 3-manifolds*, Annals of Math. **117** (1983), 109–146   {Introduction}.

Day–57. M.M. Day, *Amenable semigroups*, Illinois J. Math. **1** (1957), 509–544   {III.35, VIII.68}.

Dehn–38. M. Dehn, *Die Gruppe der Abbildungsklassen*, Acta Math. **69** (1938), 135–206   {V.7}.

Delig–78. P. Deligne, *Extensions centrales non résiduellement finies de groupes arithmétiques*, C.R. Acad. Sci. Paris, Sér. A **287** (1978), 203–208   {III.18}.

Delsa–42. J. Delsarte, *Sur le gitter fuchsien*, C.R. Acad. Sci. Paris **214** (1942), 147–149   {I.2}.

Delz–96a. T. Delzant, *Sous-groupes distingués et quotients des groupes hyperboliques*, Duke Math. J. **83** (1996), 661–682   {IV.3.viii, V.55}.

Delz–96b. T. Delzant, *Décomposition d'un groupe en produit libre ou somme amalgamée*, J. reine angew. Math. **470** (1996), 153–180   {III.13}.

Delz–99. T. Delzant, *Sur l'accessibilité acylindrique des groupes de présentation finie*, Ann. Inst. Fourier **49** (1999), 1215–1224   {II.36, III.15}.

DeuSS-95. W.A. Deuber, M. Simonovitz, and V.T. Sós, *A note on paradoxical metric spaces*, Studia Scient. Math. Hungarica **30** (1995), 17–23 {VI.51}.

Dieud-54. J. Dieudonné, *Les isomorphismes exceptionnels entre les groupes classiques finis*, Canadian J. Math. **6** (1954), 305–315 {III.32, V.36}.

Dieud-78. J. Dieudonné, *Abrégé d'histoire des mathématiques, 1700–1900 (2 volumes)*, Hermann, 1978 {Introduction}.

Dinab-71. E.I. Dinaburg, *A connection between various entropy characteristics of dynamical systems*, Math. USSR Izvestiya **5** (1971), 337–378 {VI.53}.

Dioub. A. Dioubina, *On some properties of groups not preserved by quasi-isometry*, preprint, St.Petersburg State University, July 1999 {IV.9, IV.36, IV.44}.

DiDMS-91. J.D. Dixon, M.P.F. du Sautoy, A. Mann, and D. Segal, *Analytic pro-p groups*, second edition, Cambridge Univ. Press, 1999 (first published 1991) {III.20, VII.29}.

Dixmi-60. J. Dixmier, *Opérateurs de rang fini dans les représentations unitaires*, Publ. Math. I.H.E.S. **6** (1960), 13–25 {VII, VII.26}.

DixMo-96. J.D. Dixon and B. Mortimer, *Permutation groups*, Springer, 1996 {II.39, III.47, VIII.3}.

DouNF-82. B. Doubrovine, S. Novikov, and A. Fomenko, *Géométrie contemporaine, deuxième partie*, Editions Mir, 1982 {V.40}.

DoySn-84. P.G. Doyle and J.L. Snell, *Random walks and electric networks*, Carus Mathematical Monographs **22**, Mathematical Association of America, 1984 {I.B, I.6}.

DunJo-99. M.J. Dunwoody and J.M. Jones, *A group with very strange decomposition properties*, J. Austral. Math. Soc. (Series A) **67** (1999), 185–190. {III.14}.

DunSa-99. M.J. Dunwoody and M.E. Sageev, *JSJ-splittings for finitely presented groups*, Inventiones Math. **135** (1999), 25–44 {III.5}.

Dunwo-69. M.J. Dunwoody, *The ends of finitely generated groups*, J. of Algebra **12** (1969), 339–344 {IV.50}.

Dunwo-85. M.J. Dunwoody, *The accessibility of finitely presented groups*, Inventiones Math. **81** (1985), 449–457 {III.15}.

Dunwo-93. M.J. Dunwoody, *An inaccessible group*, in "Geometry group theory, Volume 1, Sussex 1991", G.A. Niblo and M.A. Roller, Editors, Cambridge Univ. Press (1993), 75–78 {III.15}.

DynYu-67. E.B. Dynkin and A.A. Yushkevich, *Markov Processes, theorems and problems*, Plenum Press, 1969 [Russian original, Nauka Press, 1967] {I.5}.

Eberl-96. P.B. Eberlein, *Geometry of nonpositively curved manifolds*, University of Chicago Press, 1996 {V.C}.

Eckma. B. Eckmann, *Introduction to ℓ₂-homology*, preprint, 1998 {V.31}.

Eda-92. K. Eda, *Free σ-products and noncommutative slender groups*, J. of Algebra **148** (1992), 243–263 {II.23}.

Efrem-53. V.A. Efremovic, *The proximity geometry of Riemannian manifolds*, Uspekhi Math. Nauk. **8** (1953), 189 {IV.22, VI.B, VII}.

Egoro-99. A.V. Egorov, *Finitely generated groups of unitary operators*, Russian Math. Surveys **54:3** (1999), 632–633 {III.18}.

Ehrha-67. E. Ehrhart, *Sur un problème de géométrie diophantienne linéaire I. Polyèdres et réseaux*, J. reine angew. Math. **226** (1967), 1–29 {VI.2}.

ElsGM-98. J. Elstrodt, F. Grunewald, and J. Mennicke, *Groups acting on hyperbolic space (harmonic analysis and number theory)*, Springer, 1998 {II.28, III.54, V.49}.

EpsIZ-96. D.B.A. Epstein, A.R. Iano-Fletcher, and U. Wzick, *Growth functions and automatic groups*, J. Experiment. Math. **5** (1996), 297–315 {VII.39}.

EpsPe-94. D.B.A. Epstein and C. Petronio, *An exposition of Poincaré's polyhedron theorem*, l'Enseignement math. **40** (1994), 113–170 {V.40, V.48}.

Epste-61. D.B.A. Epstein, *Finite presentations of groups and 3-manifolds*, Quart. J. Math. Oxford **12** (1961), 205–212 {V.23, V.30–31}.

Epste-62. D.B.A. Epstein, *Ends*, in "Topology of 3-manifolds and related topics", M.K. Fort, Editor, Prentice Hall (1962), 110–117 {IV.25.vi}.

Epste–71. D.B.A. Epstein, *Almost all subgroups of a Lie group are free*, J. of Algebra **19** (1971), 261–262   {II.41}.

Epst6–92. D.B.A. Epstein, J.W. Cannon, D.F. Hold, S.V.F. Levy, M.S. Paterson, and W.P. Thurston, *Word processing in groups*, Jones and Bartlett, 1992
{Introduction, V.60}.

EskFa–97. A. Eskin and B. Farb, *Quasi-flats and rigidity of higher rank symmetric spaces*, J. Amer. Math. Soc. **10** (1997), 653–692   {IV.41}.

EvaMo–72. B. Evans and L. Moser, *Solvable fundamental groups of compact 3-manifolds*, Trans. Amer. Math. Soc. **168** (1972), 189–210   {V.30}.

FalZo–99. E. Falbel and V. Zocca,, *A Poincaré's polyhedron theorem for complex hyperbolic geometry*, J. reine angew. Math. **516** (1999), 133–158   {V.40}.

Farb–94. B. Farb, *The extrinsic geometry of subgroups and the generalized word problem*, Proc. London Math. Soc. **68** (1994), 577–593   {VI.43}.

Farb–97. B. Farb, *The quasi-isometry classification of lattices in semisimple Lie groups*, Math. Research Letters **4** (1997), 705–717   {IV.42}.

FarJ–89a. T. Farell and L. Jones, *A topological analogue of Mostow's rigidity theorem*, J. Amer. Math. Soc. **2** (1989), 257–370   {V.52}.

FarJ–89b. T. Farell and L. Jones, *Negatively curved manifolds with exotic smooth structures*, J. Amer. Math. Soc. **2** (1989), 899–908   {V.52}.

FarMo–98. B. Farb and L. Mosher (with an appendix by D. Cooper), *A rigidity theorem for the solvable Baumslag-Solitar groups*, Inventiones Math. **131** (1998), 419–451
{III.21, IV.43}.

FarMo–99. B. Farb and L. Mosher, *Quasi-isometric rigidity for the solvable Baumslag-Solitar groups, II*, Inventiones Math. **137** (1999), 613–649   {III.21, IV.43}.

FarMo. B. Farb and L. Mosher, *On the asymptotic geometry of abelian-by-cyclic groups, I*, preprint, University of Chicago and Rutgers University, 1998   {IV.43}.

Fatou–06. P. Fatou, *Séries trigonométriques et séries de Taylor*, Acta Math. **30** (1906), 335–400   {VIII.67}.

Feit–95. W. Feit, *On the work of Efim Zelmanov*, Proceedings of the International Congress of Mathematicians, Zürich, 1994, 1 (Birkhäuser, 1995), 17–24 {VIII.19}.

Felle–50. W. Feller, *An introduction to probability theory and its applications, Volume I*, second edition, J. Wiley, 1957 (first published 1950)   {I.3, I.7–8, VIII.61}.

FenRo–96. R. Fenn and C. Rourke, *Klyachko's methods and the solution of equations over torsion-free groups*, l'Enseignement math. **42** (1996), 49–74   {II.46}.

FinLR–88. B. Fine, F. Levin, and G. Rosenberger, *Free subgroups and decompositions of one-relator products of cyclics. Part I: The Tits alternative*, Arch. Math. **50** (1988), 97–109   {II.42}.

Flajo–87. P. Flajolet, *Analytic models and ambiguity of context-free languages*, Theoretical Computer Sciences **49** (1987), 283–309   {I.8}.

FloPa–97. W. Floyd and W. Parry, *The growth of nonpositively curved triangles of groups*, Inventiones math. **129** (1997), 289–359   {III.14, VI.2.viii}.

Følne–55. E. Følner, *On groups with full Banach mean value*, Math. Scand. **3** (1955), 243–254   {IV.50}.

ForPr–92. E. Formanek and C. Procesi, *The automorphism group of a free group is not linear*, J. of Algebra **149** (1992), 494–499   {III.20}.

Fox–49. R.H. Fox, *A remarkable simple closed curve*, Annals of Math. **50** (1949), 264–265   {V.18}.

Fox–52. R.H. Fox, *On Fenchel's conjecture about F-groups*, Mat. Tidsskr. B. **1952** (1952), 61–65   {V.20}.

Frank–70. J. Franks, *Anosov diffeomorphisms*, in "Global Analysis, Berkeley, 1968", Proc. Symp. Pure Math. **14** (Amer. Math. Soc. 1970), 61–93   {VII.23}.

Fried–70. N.A. Friedman, *Introduction to ergodic theory*, Van Nostrand, 1970 {VIII.7}.

Furm–99a. A. Furman, *Gromov's measure equivalence and rigidity of higher rank lattices*, Annals of Math. **150** (1999), 1059–1081   {IV.27, IV.35}.

Furm–99b. A. Furman, *Orbit equivalence rigidity*, Annals of Math. **150** (1999), 1083–1108
{IV.27, IV.35}.

GabLP–94. D. Gaboriau, G. Levitt, and F. Paulin, *Pseudogroups of isometries of $\mathbb{R}$ and Rips' theorem on free actions on $\mathbb{R}$-trees*, Israel J. Math. **87** (1994), 403–428 {II.18}.

Gabor–00. D. Gaboriau, *Coût des relations d'équivalence et des groupes*, Inventiones Math. **139** (2000), 41–98 {III.44, IV.47}.

GalHL–87. S. Gallot, D. Hulin, and J. Lafontaine, *Riemannian geometry*, Universitext, Springer, 1987 {VII.6}.

Gers–92a. S.M. Gersten, *Bounded cocycles and combings of groups (with an appendix by Domingo Toledo)*, Internat. J. Algebra Comput. **2** (1992), 307–326 {IV.48}.

Gers–92b. S.M. Gersten, *Dehn functions and ℓ_1-norms of finite presentations*, in "Algorithms and classification in combinatorial group theory", G. Baumslag and C.F. Miller III, Editors, M.S.R.I. Publ. **23**, Springer (1992), 195–224 {V.57}.

Gers–93. S.M. Gersten, *Quasi-isometry invariance of cohomological dimension*, C.R. Acad. Sci. Paris **316** (1993), 411–416 {IV.50}.

Gers–94. S.M. Gersten, *Divergence in 3-manifold groups*, GAFA Geom. Funct. Anal. **4** (1994), 633–647 {V.31}.

GerSh–91. S.M. Gersten and H.B. Short, *Rational subgroups of biautomatic groups*, Annals of Math. **134** (1991), 125–158 {IV.3.viii}.

GhyHf–90. E. Ghys and A. Haefliger, *Groupes de torsion*, in "Sur les groupes hyperboliques d'après Mikhael Gromov", E. Ghys and P. de la Harpe, Editors, Birkhäuser (1990), 215–226 {VIII.19}.

GhyHp–90. E. Ghys and P. de la Harpe, *Sur les groupes hyperboliques d'après Mikhael Gromov*, Birkhäuser, 1990 {II.42, IV.3.vii–viii, IV.40, IV.50, V.54–56, V.58, V.60, VI.8, VI.19}.

Ghys–90. E. Ghys, *Les groupes hyperboliques*, Séminaire Bourbaki, vol. 1989/90, Astérisque **189–190**, Soc. Math. France (1990), 203–238 {IV.48, V.60}.

Ghys–99. E. Ghys, *Group acting on the circle*, Monografias del IMCA, Universidad Nacional de Ingenieria, Lima, 1999 {II.18, II.48, IV.48}.

Gilma–95. J. Gilman, *Two-generator discrete subgroups of $PSL(2, \mathbb{R})$*, Mem. Amer. Math. Soc. **117**, 1995 {II.25}.

Godbi–71. C. Godbillon, *Eléments de topologie algébrique*, Hermann, 1971 {V.7}.

Golod–64. E.S. Golod, *On nil-algebras and finitely approximable p-groups*, Amer. Math. Soc. Transl. (2) **48** (1965), 103–106 [Russian original: Izv. Akad. Nauk. SSSR Ser. Mat. **28** (1964) pp. 273–276] {VII.14, VII.19}.

Gonci. C. Gonciulea, *Non virtually abelian Coxeter groups virtually surject onto $\mathbb{Z} \star \mathbb{Z}$*, preprint, Ohio State University, 1998 {VII.14}.

GooHJ–89. F.M. Goodman, P. de la Harpe, and V.F.R. Jones, *Coxeter graphs and towers of algebras*, M.S.R.I. Publ. **14**, Springer, 1989 {II.25, II.28}.

Goren–68. D. Gorenstein, *Finite groups*, Harper and Row, 1968 {III.47}.

Graev–48. M.I. Graev, *Free topological groups*, Transl. Amer. Math. Soc. Ser. 1 **8** (1962) 305–364 [Russian original: Izvestiya Akad. Nauk SSSR. Ser. Mat. **12** (1948) pp. 279–324] {II.23}.

GraKP–91. R.L. Graham, D.E. Knuth, and O. Patashnik, *Concrete mathematics*, Addison-Wesley, 1991 {VI.7, VI.14, VIII.67}.

Grigo–78. R.I. Grigorchuk, *Symmetrical random walks on discrete groups*, in "Multicomponent random systems", R.L. Dobrushin. Ya. G. Sinai, and D. Griffeath, Editors, Advances in Probability and Related Topics **6** (Dekker 1980), 285–325 [Russian original: Nauka, Moscow (1978) pp. 132–152] {VII.35, VII.37}.

Grigo–80. R.I. Grigorchuk, *Bernside's problem on periodic groups*, Functional Anal. Appl. **14** (1980), 41–43 {VIII, VIII.8}.

Grigo–83. R.I. Grigorchuk, *On Milnor's problem of group growth*, Soviet Math. Dokl. **28:1** (1983), 23–26 [Russian original: Dokl. Akad. Nauk. SSSR Ser. Mat. 271 (1983), no 1, pp. 30–33] {V.10}.

Grigo–84. R.I. Grigorchuk, *Degrees of growth of finitely generated groups, and the theory of invariant means*, Math. USSR Izv. **25:2** (1985), 259–300 [Russian original: Izv. Akad. Nauk. SSSR Ser. Mat. 48 (1984), no. 5, pp. 939–985]
{III.45, V.10, VI.B, VI.21, VI.35, VI.53, VII.24, VIII.20}.

Grigo–85. R.I. Grigorchuk, *On the growth degrees of p-groups and torsion-free groups*, Math. USSR Sbornik **54** (1986), 185–205 [Russian original: Mat. Sb. (N.S.) 126(168) (1985), no. 2, pp. 194–214, 286]   {III.45, VIII.18, VIII.59, VIII.66}.

Grigo–89. R.I. Grigorchuk, *On the Hilbert-Poincaré series of graded algebras that are associated with groups*, Math. USSR Sbornik **66** (1990), 211–229 [Russian original: Mat. Sb. (N.S.) 180 (1989), no.2, pp. 211–229]   {VII.29, VII.39}.

Grigo–91. R.I. Grigorchuk, *On growth in group theory*, Proceedings of the International Congress of Mathematicians, Kyoto, 1990, **I** (Math. Soc. of Japan, 1991), 325–338   {VI.35, VI.53, VIII}.

Grigo–96. R.I. Grigorchuk, *On the problem of M. Day about nonelementary amenable groups in the class of finitely presented groups*, Math. Notes **60** (1996), 774–775   {VIII, VIII.68}.

Grigo–98. R.I. Grigorchuk, *An example of a finitely presented amenable group not belonging to the class EG*, Sbornik Math. **189:1** (1998), 75–95 {VIII, VIII.68}.

Grigo–99. R.I. Grigorchuk, *On the system of defining relations and the Schur multiplier of periodic groups generated by finite automata*, in "Groups St Andrews 1997 in Bath, I", C.M. Campbell, E.F. Robertson, N. Ruskuc, and G.C. Smith, Editors, Cambridge Univ. Press (1999), 290–317   {VIII, VIII.56}.

Grigo. R.I. Grigorchuk, *Just infinite branch groups*, in "Horizons in profinite groups", D. Segal, Editor, Birkhäuser, to appear   {VIII.31, VIII.41}.

GriHa–97. R.I. Grigorchuk and P. de la Harpe, *On problems related to growth, entropy and spectrum in group theory*, J. of Dynamical and Control Systems **3:1** (1997), 51–89   {I.8, VI.8, VI.18, VI.53, VI.62, VII.12, VII.17, VII.35–36, VII.39}.

GriHa–98. R.I. Grigorchuk and P. de la Harpe, *On the regularity of growth series for finitely generated groups and a problem of the Kourovka Notebook (in Russian)*, Algebra i Logika **37:6** (1998), 621–626   {VI.9}.

GriHa. R.I. Grigorchuk and P. de la Harpe, *Limit behaviour of exponential growth rates for finitely generated groups*, preprint, University of Geneva, 1999   {VI.61}.

GriKu–93. R.I. Grigorchuk and P.F. Kurchanov, *Some questions of group theory related to geometry*, in "Algebra VII", A.N. Parshin and I.R. Shafarevich, Editors, Encyclopaedia of Math. Sciences **58**, Springer (1993), 167–232   {II.45}.

GriKZ–92. R.I. Grigorchuk, P.F. Kurchanov, and H. Zieschang, *Equivalence of homomorphisms of surface groups to free groups and some properties of 3-dimensional hanldlebodies*, in "Proceedings of the International Conference on Algebra Dedicated to the Memory of A.I. Mal'cev", L.A. Bokut', Yu. L. Ershov and A.I. Kostrikin, Editors, Contemporary Mathematics **131.1** (1992), 521–530 {II.47}.

GriMa–97. G.I. Grigorchuk and M.J. Mamaghani, *On use of iterates of endomorphisms for constructing groups with specific properties*, Mat. Stud. **8** (1997), 198–206   {VII.19}.

GriNa–97. R.I. Grigorchuk and T. Nagnibeda, *Complete growth functions of hyperbolic groups*, Inventiones Math. **130** (1997), 159–188   {VI.18}.

GriPa–94. R. Grimaldi and P. Pansu, *Sur la régularité de la fonctions croissance d'une variété riemannienne*, Geometriae Dedicata **50** (1994), 301–307   {VI.40}.

GroLP–81. M. Gromov, J. Lafontaine and P. Pansu, *Structures métriques pour les variétés riemanniennes*, Cedic/F. Nathan, 1981   {IV.17, VI.53, VII.12–13, VII.19}.

GroMe–69. D. Gromoll and W. Meyer, *On complete open manifolds of positive curvature*, Annals of Math. **90** (1968), 75–90   {V.51}.

Gromo–81. M. Gromov, *Groups of polynomial growth and expanding maps*, Publ. Math. I.H.E.S. **53** (1981), 53–73
{Introduction, IV.50, VI.B, VI.38–39, VII.23, VII.29–30}.

Gromo–82. M. Gromov, *Volume and bounded cohomology*, Publ. Math. I.H.E.S. **56** (1982), 5–100   {V.31}.

Gromo–84. M. Gromov, *Infinite groups as geometric objects*, Proceedings of the International Congress of Mathematicians, Warsaw, 1983 **1** (1984), 385–392
{IV, IV.42, IV.46}.

Gromo–87. M. Gromov, *Hyperbolic groups*, in "Essays in Group Theory", S.M. Gersten, Editor, M.S.R.I. Publ. **8**, Springer (1987), 75–263 {II.42, III.22, IV.3, IV.26, IV.37, IV.49, V.53, V.55–58, V.60, VI.8, VII.39, VIII.19}.

Gromo–91. M. Gromov, *Sign and geometric meaning of curvature*, Rend. Sem. Math. Fis. Milano **61** (1991), 9–123 {V.51, V.53}.

Gromo–93. M. Gromov, *Asymptotic invariants of infinite groups*, Volume 2 of "Geometry group theory, Sussex 1991", G.A. Niblo and M.A. Roller, Editors, Cambridge Univ. Press, 1993 {Introduction, IV, IV.33, IV.35, IV.37, IV.46, IV.48, IV.50, V.33, V.55, VI.26, VI.43}.

GroPa–91. M. Gromov and P. Pansu, *Rigidity of lattices: an introduction*, in "Geometric topology: recent developments (Montecatini Terme, 1990)", Springer Lecture Notes in Mathematics **1504** (1991), 39–137 {II.33, IV.41}.

Grune–78. F.J. Grunewald, *On some groups which cannot be finitely presented*, J. London Math. Soc. (2) **17** (1978), 427–436 {III.3}.

GruSS–88. F.J. Grunewald, D. Segal, and G.C. Smith, *Subgroups of finite index in nilpotent groups*, Inventiones Math. **93** (1988), 185–223 {II.21}.

Guba–86. V.S. Guba, *A finitely generated simple group with free 2-generated subgroups*, Siberian Mathematical J. **27** (1986), 670–684 {III.46}.

Guiva–70. Y. Guivarc'h, *Groupes de Lie à croissance polynomiale*, C.R. Acad. Sc. Paris, Sér. A **271** (1970), 237–239 {VII.26}.

Guiva–71. Y. Guivarc'h, *Groupes de Lie à croissance polynomiale*, C.R. Acad. Sc. Paris, Sér. A **272** (1971), 1695–1696 {VII.26}.

Guiva–73. Y. Guivarc'h, *Croissance polynomiale et périodes des fonctions harmoniques*, Bull. Soc. Math. France **101** (1973), 333–379 {VII.26}.

Guiva–76. Y. Guivarc'h, *Equirépartition dans les espaces homogènes*, in "Théorie ergodique, Rennes 1973/1974", J.P. Conze and M.S. Keane, Editors, Lecture Notes in Mathematics **532**, Springer (1976), 131–142 {II.41}.

Gunth–60. P. Günther, *Einige Sätze über das Volumenelement eines Riemannschen Raumes*, Publ. Math. Debrecen **7** (1960), 78–93 {VII.6}.

GupS–83a. N. Gupta and S. Sidki, *Some infinite p-groups*, Algebra i Logika **22:5** (1983), 584–589 {VIII.1}.

GupS–83b. N. Gupta and S. Sidki, *On the Burnside problem for periodic groups*, Math. Z. **182** (1983), 385–388 {II.42, VIII.1, VIII.19}.

Gupta–89. N. Gupta, *On groups in which every element has finite order*, Amer. Math. Monthly **96** (1989), 297–308 {VIII.1}.

Haefl–91. A. Haefliger, *Complexes of groups and orbihedra*, in "Group theory from a geometrical viewpoint (26 March — 6 April 1990, Trieste)", E. Ghys, A. Haefliger and A. Verjovsky, Editors, World Scientific (1991), 504–540 {III.14}.

HallM–49. M. Hall, Jr, *Subgroups of finite index in free groups*, Canadian J. Math. **1** (1949), 187–190 {II.21, VII.38}.

HallP–54. P. Hall, *Finiteness conditions for soluble groups*, Proc. London Math. Soc. **4** (1954), 419–436 {III.41–42, V.19}.

Harpe–83. P. de la Harpe, *Free groups in linear groups*, l'Enseignement math. **29** (1983), 129–144 {II.B, II.37}.

Harpe–91. P. de la Harpe, *An invitation to Coxeter groups*, in "Group theory from a geometrical viewpoint (26 March — 6 April 1990, Trieste)", E. Ghys, A. Haefliger and A. Verjovsky, Editors, World Scientific (1991), 193–253 {V.40, V.49, VI.10}.

Harpe–95. P. de la Harpe, *Operator algebras, free groups and other groups*, in "Recent advances in operator algebras, Orléans, 1992", Astérisque **232**, Soc. Math. France (1995), 121–153 {II.23, III.46}.

HarRV–93. P. de la Harpe, G. Robertson, and A. Valette, *On the spectrum of the sum of generators for a finitely generated group*, Israel J. Math. **81** (1993), 65–96 {VII.39}.

HarVa–89. P. de la Hárpe and A. Valette, *La propriété (T) de Kazhdan pour les groupes localement compacts*, Astérisque **175**, Soc. Math. France, 1989 {III.4, V.20.iv}.

HatTh–80. A. Hatcher and W. Thurston, *A presentation for the mapping class group of a closed surface*, Topology **19** (1980), 221–237 {V.7}.

Hausm–81. J.C. Hausmann, *Sur l'usage de critères pour reconnaître un groupe libre, un produit amalgamé ou une HNN-extension*, l'Enseignement math. **27** (1981), 221–242 {II.B, II.36}.

HauWe–85. J.C. Hausmann and S. Weinberger, *Caractéristiques d'Euler et groupes fondamentaux des variétés de dimension 4*, Comment. Math. Helvetici **60** (1985), 139–144 {V.31}.

HecHi–83. G. Hector and U. Hirsch, *Introduction to the geometry of foliations, Part B, foliations of codimension one*, Vieweg, 1983 {II.18}.

Hecke–36. E. Hecke, *Über die Bestimmung Dirichletscher Reihen durch ihre Funktionalgleichung*, Math. Ann. **112** (1936), 664–699 [Werke: pages 591–626] {II.28}.

Held–66. D. Held, *On abelian subgroups of infinite 2-groups*, Acta Sci. Math. (Szeged) **27** (1966), 97–98 {VIII.84}.

Hempe–76. J. Hempel, *3-manifolds*, Annals of Math. Studies **86**, Princeton Univ. Press, 1976 {II.47}.

Herma–79. M. Herman, *Sur la conjugaison différentiable des difféomorphismes du cercle à des rotations*, Publ. Math. I.H.E.S **49** (1979), 5–234 {IV.48}.

Herst–68. I.N. Herstein, *Noncommutative rings*, Carus Mathematical Monographs **15**, Mathematical Association of America, J. Wiley, 1968 {VIII.19}.

Higma–51. G. Higman, *A finitely generated infinite simple group*, J. London Math. Soc. **26** (1951), 61–64 {V.26}.

Higma–52. G. Higman, *Unrestricted free products, and varieties of topological groups*, J. London Math. Soc. **27** (1952), 73–81 {II.23}.

Higma–61. G. Higman, *Subgroups of finitely presented groups*, Proc. Royal Soc. London Ser. A **262** (1961), 455–475 {III.17}.

Higma–74. G. Higman, *Finitely presented infinite simple groups*, Notes on pure mathematics **8**, Univ. of Canberra, 1974 {V.26}.

HigNN–49. G. Higman, B.H. Neumann, and H. Neumann, *Embedding theorems for groups*, J. London Math. Soc. **24** (1949), 247–254 {III.14, III.16, III.44}.

Hirsh–77. R. Hirshon, *Some properties of endomorphisms in residually finite groups*, J. Austral. Math. Soc. **24** (1977), 117–120 {III.19}.

HirTh–75. M.W. Hirsch and W.P. Thurston, *Foliated bundles, invariant measures and flat manifolds*, Annals of Math. **101** (1975), 369–390 {III.13, VII.34}.

HogMe–93. C. Hog-Angeloni and W. Metzler, *The Andrews-Curtis conjecture and its generalizations*, in "Two-dimensional homotopy and combinatorial group theory", London Math. Soc. Lecture Note Ser., Cambridge Univ. Press **197** (1993), 365–380 {II.45, V.31}.

HopUl–79. J.E. Hopcroft and J.D. Ullman, *Introduction to automata theory, languages and computation*, Addison-Wesley, 1979 {VIII.10}.

Hughe–95. B.D. Hughes, *Random walks and random environments, Volume 1: Random walks*, Oxford Science Publications, 1995 {I.5, I.12}.

Huppe–67. B. Huppert, *Endliche Gruppen I*, Die Grundlehren der mathematischen Wissenschaften, Band 134, Springer, 1967 {III.52}.

Huppe–98. B. Huppert, *Character theory of finite groups*, de Gruyter, 1998 {VIII.3}.

HyeUl–47. D.H Hyers and S.M. Ulam, *On approximate isometries on the space of continuous functions*, Annals of Math. **48** (1947), 285–289 {IV.45}.

ImaTa–72. Y. Imayoshi and M. Tanguchi, *An introduction to Teichmüller spaces*, Springer, 1992 {II.33}.

Ivano–84. N. Ivanov, *Algebraic properties of the Teichmüller modular group*, Dokl. Akad. Nauk. SSSR **275** (1984), 786–789 {II.42}.

Ivano–94. S.V. Ivanov, *The free Burnside groups of sufficiently large exponents*, Internat. J. Algebra Comput. **4** (1994), 1–308 [see also *On the Burnside problem on periodic groups*, Bull. Amer. Math. Soc. **27** (1992), 257–260]   {VIII.19}.

Ivano–98. S.V. Ivanov, *On the Burnside problem for groups of even exponent*, Proceedings of the International Congress of Mathematicians, Berlin 1998, **II** (Doc. Math. J. DMV, 1998), 67–75   {VIII.19}.

IvaOl–96. S.V. Ivanov and A.Yu. Ol'shanskii, *Hyperbolic groups and their quotients of bounded exponents*, Trans. Amer. Math. Soc. **348** (1996), 2091–2138   {III.18, VIII.19}.

Ivers–92. B. Iversen, *Hyperbolic geometry*, London Math. Soc., Student Texts **25**, Cambridge Univ. Press, 1992   {II.16, V.35, V.39–40}.

Jaco–69. W Jaco, *Heegard splittings and splitting homomorphisms*, Trans. Amer. Math. Soc. **144** (1969), 365–379   {II.47}.

Jacob–92. N. Jacobson, *Magnus' method in the theory of free groups*, Ulam Quarterly **1:1** (1992), 11 pages [electronic journal]   {VIII.31}.

Jenki–73. J.W. Jenkins, *Growth of connected locally compact groups*, J. Functional Analysis **12** (1973), 113–127   {VII.26}.

JohKS–95. D.L. Johnson, A.C. Kim, and H.J. Song, *The growth of the trefoil group*, in "Groups — Korea '94", A.C. Kim and D.L. Johnson, Editors, de Gruyter (1995), 157–161   {VI.20}.

Johns–83. D. Johnson, *The structure of the Torelli group I: a finite set of generators of T*, Annals of Math. **118** (1983), 423–442   {III.16}.

Johns–90. D.L. Johnson, *Presentations of groups*, London Math. Soc., Student Texts **15**, Cambridge Univ. Press, 1990   {II.17, V.15, V.36}.

Johns–91. D.L. Johnson, *Rational growth of wreath products*, in "Groups St-Andrews 1989, Volume 2", London Math. Soc. Lecture Note Ser., Cambridge Univ. Press **160** (1991), 309–315   {VI.3}.

Jolis–90. P. Jolissaint, *Rapidly decreasing functions in reduced C^*-algebras of groups*, Trans. Amer. Math. Soc. **317** (1990), 167–196   {VI.44}.

Jon–1856. O. Jones, *The grammar of ornament*, Van Nostrand, 1982 [first published in 1856]   {Introduction}.

Jones–86. G.A. Jones, *Congruence and non-congruence subgroups of the modular group: a survey*, in "Proceedings of Groups — St Andrews 1985", E.F. Robertson and C.M. Campbell, Editors, Cambridge Univ. Presss (1986), 223–234 {II.21, III.D}.

Jones. V.F.R. Jones, *Ten problems*, preprint, February 1999, to appear in a book "Mathematics tomorrow" of the IMU   {III.20}.

KacPe–85. V.G. Kac and D.H. Peterson, *Defining relations of certain infinite dimensional groups*, in "Élie Cartan et les mathématiques d'aujourd'hui, Lyon, 25–29 juin 1984", Astérisque (numéro hors série), Soc. Math. France (1985), 165–208   {V.2}.

KaiVe–83. V.A. Kaimanovich and A.M. Vershik, *Random walks on discrete groups: boundary and entropy*, Ann. of Probability **11** (1983), 457–490   {VII.40}.

KapLe–95. M. Kapovich and B. Leeb, *Asymptotic cones and quasi-isometry classes of fundamental groups of 3-manifolds*, GAFA Geom. Funct. Anal. **5** (1995), 582–603   {Introduction}.

Kapov–99. I. Kapovich, *Howson property and one-relator groups*, Comm. in Algebra **27** (1999), 1057–1072   {IV.49}.

KapWi–00. I. Kapovich and D.T. Wise, *The equivalence of some residual properties of word-hyperbolic groups*, J. of Algebra **223** (2000), 562–583   {V.60}.

KarMe–85. M. Kargapolov and Iou. Merzliakov, *Eléments de la théorie des groupes*, Editions Mir, 1985   {Introduction, II.17, VIII.3–4, VIII.10, VIII.76}.

KarSo–71. A. Karras and D. Solitar, *Subgroups of HNN groups and groups with one defining relation*, Canadian J. Math. **23** (1971), 627–643   {II.42}.

Katok–92. S. Katok, *Fuchsian groups*, University of Chicago Press, 1992   {II.28, III.25}.

Kazhd–65. D.A. Kazhdan, *Uniform distribution in the plane*, Trans. Moscow Math. Soc. **14** (1965), 325–332   {II.41}.

Kazhd–67. D. Kazhdan, *Connection of the dual space of a group with the structure of its closed subgroups*, Funct. Anal. and its Appl. **1** (1967), 63–65
{III.2, III.4, V.20.iv}.

KegWe–73. O. Kegel and B. Wehrfritz, *Locally finite groups*, North Holland, 1973 {III.4}.

Kerva–95. M. Kervaire, *On higher dimensional knots*, in "Differential and combinatorial topology — a symposium in honour of Marston Morse", S. Cairns, Editor, Princeton Univ. Press (1965), 105–120   {II.46}.

Keste–59. H. Kesten, *Symmetric random walks on groups*, Trans. Amer. Math. Soc. **92** (1959), 336–354   {I.4, I.9, VII.37}.

Kirby–97. R.C. Kirby, *Problems in low-dimensional topology*, in "Geometric topology", W.H. Kazez Editor, Studies in Adv. Math., Vol 2, Part 2 (1997), 35–473
{II.45–46, III.5}.

KirSi–77. R.C. Kirby and L.C. Siebenmann, *Foundational essays on topological manifolds, smoothings, and triangulations*, Princeton Univ. Press, 1977   {V.7}.

KlaLP–97. G. Klaas, C.R. Leedham-Green, and W. Plesken, *Linear pro-p-groups of finite width*, Lecture Notes in Mathematics **1674**, Springer, 1997   {VIII.31}.

Klarn–81. D.A. Klarner, *Mathematical crystal growth II*, Discrete Appl. Math. **3** (1981), 113–117   {VI.2}

Klein–80. F. Klein, *Zur Theorie der elliptischen Modulfunctionen*, Math. Ann. **17** (1880), 62–70 [Gesamm. Math. Abh., Bd. 3, p. 169–178]   {III.D}.

Klein–83. F. Klein, *Neue Beiträge zur Riemann'schen Functionentheorie*, Math. Annalen **21** (1883), 141–218   {II.26}.

KleLe–98. B. Kleiner and B. Leeb, *Rigidity of quasi-isometries for symmetric spaces and Euclidean buildings*, Publ. Math. I.H.E.S. **86** (1998), 115–197
{Introduction, IV.41}.

Klyac–93. A. Klyachko, *A funny property of sphere and equations over groups*, Comm. in Algebra **21** (1993), 2555–2575   {II.46}.

KosMa–89. A.I. Kostrikin and Yu.I. Manin, *Linear algebra and geometry*, Gordon and Breach, 1989   {VI.2}.

Koubi–98. M. Koubi, *Croissance uniforme dans les groupes hyperboliques*, Ann. Inst. Fourier **48** (1998), 1441–1453   {VII.17}.

Kouks–98. D. Kouksov, *On rationality of the cogrowth series*, Proc. Amer. Math. Soc. **126** (1998), 2845–2847   {I.11}.

Kouro–95. V.D. Mazurov and E.I. Khakhro (Editors), *Unsolved problems in group theory — The Kourovka notebook*, thirteenth augmented edition, Russian Academy of Sciences, Siberian Division, Novosibirsk, 1995   {III.5, VI.9, VI.53}.

KraLe–85. G.R. Krause and T.H. Lenagan, *Growth of algebras and Gelfand-Kirillov dimension*, revised edition, Graduate Studies in Math. **22**, Amer. Math. Soc., 2000 (first published 1985)   {VII.28}.

Kramm–a. D. Krammer, *The braid group B_4 is linear*, Inventiones Math., to appear
{III.20}.

Kramm–b. D. Krammer, *Braid groups are linear*, preprint, 2000   {III.20}.

Kulka–83. R.S. Kulkarni, *An extension of a theorem of Kurosh and applications to Fuchsian groups*, Michigan Math. J. **30** (1983), 259–272   {II.22}.

Kuran–51. M. Kuranishi, *On everywhere dense imbedding of free groups in Lie groups*, Nagoya Math. J. **2** (1951), 63–71   {II.41}.

Kuros–56. A.G. Kurosh, *The theory of groups (2 volumes)*, translated from the Russian and edited by K.A. Hirsch, Chelsea, 1955–56   {II.22, III.B}.

Labou–96. F. Labourie, *Quelques résultats récents sur les espaces localement homogènes compacts*, in "Manifolds and geometry", P. de Bartolomeis, F. Tricerri, and E. Vesentini, Editors, Symposia Math. **XXXVI**, Cambridge Univ. Press (1996), 267–283   {IV.25.ix}.

Lam–91. T.Y. Lam, *A first course in noncommutative rings*, Graduate Texts in Mathematics **131**, Springer, 1991   {VIII.19}.

Lamy–98. S. Lamy, *L'alternative de Tits pour Aut* $[\mathbb{C}^2]$, C.R. Acad. Sci. Paris **327** (1998), 537–540   {II.42}.

Lamy. S. Lamy, *L'alternative de Tits pour Aut* $[\mathbb{C}^2]$, preprint, University of Rennes I, 1998   {II.42}.

Ledra. F. Ledrappier, *Some asymptotic properties of random walks on free groups*, preprint, 1992   {II.23}.

Lehne–64. J. Lehner, *Discontinuous groups and automorphic functions*, Math. Surveys VIII, Amer. Math. Soc., 1964   {II.26, III.D}.

Leon–98a. Yu.G. Leonov, *The conjugacy problem in a class of 2-groups*, Math. Notes **64** (1998), 496–505   {VIII.49}.

Leon–98b. Yu.G. Leonov, *On precisement of estimation of period's growth for Grigorchuk's 2-groups*, Voprosi algebri **13** (1998), 58–67   {VIII.66, VIII.70}.

Leon. Yu.G. Leonov, *On growth function for some torsion residually finite groups*, International conference dedicated to the 90th anniversary of L.S. Pontryagin, Moscow, August 31 — September 6, 1998   {VIII.66}.

Levin–78. J. Levine, *Some results on higher dimensional knot groups (with an appendix of C. Weber)*, in "Knot theory, Proceedings, Plans-sur-Bex 1977", J.C. Hausmann, Editor, Springer Lecture Notes in Mathematics **685** (1978), 243–273   {V.31}.

Levit–95. G. Levitt, *On the cost of generating an equivalence relation*, Ergod. Th. & Dynam. Sys. **15** (1995), 1173–1181   {IV.47}.

Lewin–67. J. Lewin, *A finitely presented group whose group of automorphisms is infinitely generated*, J. London Math. Soc. **42** (1967), 610–613   {V.22}.

Liard–96. F. Liardet, *Croissance dans les groupes virtuellement abéliens*, Thèse, University of Genève, 1996   {VI.18}.

Licko–64. W.B.R. Lickorish, *A finite set of generators for the homeotopy group of a 2-manifold*, Proc. Camb. Phil. Soc. **60** (1964), 769–778 [see also **62** (1966), 679–681]   {V.7}.

LidNi–83. R. Lidl and H. Niederreiter, *Finite fields*, Encyclopedia of Mathematics and its Applications, **20**, Cambridge Univ. Press, 1997 [first edition 1983]   {III.4}.

LieSh–96. M.W. Liebeck and A. Shalev, *Classical groups, probabilistic methods and the* $(2,3)$-*generator problem*, Annals of Math. **144** (1996), 77–125   {III.46}.

Loser–87. V. Losert, *On the structure of groups with polynomial growth*, Math. Z. **195** (1987), 109–117   {VII.29}.

Lotha–83. M. Lothaire, *Combinatorics on words*, Cambridge Univ. Press, 1997 (first published 1983)   {II.9}.

Lott–99. J. Lott, *Deficiencies of lattice subgroups of Lie groups*, Bull. London Math. Soc. **31** (1999), 191–195   {V.31}.

Lovás–79. L. Lovász, *Combinatorial problems and exercises*, North Holland, 1979 {IV.11}.

LubMa–91. A. Lubotzky and A. Mann, *On groups of polynomial subgroup growth*, Invetiones Math. **104** (1991), 521–533   {VII.29}.

LubMR–93. A. Lubotzky, S. Mozes, and M.S. Raghunathan, *Cyclic subgroups of exponential growth and metrics on discrete groups*, C.R. Acad. Sci. Paris **317** (1993), 735–740   {VI.43, VI.45}.

LubMZ–94. A. Lubotzky, S. Mozes, and R.J. Zimmer, *Superrigidity for the commensurability group of tree lattices*, Comment. Math. Helvetici **69** (1994), 523–548 {III.20}.

Lubo–87. A. Lubotzky, *On finite index subgroups of linear groups*, Bull. London Math. Soc. **19** (1987), 325–328   {III.20}.

Lubo–88. A. Lubotzky, *A group theoretic characterization of linear groups*, J. of Algebra **113** (1988), 207–214   {III.20, IV.48}.

Lubo–89. A. Lubotzky, *Trees and discrete subgroups of Lie groups over local fields*, Bull. Amer. Math. Soc. **20** (1989), 27–30   {III.4}.

Lubo–91. A. Lubotzky, *Lattices in rank one Lie groups over local fields*, GAFA Geom. Funct. Anal. **1** (1991), 405–431   {II.26, III.4}.

Lubo–94. A. Lubotzky, *Discrete groups, expanding graphs and invariant measures*, Birkhäuser, 1994   {II.37, II.41}.

Lubo–95a. A. Lubotzky, *Counting finite index subgroups*, in "Groups '93 Galway / St Andrews, vol. 2", C.M. Campbell et al., Editors, Cambridge Univ. Press (1995), 368–404   {II.21}.

Lubo–95b. A. Lubotzky, *Cayley graphs: eigenvalues, expanders and random walks*, in "Surveys in Combinatorics, 1995", P. Rowlinson, Editor, Cambridge Univ. Press (1995), 155–189   {IV.15}.

Lubo–95c. A. Lubotzky, *Subgroup growth*, Proceedings of the International Congress of Mathematicians, Zürich, 1994, 1 (Birkhäuser, 1995), 309–317   {II.21}.

Lubo–95d. A. Lubotzky, *Tree-lattices and lattices in Lie groups*, in "Combinatorial and geometric group theory, Edinburgh 1993", A.J. Duncan, N.D. Gilbert, and J. Howie, Editors, London Math. Soc. Lecture Note Series **204** (1995), 217–232
{V.20.vii}.

Lubo–95e. A. Lubotzky, *Subgroup growth and congruence subgroups*, Inventiones Math. **119** (1995), 267–295   {II.21}.

Lubo–96. A. Lubotzky, *Free quotients and the first Betti number of some hyperbolic manifolds*, Transformation Groups 1 (1996), 71–82   {VII.18}.

LubPS–96. A. Lubotzky, L. Pyber, and A. Shalev, *Discrete groups of slow subgroup growth*, Israel J. Math. **96** (1996), 399–418   {III.36}.

LubWe–93. A. Lubotzky and B. Weiss, *Groups and expanders*, DIMACS Series in Discrete Mathematics and Theoretical Computer Science **10** (1993), 95–109   {III.36}.

LucTW. A. Lucchini, M.C. Tamburini, and J.S. Wilson, *Hurwitz groups of large rank*, J. London Math. Soc., to appear   {III.39}.

Lusti–93. M. Lustig, *Fox ideals, $\mathcal{N}$-torsion and applications to groups and 3-manifolds*, in "Two-dimensional homotopy and combinatorial group theory", London Math. Soc. Lecture Note Ser., Cambridge Univ. Press **197** (1993), 219–249   {V.31}.

Lusti–95. M. Lustig, *Non-efficient torsion-free groups exist*, Comm. in Algebra **23** (1995), 215–218   {V.31}.

Lyndo–50. R.C. Lyndon, *Cohomology theory of groups with a single defining relation*, Annals of Math. **52** (1950), 650–665   {V.13}.

Lyndo–73. R. Lyndon, *Two notes on Rankin's book on the modular group*, J. Austral. Math. Soc. **46** (1973), 454–457   {II.16}.

LynSc–77. R.C. Lyndon and P.E. Schupp, *Combinatorial group theory*, Ergebnisse der Mathematik und ihrer Grenzgebiete, Band 89, Springer, 1977
{II.B, II.25, II.47, III.37}.

LynUl–69. R.C. Lyndon and J.L. Ullman, *Groups generated by two parabolic linear fractional transformations*, Canad. J. Math. **21** (1969), 1388–1403   {II.34}.

Lyons–95. R.Lyons, *Random walks and the growth of groups*, C.R. Acad. Sci. Paris **320** (1995), 1361–1366   {VI.53.vii}.

LyoPP–96. R. Lyons, R. Pemantle, and Y. Peres, *Random walks on the lamplighter group*, Annals of Prob. **24** (1996), 1993–2006   {VI.53.vii}.

Lysen–85. I.G. Lysionok, *A system of defining relations for the Grigorchuk group*, Math. Notes **38** (1985), 784–792 [Russian original: pp. 503–516]   {VIII.56}.

Lysen–96. I.G. Lysionok, *Infinite Burnside groups of even period*, Izv. Ross. Akad. Nauk Ser. Mat. **60** (1996), 3–224 [see also *The infinitude of Burnside groups of period 2^k for $k \geq 13$*, Russian Math. Surveys **47**:2 (1992), pp. 229–230]   {VIII.19}.

Lysen. I.G. Lysionok, *Growth of orders of elements in the Grigorchuk 2-group*, preprint, University of Geneva, 1998   {VIII.70–71}.

Macbe–64. A.M. Macbeath, *Groups of homeomorphisms of a simply connected space*, Annals of Math. **79** (1964), 473–488   {IV.23}.

Macbe–83. A.M. Macbeath, *Commensurability of co-compact three-dimensional hyperbolic groups*, Duke Math. J. **50** (1983), 1245–1253 [and *Erratum*, **56** (1988) 219]
{IV.29}.

MacDo–88. I.D. Macdonald, *The theory of groups*, R.E. Krieger Publ. Comp., 1988
{III.5, III.26, V.31, VII.26, VIII.66}.

MacDu–46. C.C. MacDuffee, *The theory of matrices*, Chelsea Publ. Comp., 1946   {III.2}.

MadSl–93. N. Madras and G. Slade, *The self-avoiding walk*, Birkhäuser, 1993   {I.12}.

MagKS–66. W. Magnus, A. Karras, and D. Solitar, *Combinatorial group theory*, J. Wiley, 1966   {II.16–17, II.44, II.46, III.3, III.8, III.13, III.19, III.21, IV.3, V.5–6, V.22, VIII.19, VIII.60}.

Magnu–73. W. Magnus, *Rational representations of Fuchsian groups and non-parabolic subgroups of the modular group*, Nachr. Akad. Wiss. Göttingen Math.-Phys. Kl. II (1973), 179–189   {V.46}.

Magnu–74. W. Magnus, *Noneuclidean tesselations and their groups*, Academic Press, 1974   {II.26, II.32, III.D, V.36}.

Mal'c–40. A.I. Mal'cev, *On the faithful representation of infinite groups by matrices*, Amer. Math. Soc. Transl (2) **45** (1965), 1–18 [Russian original: Mat. SS.(N.S.) 8(50) (1940), pp. 405–422]   {III.4, III.19–20, V.20}.

Mal'c–49. A.I. Mal'cev, *On a class of homogeneous spaces*, Amer. Math. Soc. Transl, Ser 1 **9** (1962), 276–307 [Russian original: Izv. Akad. Nauk. SSSR **13** (1949), pp. 9–32]   {VII.26}.

Manni–79. A. Manning, *Topological entropy for geodesic flows*, Annals of Math. **110** (1979), 567–573   {VI.53}.

Margu–91. G.A. Margulis, *Discrete subgroups of semisimple Lie groups*, Ergebnisse der Mathematik und ihrer Grenzgebiete, Band 17, Springer, 1991   {II.28, IV.24, V.20.vii}.

Margu–97. G. Margulis, *Existence of compact quotients of homogeneous spaces, measurably proper actions, and decay of matrix coefficients*, Bull. Soc. Math. France **125** (1997), 447–456   {IV.25.ix}.

MarSo–81. G.A. Margulis and G.A. Soifer, *Maximal subgroups of infinite index in finitely generated linear groups*, J. of Algebra **69** (1981), 1–23   {III.24}.

MarVi. G.A. Margulis and E.B. Vinberg, *Some linear groups virtually having a free quotient*, preprint, 1998   {VII.14}.

Marko–45. A. Markoff, *On free topological groups*, Transl. Amer. Math. Soc. Ser. 1 **8** (1962) 195–272 [Russian original: Izvestiya Akad. Nauk SSSR. Ser. Mat. **9** (1945) pp. 3–64]   {II.23}.

Maski–71. B. Maskit, *On Poincaré's theorem for fundamental polygons*, Adv. in Math. **7** (1971), 219–230   {V.40}.

Maski–94. B. Maskit, *Explicit matrices for Fuchsian groups*, in "The mathematical legacy of Wilhelm Magnus — Groups, geometry and special functions", W. Abikoff, J.S. Birman, and K. Kuiken Editors, Contemporary Mathematics **169** (1994), 451–466   {V.46}.

Masse–67. W.S. Massey, *Algebraic topology: an introduction*, Harcourt, Brace & World, 1967   {Introduction, II.4, II.11, II.17, II.22, V.7, V.29}.

McCar–68. D. McCarthy, *Infinite groups whose proper quotient groups are finite, I*, Comm. Pure Appl. Math. **21** (1968), 545–562   {VIII.31.f}.

McCar–70. D. McCarthy, *Infinite groups whose proper quotient groups are finite, II*, Comm. Pure Appl. Math. **23** (1970), 767–789   {VIII.31.f}.

McCar–85. J. McCarthy, *A "Tits-alternative" for subgroups of mapping class groups*, Trans. Amer. Math. Soc. **291** (1985), 583–612   {II.42}.

McCul–80. D. McCullough, *Finite aspherical complexes with infinitely-generated groups of self-homotopy equivalences*, Proc. Amer. Math. Soc. **80** (1980), 337–340   {V.7, V.22}.

McCul–84. D. McCullough, *Compact 3-manifolds with infinitely-generated groups of self-homotopy equivalences*, Proc. Amer. Math. Soc. **91** (1984), 625–629   {V.7}.

McCuM–86. D. McCullough and A. Miller, *The genus 2 Torelli group is not finitely generated*, Topology and its Applications **22** (1986), 43–49   {III.16}.

McKTh–73. R. McKenzie and R.J. Thompson, *An elementary construction of unsolvable problems in group theory*, in "Word Problem", W.W. Boone, F.B. Cannonito, and R.C. Lyndon Editors, North Holland (1973), 457–478   {V.26}.

McMul–98. C.T. McMullen, *Lipschitz maps and nets in Euclidean space*, GAFA Geom. Funct. Anal. **8** (1998), 304–314   {IV.46}.

Medny–79. A.D. Mednyh, *On unramified coverings of compact Riemann surfaces*, Soviet Math. Dokl. **20**:1 (1979), 85–88   {II.21}.

Meier–82. D. Meier, *Non-Hopfian groups*, J. London Math. Soc. **26** (1982), 265–270
{III.20, VIII.64}.

Menni–65. J. Mennicke, *Finite factor groups of the unimodular group*, Annals of Math. **81** (1965), 31–37   {III.D}.

Menni–67. J. Mennicke, *Eine Bemerkung über Fuchsshe Gruppen*, Inventiones math. **2** (1967), 301-305 [see also **3** (1968) page 106]   {V.20}.

Merzl–87. Yu. I. Merzlyakov, *Rational groups*, Nauka, 1987 [in Russian]   {V.20}.

MeSWZ–93. J.-F. Mestre, R. Schoof, L. Washington, and D. Zagier, *Quotients homophones des groupes libres*, Experiment. Math. **2** (1993), 153–155   {V.6}.

MihTo–94. M.L. Mihalik and W. Towle, *Quasiconvex subgroups of negatively curved groups*, J. of Pure and Appl. Algebra **95** (1994), 297–301   {IV.49}.

Mille–52. C. Miller, *The second homology group of a group; relations among commutators*, Proc. Amer. Math. Soc. **3** (1952), 588–595   {V.23}.

Mille–71. C.F. Miller, III, *On group-theoretic problems and their classification*, Annals of Math. Studies **68**, Princeton Univ. Press, 1971   {III.18}.

Mille–92. C.F. Miller III, *Decision problems for groups — survey and reflections*, in "Algorithms and classification in combinatorial group theory", G. Baumslag and C.F. Miller III, Editors, M.S.R.I. Publ. **23**, Springer (1992), 1–59 {III.20, V.26}.

MillS–71. C.F. Miller III and P.E. Schupp, *Embeddings into Hopfian groups*, J. of Algebra **17** (1971), 171–176   {III.22}.

Miln–57. J. Milnor, *Groups which act on S^n without fixed points*, Amer. J. Math. **79** (1957), 623–630 [Collected Papers, vol. 2, pages 93–95 and 97–104]   {V.30}.

Miln–58. J. Milnor, *On the existence of a connection with curvature zero*, Comment. Math. Helvetici **32** (1958), 215–223 [Collected Papers, vol. 1, pages 37–38 and 39–47]   {IV.48}.

Miln–62. J. Milnor, *A unique decomposition theorem for 3-manifolds*, Amer. J. Math. **84** (1962), 1–7 [Collected Papers, vol. 2, pages 235 and 237–243]   {III.13}.

Miln–63. J. Milnor, *Morse theory*, Annals of Mathematics Studies **51**, Princeton Univ. Press, 1963   {IV.18}.

Miln–68a. J. Milnor, *Problem 5603*, Amer. Math. Monthly **75**[6] (1968), 685–686 {VI.53}.

Miln–68b. J. Milnor, *A note on curvature and fundamental group*, J. Diff. Geom. **2** (1968), 1–7 [Collected Papers, vol. 1, pages 53 and 55–61]
{IV.22, VI.6, VI.B, VII, VII.6, VII.31}.

Miln–68c. J. Milnor, *Growth of finitely generated solvable groups*, J. Diff. Geom. **2** (1968), 447–449   {VII.27}.

Miln–71. J. Milnor, *Introduction to algebraic K-theory*, Annals of Mathematics Studies **72**, Princeton Univ. Press, 1971   {V.9, V.16}.

Miln–75. J. Milnor, *On the 3-dimensional Brieskorn manifolds M(p,q,r)*, in "Knots, groups, and 3-manifolds", Papers dedicated to the memory of R.H. Fox, L.P. Neuwirth, Editor, Annals of Math. Studies **84** (Princeton Univ. Press 1975), 175–225 [Collected Papers, vol. 2, pages 235 and 245–295]   {V.35}.

Miln–76. J. Milnor, *Curvatures of left invariant metrics on Lie groups*, Adv. in Math. **21** (1976), 293–329 [Collected Papers, vol. 1, pages 73 and 75–111] {V.53, VII.7}.

MilSc–99. C.F. Miller III and P.E. Schupp, *Some presentations of the trivial group*, in "Groups, languages and geometry", R.H. Gilman, Editor, Contemporary Math. **250** (1999), 113–115   {V.6}.

MilSh–98. C.F. Miller III and M. Shapiro, *Solvable Baumslag-Solitar groups are not almost convex*, Geometriae Dedicata **72** (1988), 123–127   {III.21, IV.49}.

Mirsk–71. L. Mirsky, *Transversal theory*, Academic Press, 1971   {IV.46}.

MohWo–89. B. Mohar and W. Woess, *A survey on spectra of infinite graphs*, Bull. London Math. Soc. **21** (1989), 209–234   {I.4}.

Monod–97. N. Monod, *Eléments de géométrie grossière*, "Travail de diplôme" with M. Burger, University of Lausanne, Juillet 1997   {IV.45}.

Morga–92. J.W. Morgan, Λ-*trees and their applications*, Bull. Amer. Math. Soc. **26** (1992), 87–112 {Introduction, II.22}.

Mosto–68. G.D. Mostow, *Quasi-conformal mappings in n-space and the rigidity of hyperbolic space forms*, Publ. Math. I.H.E.S. **34** (1968), 53–104 {Introduction}.

Mosto–73. G.D. Mostow, *Strong rigidity of locally symmetric spaces*, Annals of Mathematics Studies **78**, Princeton Univ. Press, 1973 {Introduction, II.33}.

MouPe–74. R. Moussu and F. Pelletier, *Sur le théorème de Poincaré-Bendixon*, Ann. Inst. Fourier **24,1** (1974), 131–148 {VI.36}.

Mozes–98. S. Mozes, *Products of trees, lattices and simple groups*, Proceedings of the International Congress of Mathematicians, Berlin 1998, **II** (Doc. Math. J. DMV, 1998), 571–582 {V.26}.

MulSc–81. D.E. Muller and P.E. Schupp, *Context-free languages, groups, the theory of ends, second-order logic, tiling problems, cellular automata and vector addition systems*, Bull. Amer. Math. Soc. **4** (1981), 331–334 {VIII.10}.

MulSc–83. D.E. Muller and P.E. Schupp, *Groups, the theory of ends, and context-free languages*, J. of Computer and System Sciences **26** (1983), 295–310 {V.17, VIII.10}.

Mycie–73. J. Mycielski, *Almost every function is independent*, Fund. Math. **81** (1973), 43–48 [see also [Ulam–74], page 681] {II.41}.

Nadka–95. M.G. Nadkarni, *Basic ergodic theory*, second edition, Birkhäuser, 1998 (first published 1995) {VIII.7}.

Natha–99. M.B. Nathanson, *Number theory and semigroups of intermediate growth*, Amer. Math. Monthly **106** (1999), 666–669 {VI.53}.

Nekra. V. Nekrashevych, *Nets of metric spaces*, preprint, State University of Kiev, 1996 {IV.46}.

NeuBH–37. B.H. Neumann, *Some remarks on infinite groups*, J. London Math. Soc. **12** (1937), 120–127 {III.24, III.B, III.29, III.35, V.8, V.11}.

NeumW–99. W.D. Neumann, *Notes on geometry and 3-manifolds*, in "Low dimensional topology", K. B Böröczky, Jr, W. Neumann, and A. Stipsicz, Editors, Bolyai Society Mathematical Studies **9**, Budapest (1999), 191–267 {Introduction, II.28, III.15}.

NeumP–99. P.M. Neumann, *What groups were: a study of the development of the axiomatics of group theory*, Bull. Austral. Math. Soc. **60** (1999), 285–301 {Introduction}.

NeuPM–73. P.M. Neumann, *The SQ-universality of some finitely presented groups*, J. Austral. Math. Soc **16** (1973), 1–6 {III.37}.

Niels–24. J. Nielsen, *Die Isomorphismengruppe der freien Gruppen*, Math. Ann. **91** (1924), 169–209 {V.22}.

Nori–96. M.V. Nori, *The Schottky groups in higher dimensions*, Contemporary Mathematics **58, Part I** (1996), 195–197 {II.27}.

Obraz–93. V.N. Obraztsov, *Growth sequences of 2-generator simple groups*, Proc. Royal Soc. Edinburgh **123A** (1993), 839–855 {III.46}.

OdlPo–93. A.M. Odlyzko and B. Poonen, *Zeros of polynomials with* 0, 1 *coefficients*, l'Enseignement math. **39** (1993), 317–348 {VII.4}.

Oh–98. Hee Oh, *Discrete subgroups generated by lattices in opposite horospherical subgroups*, J. of Algebra **203** (1998), 621–676 {II.25}.

OknSa–95. J. Okniński and A. Salwa, *Generalised Tits alternative for linear semigroups*, J. of Pure and Appl. Algebra **103** (1995), 211–220 {VII.28}.

Olijn. A. Olijnyk, *Free products of C_2 as groups of finitely automatic permutations*, preprint, University of Kiev {VIII.7}.

Ol's–80. A. Yu. Ol'shanskii, *On the problem of the existence of an invariant mean on a group*, Russian Math. Surveys **35:4** (1980), 180–181 [Russian original: Uspekhi Mat. Nauk. 35:4 (1980) pp. 199–200] {VII.35}.

Ol's–89. A.Yu. Ol'shanskii, *Geometry of defining relations in groups*, Kluwer Academic Publ., 1991 [Russian original, Nauka Publ., 1989] {Introduction, III.5, III.18, III.24}.

Ol's–91a. A.Yu. Ol'shanskii, *Periodic quotient groups of hyperbolic groups*, Mat. Sb. **182** (1991), 543–567 {VIII.19}.

Ol's–91b. A.Yu. Ol'shanskii, *Hyperbolicity of groups with subquadratic isoperimetric inequality*, Internat. J. Algebra Comput. **1** (1991), 281–289   {V.57}.

Ol's–92. A.Yu. Ol'shanskii, *Almost every group is hyperbolic*, Internat. J. Algebra Comput. **2** (1992), 1–17   {V.55}.

Ol's–93. A.Yu. Ol'shanskii, *On residualing homomorphisms and G-subgroups of hyperbolic groups*, Internat. J. Algebra Comput. **3** (1993), 365–409   {III.5}.

Ol's–95a. A.Yu. Ol'shanskii, *A simplification of Golod's example*, in "Groups — Korea '94", A.C. Kim and D.L. Johnson, Editors, de Gruyter (1995), 263–265   {VIII.19}.

Ol's–95b. A.Yu. Ol'shanskii, *The SQ-universality of hyperbolic groups*, Matem. Sbornik **186:8** (1995), 119–132   {III.37}.

Ol's–97. A.Yu. Ol'shanskii, *On subgroup distortion in finitely presented groups*, Sbornik Math. **188:11** (1997), 1617–1664   {IV.3.vii, VI.43}.

Ol's–99a. A.Yu. Ol'shanskii, *The growth of finite subgroups in p-groups*, in "Groups St Andrews 1997 in Bath, II", C.M. Campbell, E.F. Robertson, N. Ruskuc, and G.C. Smith, Editors, Cambridge Univ. Press (1999), 579–595   {VII.39}.

Ol's–99b. A.Yu. Ol'shanskii, *Distortion functions for subgroups*, in "Geometric group theory down under", J. Cossey, C.F. Miller III, W.D. Neumann and M. Shapiro, Editors, de Gruyter (1999), 281–291   {IV.3.vii, VI.43}.

Osin–99a. D.V. Osin, *On the generating set's growth function for subgroups of free groups*, Internat. J. Algebra Comput. **9** (1999), 41–50   {VI.41}.

Osin–99b. D.V. Osin, *On the growth of rank for subgroups of finitely generated groups*, Sbornik Math. **190:8** (1999), 1151–1172   {VI.41}.

Osin. D.V. Osin, *Distortions of subgroups in nilpotent groups*, preprint, February 2000   {VI.43}.

Osofs–78. B. Osofsky, *Problem 6102*, Amer. Math. Monthly **85** (1978), 504–505   {II.37}.

Pansu–83. P. Pansu, *Croissance des boules et des géodésiques fermées dans les nilvariétés*, Ergod. Th. & Dynam. Sys. **3** (1983), 415–445   {VII.33–34}.

Pansu–89. P. Pansu, *Métriques de Carnot-Carathéodory et quasiisométries des espaces symétriques de rang un*, Annals of Math. **129** (1989), 1–60   {IV.25.vii, IV.41}.

Papas–95. P. Papasoglu, *Homogeneous trees are bi-Lipschitz equivalent*, Geometriae Dedicata **54** (1995), 301–306   {IV.46}.

PapWh. P. Papasoglu and K. Whyte, *Quasi-isometries between groups with many ends*, preprint, 1999   {IV.46}.

Paris–91. L. Paris, *Growth series of Coxeter groups*, in "Group theory from a geometrical viewpoint (26 March — 6 April 1990, Trieste)", E. Ghys, A. Haefliger and A. Verjovsky, Editors, World Scientific (1991), 302–310   {VI.10}.

Parr–88. W. Parry, *Counter-examples involving growth series and Euler characteristics*, Proc. Amer. Math. Soc. **102** (1988), 49–51   {VI.15}.

Parr–92a. W. Parry, *A sharper Tits alternative for 3-manifold groups*, Israel J. Math. **77** (1992), 265–271   {II.42}.

Parr–92b. W. Parry, *Growth series of some wreath products*, Trans. Amer. Math. Soc. **331** (1992), 751–759   {IV.44, VI.8}.

PasTe–96. D.S. Passman and W.V. Temple, *Representations of the Gupta-Sidki group*, Proc. Amer. Math. Soc. **124** (1996), 1403–1410   {VII.39, VIII.87}.

Pater–99. G.P. Paternain, *Geodesic flows*, Birkhäuser, 1999   {VI.53}.

Patte–87. S.J. Patterson, *Lectures on measures on limit sets of Kleinian groups*, in "Analytical and geometric aspects of hyperbolic space, Warwick and Durham", D.B.A. Epstein, Editor, Cambridge Univ. Press (1987), 281–323   {I.2}.

Pauli–97. F. Paulin, *Actions de groupes sur les arbres*, Séminaire Bourbaki, vol. 1995/96, Astérisque **241**, Soc. Math. France (1997), 97–137   {Introduction, III.15}.

Pauli. F. Paulin, *Un groupe hyperbolique est déterminé par son bord*, J. London Math. Soc., to appear   {V.58}.

Pawli–98. J. Pawlikowski, *The fundamental group of a compact metric space*, Proc. Amer. Math. Soc. **126** (1998), 3083–3087   {III.6}.

PayVa–91. I. Pays and A. Valette, *Sous-groupes libres dans les groupes d'automorphismes d'arbres*, l'Enseignement math. **37** (1991), 151–174   {II.24, II.42}.

PitSa–99. C. Pittet and L. Saloff–Coste, *Amenable groups, isoperimetric profiles and random walks*, in "Geometric group theory down under", J. Cossey, C.F. Miller III, W.D. Neumann and M. Shapiro, Editors, de Gruyter (1999), 293–316 {VII.32}.

Pitte–93. C. Pittet, *Surface groups and quasi-convexity*, Volume 1 of "Geometry group theory, Sussex 1991", G.A. Niblo and M.A. Roller, Editors, Cambridge Univ. Press (1993), 169–175   {IV.49}.

Pitte–97. C. Pittet, *Isoperimetric inequalities in nilpotent groups*, J. London Math. Soc. **55** (1997), 588–600   {V.57}.

Pitte. C. Pittet, *The isoperimetric profile of homogeneous Riemannian manifolds*, preprint, University of Toulouse, 2000   {VII.34}.

PlaRa–91. V. Platonov and A. Rapinchuk, *Algebraic groups and number theory*, Academic Press, 1994 [Russian original: 1991]   {V.20.viii}.

Poinc–82. H. Poincaré, *Théorie des groupes fuchsiens*, Acta Math. **1** (1882), 1–62 [Oeuvres, tome II, pages 108–168]   {V.40}.

Poinc–85. H. Poincaré, *Sur les courbes définies par les équations différentielles*, J. de Math. Pures et Appl. **1** (1885), 167–244 [Oeuvres, tome I, pages 90–161, in particular 137–158]   {IV.48}.

Poinc–95. H. Poincaré, *Analysis situs*, Journal de l'École Polytechnique **1** (1895), 1–121 [Oeuvres, tome VI, pages 193–288]   {IV.29, V.27}.

PolSh–98. M. Pollicott and R. Sharp, *Comparison theorems and orbit counting in hyperbolic geometry*, Trans. Amer. Math. Soc. **350** (1998), 473–499   {VI.46}.

PólSz–76. G. Pólya and G. Szegö, *Problems and theorems in analysis, vol. I & vol. II*, Die Grundlehren der mathematischen Wissenschaften, Band 193 & Band 216, Springer, 1972 & 1976   {VI.55, VIII.67}.

Pólya–21. G. Pólya, *Ueber eine Aufgabe der Wahrscheinlichkeitsrechnung betreffend die Irrfahrt im Strassennetz*, Math. Ann. **84** (1921), 149–160   {I.B}.

PraRa–96. G. Prasad and A.S. Rapinchuk, *Computation of the metaplectic kernel*, Publ. Math. I.H.E.S. **84** (1996), 91–187   {III.18}.

PrasG–76. G. Prasad, *Discrete subgroups isomorphic to lattices in semisimple Lie groups*, American J. Math. **98** (1976), 241–261   {III.22}.

PrasG–91. G. Prasad, *Semi-simple groups and arithmetic subgroups*, Proceedings of the International Congress of Mathematicians, Kyoto, 1990, **II** (Math. Soc. of Japan, 1991), 821–832   {Introduction}.

PrasV–81. V.S. Prasad, *Generating dense subgroups of measure preserving transformations*, Proc. Amer. Math. Soc. **83** (1981), 286–288   {III.46}.

Preis–42. A. Preissmann, *Quelques propriétés globales des espaces de Riemann*, Comment. Math. Helvetici. **15** (1942), 175–216   {V.53}.

Pride–80. S.J. Pride, *The concept of "largeness" in group theory*, in "Word Problem II", S.D. Adian, W.W. Boone, and G. Higman, Editors, North Holland (1980), 299–335   {VII.14}.

Raghu–72. M.S. Raghunathan, *Discrete subgroups of Lie groups*, Ergebnisse der Mathematik und ihrer Grenzgebiete, Band 68, Springer, 1972
{III.5, IV.25.viii, IV.29, IV.42, V.20, VII.26}.

Raghu–84. M.S. Raghunathan, *Torsion in cocompact lattices in coverings of Spin(2, n)*, Math. Ann. **266** (1984), 403–419   {III.18}.

Ranki–69. R.A. Rankin, *The modular group and its subgroups*, Ramanujan Institute, University of Madras, 1969   {II.28}.

Rapin–92. A. Rapinchuk, *Congruence subgroup problem for algebraic groups: old and new*, in "Journées arithmétiques de Genève, 9–13 septembre 1991", D.F. Coray and Y.-F. S. Pétermann, Editors, Astérisque **209**, Soc. Math. France (1992), 73–84
{Introduction}.

Rapin–99. A. Rapinchuk, *On the finite-dimensional unitary representations of Kazhdan groups*, Proc. Amer. Math. Soc. **127** (1999), 1557–1562   {VII.39}.

Rapin-99. A.S. Rapinchuk, *The congruence subgroup problem*, in "Algebra, K-theory, groups and education, on the occasion of Hyman Bass's 65th birthday", T.Y. Lam and A.R. Magid, Editors, Contemporary Math. **243** (1999), 175–188 {Introduction}.

RehSo-76. U. Rehmann and C. Soulé, *Finitely presented groups of matrices*, in "Algebraic K-theory, Evanston 1976", Lecture Notes in Mathematics **551**, Springer (1976), 164–169   {V.21}.

Reid-90. A.W. Reid, *A note on trace-fields of Kleinian groups*, Bull. London Math. Soc. **22** (1990), 349–352   {IV.29}.

Reide-32. K. Reidemeister, *Einführung in die kombinatorische Topologie*, Vieweg, 1932 {II.18}.

Remes-74. V.N. Remeslennikov, *Finitely presented group whose center is not finitely generated*, Algebra and Logic **13:4** (1974), 258–264   {III.5}.

Rham-69. G. de Rham, *Lectures on introduction to algebraic topology*, Tata Inst. of Fund. Research, Bombay, 1969   {Introduction, II.12–15, III.6}.

Rham-71. G. de Rham, *Sur les polygones générateurs des groupes fuchsiens*, l'Enseignement math. **17** (1971), 49–61 [Oeuvres mathématiques : 610–622]   {V.40}.

Rips-82. E. Rips, *Subgroups of small cancellation groups*, Bull. London Math. Soc. **14** (1982), 45–47   {IV.49, V.55}.

Rips-95. E. Rips, *Cyclic splittings of finitely presented groups and the canonical JSJ-decomposition*, Proceedings of the International Congress of Mathematicians, Zürich, 1994, **1** (Birkhäuser, 1995), 595–600   {III.15}.

RipSe-97. E. Rips and Z. Sela, *Cyclic splittings of finitely presented groups and the canonical JSJ decomposition*, Annals of Math. **146** (1997), 53–109   {III.15}.

Rivin. I. Rivin, *Growth in free groups (and other stories)*, preprint, University of Warwick, 1998   {VII.39}.

Roe-93. J. Roe, *Coarse cohomology and index theory on complete Riemannian manifolds*, Mem. Amer. Math. Soc. **104**, 1993   {V.23}.

Roman-92. V.A. Roman'kov, *Automorphisms of groups*, Acta Applicandae Math. **29** (1992), 241–280   {V.22}.

Rong-94. Xiaochun Rong, *On the fundamental groups of manifolds of positive sectional curvature*, Annals of Math. **143** (1996), 397–411   {V.51}.

Rosbl-74. J.M. Rosenblatt, *Invariant measures and growth conditions*, Trans. Amer. Math. Soc. **193** (1974), 33–53   {VII.8, VII.28}.

Rosen-94. J. Rosenberg, *Algebraic K-theory and its applications*, Graduate Texts in Mathematics **147**, Springer, 1994   {III.2}.

Rosse-76. S. Rosset, *A property of groups of nonexponential growth*, Proc. Amer. Math. Soc. **54** (1976), 24–26   {VII.27}.

Rotma-95. J.J. Rotman, *An introduction to the theory of groups*, fourth edition, Graduate Texts in Mathematics **148**, Springer, 1995 [first edition 1965]   {Introduction, III.14, III.16–17, III.24, III.47, III.51, V.24–26, VIII.3, VIII.5, VIII.49}.

Röver-99. C.E. Röver, *Constructing finitely presented simple groups that contain Grigorchuk groups*, J. of Algebra **220** (1999), 284–313   {VIII.49}.

Rozhk-94. A.V. Rozhkov, *Centralizers of elements in a group of tree automorphisms*, Russian Acad. Sci. Izv. Math. **43:3** (1994), 471–492   {VIII.21, VIII.24–25}.

Rozhk-96. A.V. Rozhkov, *Lower central series of a group of tree automorphisms*, Mat. Zameski **60:2** (1996), 225–237   {VIII.31}.

Rubel-96. L.A. Rubel (with J.E. Colliander), *Entire and meromorphic functions*, Springer, 1996   {VI.40}.

Rudin-53. W. Rudin, *Principles of mathematical analysis*, third edition, McGraw-Hill, 1976 [first edition 1953]   {VII.34}.

Saitô-68. T. Saitô, *Generators of certain von Neumann algebras*, Tôhoku Math. J. **20** (1968), 101–105   {III.46}.

Sakai-71. S. Sakai, *C*-algebras and W*-algebras*, Ergebnisse der Mathematik und ihrer Grenzgebiete, Band 60, Springer, 1971   {III.45}.

SakZy–65. S. Saks and A. Zygmund, *Analytic functions*, second edition enlarged, PWN — Polish Scientific Publishers, 1965   {I.B, I.7}.

SalSo–78. A. Salomaa and M. Soittola, *Automata-theoretic aspects of formal power series*, Texts and Monographs in Computer Science, Springer, 1978   {VI.17}.

Sambu–99. A. Sambusetti, *Minimal growth of non-Hopfian free products*, C.R. Acad. Sci. Paris **329** (1999), 943-946   {VII.19}.

Sarna–90. P. Sarnak, *Some applications of modular forms*, Cambridge Univ. Press, 1990   {II.41}.

Satô–95. K. Satô, *A free group acting without fixed points on the rational unit sphere*, Fund. Math. **148** (1995), 63–69   {II.38}.

Satô–98. K. Satô, *Free groups acting without fixed points on rational spheres*, Acta Arithmetica **85** (1998), 135–140   {II.38}.

Satô–99. K. Satô, *A free group acting on $\mathbb{Z}^2$ without fixed point*, l'Enseignement math. **45** (1999), 189–194   {II.25}.

Schmu–99. P. Schmutz Schaller, *Teichmüller space and fundamental domains of Fuchsian groups*, l'Enseignement math. **45** (1999), 169–187   {II.33.viii}.

Schre–27. O. Schreier, *Die Untergruppen der freien Gruppen*, Abh. Math. Sem. Univ. Hamburg **5** (1927), 161–183   {III.14}.

Schre–28. O. Schreier, *Über den Jordan-Hölderschen Satz*, Abh. Math. Sem. Univ. Hamburg **6** (1928), 300–302   {III.47}.

SchUl–33. J. Schreier and S. Ulam, *Über die Permutationsgruppe der natürlichen Zahlenfolge*, Studia Math. **4** (1933), 134–141   {III.35}.

SchUl–35. J. Schreier and S. Ulam, *Sur le nombre des générateurs d'un groupe topologique compact et connexe*, Fund. Math. **24** (1935), 302–304 [see also [Ulam–74], pages 141–143 and 681]   {II.41, III.46}.

Schwa–55. A.S. Schwarzc, *A volume invariant of coverings (in Russian)*, Dokl. Ak. Nauk. **105** (1955), 32–34   {IV.22, VII, VII.5}.

Scott–73. P. Scott, *Finitely generated 3-manifold groups are finitely presented*, J. London Math. Soc. **2** (1973), 437–440   {V.31}.

Scott–83. P. Scott, *The geometries of 3-manifolds*, Bull. London Math. Soc. **15** (1983), 401–487   {Introduction, IV.48}.

Scott–92. E.A. Scott, *A tour around finitely presented infinite simple groups*, in "Algorithms and classification in combinatorial group theory", G. Baumslag and C.F. Miller III, Editors, M.S.R.I. Publ. **23**, Springer (1992), 83–119 {V.26}.

ScoWa–79. P. Scott and T. Wall, *Topological methods in group theory*, in "Homological group theory, Durham 1977", C.T.C. Wall Editor, Cambridge Univ. Press (1979), 137–203   {II.22, III.3, III.13–15, IV.25.vi, VII.18}.

Segal–83. D. Segal, *Polycyclic groups*, Cambridge Univ. Press, 1983   {III.5, V.22}.

SeiTr–34. H. Seifert and W. Threlfall, *Lehrbuch der Topologie*, Teubner, 1934   {V.29}.

SeiTr–80. H. Seifert and W. Threlfall, *A textbook of topology*, Academic Press, 1980   {V.29}.

SeiTr–97. N. Seifter and V.I. Trofimov, *Automorphism groups of graphs with quadratic growth*, J. Combinatorial Theory, Series B **71** (1997), 205–210 {IV.40}.

Sela–97a. Z. Sela, *Acylindrical accessibility for groups*, Inventiones Math. **129** (1997), 527–565   {II.36}.

Sela–97b. Z. Sela, *Structure and rigidity in (Gromov) hyperbolic groups and discrete groups in rank 1 Lie groups II*, GAFA Geom. Funct. Anal. **7** (1997), 561–593 {V.22}.

Sela–99. Z. Sela, *Endomorphisms of hyperbolic groups I: the Hopf property*, Topology **38**, 301–321   {III.19}.

Selbe–60. A. Selberg, *On discontinuous groups in higher-dimensional symmetric spaces*, in "Contributions to function theory", Tata Inst. of Fund. Research, K. Chandrasekharan, Editor (1960), 147–164 [= Collected Papers I, pages 475–492] {III.20, IV.42, V.20.iii}.

Sergi–91. V. Sergiescu, *A quick introduction to Burnside's problem*, in "Group theory from a geometrical viewpoint (26 March — 6 April 1990, Trieste)", E. Ghys, A. Haefliger and A. Verjovsky, Editors, World Scientific (1991), 622–629 {VIII.19}.

Serr–70a. J-P. Serre, *Cours d'arithmétique*, Presses univ. de France, 1970   {II.28}.

Serr–70b. J-P. Serre, *Le problème des groupes de congruence pour SL_2*, Annals of Math. **92** (1970), 489–527   {III.4, III.54}.

Serr–71. J-P. Serre, *Cohomologie des groupes discrets*, in "Prospects in mathematics", Annals of Math. Studies **70**, Princeton Univ. Press (1971), 77–169
                                                                     {II.19, IV.46–47}.

Serr–73. J-P. Serre, *Problems*, in "Proc. Conf. Canberra 1973", Lecture Notes in Mathematics **372**, Springer (1973), 734   {V.31}.

Serr–77. J-P. Serre, *Arbres, amalgames, SL_2*, Astérisque, **46**, Soc. math. France, 1977
          {II.16, II.18, II.22, II.36, III.14, IV.3, IV.23, V.2, V.26, VI.18, VII.9, VII.18}.

Shale–99. P.B. Shalen, *Three-manifold topology and the tree for PSL_2: the Smith conjecture and beyond*, in "Algebra, K-theory, groups and education, on the occasion of Hyman Bass's 65th birthday", T.Y. Lam and A.R. Magid, Editors, Contemporary Math. **243** (1999), 189–209   {III.20}.

Shalo–98. Y. Shalom, *The growth of linear groups*, J. of Algebra **199** (1998), 169–174
                                                                             {VII.27}.

Shalo–a. Y. Shalom, *Explicit Kazhdan constants for representations of semisimple and arithmetic groups*, Ann. Inst. Fourier, to appear   {III.2, VII.12, VII.17}.

Shalo–b. Y. Shalom, *Bounded generation and Kazhdan's property (T)*, Publ. Math. I.H.E.S., to appear   {III.2}.

Shapi–89. M. Shapiro, *A geometric approach to the almost convexity and growth of some nilpotent groups*, Math. Ann. **285** (1989), 601–624   {VI.6}.

Shapi–97. M. Shapiro, *Pascal's triangles in abelian and hyperbolic groups*, J. Austral. Math. Soc. **63** (1997), 281–288   {VII.39}.

Sharp–98. R. Sharp, *Relative growth series in some hyperbolic groups*, Math. Ann. **312** (1998), 125–132   {VII.35}.

ShaWa–92. P.B. Shalen and P. Wagreich, *Growth rates, $\mathbb{Z}_p$-homolgy, and volumes of hyperbolic 3-manifolds*, Trans. Amer. Math. Soc. **331** (1992), 895–917
                                                                     {VII.12, VII.14}.

Shor8–91. J.M. Alonso, T. Brady, D. Cooper, V. Ferlini, M. Lustig, M. Mihalik, M. Shapiro, H. Short; edited by H. Short, *Notes on word hyperbolic groups*, in "Group theory from a geometrical viewpoint (26 March — 6 April 1990, Trieste)", E. Ghys, A. Haefliger and A. Verjovsky, Editors, World Scientific (1991), 3–63   {IV.49, V.60}.

Short–91. H. Short, *Quasiconvexity and a theorem of Howson's*, in "Group theory from a geometrical viewpoint (26 March — 6 April 1990, Trieste)", E. Ghys, A. Haefliger and A. Verjovsky, Editors, World Scientific (1991), 168–176 {IV.49}.

Shubi–86. M.A. Shubin, *Pseudodifference operators and their Green's functions*, Math. USSR Izvestiya **26** (1986), 605–622   {VI.21, VI.26}.

Siege–71. C.L. Siegel, *Topics in complex function theory, vol. II, automorphic functions and abelian integrals*, Wiley-Interscience, 1971   {V.36}.

Smit–92. B. de Smit, *The fundamental group of the Hawaiian earring is not free*, Internat. J. Algebra Comput. **2** (1992), 33–37   {III.6}.

Solom–66. L. Solomon, *The orders of the finite Chevalley groups*, J. of Algebra **3** (1966), 376–393   {VI.10}.

Soulé–78. C. Soulé, *The cohomology of $SL_3(\mathbb{Z})$*, Topology **17** (1978), 1–22   {V.9}.

Stall–63. J. Stallings, *A finitely presented group whose 3-dimensional integral homolgy is not finitely generated*, Amer. J. Math. **85** (1963), 541–543   {V.23}.

Stall–66. J. Stallings, *How not to prove the Poincaré conjecture*, in "Topology Seminar Wisconsin, 1965", R.H. Bing and R.J. Bean, Editors, Ann. Math. Studies **60**, Princeton Univ. Press (1966), 83–88   {II.47}.

Stall–68. J. Stallings, *On torsion free groups with infinitely many ends*, Annals of Math. **88** (1968), 312–334   {II.19, IV.50}.

Stall–71. J. Stallings, *Group theory and three-dimensional manifolds*, Yale University
Press, 1971								{III.13, III.15}.

Stall–91. J. Stallings, *Non-positively curved triangles of groups*, in "Group theory from
a geometrical viewpoint (26 March — 6 April 1990, Trieste)", E. Ghys, A.
Haefliger and A. Verjovsky, Editors, World Scientific (1991), 491–503 {III.14}.

Stall–95. J. Stallings, *Problems about free quotients of groups*, in "Geometric group
theory, Ohio State University 1992", R. Charney, M. Davis, and M. Shapiro,
Editors, Ohio State University Math. Res. Inst. Publ. **3** (1995), 165–182 {II.44}.

Stanl–80. R.P. Stanley, *Differentiably finite power series*, Europ. J. Combinatorics **1**
(1980), 175–188								{I.8}.

Stanl–86. R.P. Stanley, *Enumerative combinatorics, Volume I*, Wadsworth and Brooks,
1986								{VI.14, VI.17}.

Stark–95. A.N. Starkov, *Fuchsian groups from the dynamical viewpoint*, J. of Dynamical
and Control Systems **1:3** (1995), 427–445					{III.25}.

Stein–62. R. Steinberg, *Generators for simple groups*, Canadian J. Math. **14** (1962),
277–283								{III.46}.

Stein–85. R. Steinberg, *Some consequences of the elementary relations in SL_n*, in "Finite
groups — coming of age", Contemporary Math. **45** Amer. Math. Soc. (1985),
335–350								{III.22, V.9}.

Stepi–83. A.M. Stepin, *Approximability of groups and group actions*, Russian Math.
Surveys **38:6** (1983), 131–132 [Russian original: Uspekhi Math. Nauk. 38:6
(1983) pp. 123–134]							{III.18}.

Stepi–84. A.M. Stepin, *A remark on the approximability of groups*, Vestnik Moskov. Univ.
Ser. I Math. Mekh **4** (1984), 85–87					{III.18}.

Stepi–96. A.M. Stepin, *Approximation of groups and group actions, the Cayley topology*, in "Ergodic theory of $\mathbb{Z}^d$-actions", M. Pollicott and K. Schmidt, Editors,
Cambridge Univ. Press (1996), 475–484				{III.18, V.10}.

Still–80. J. Stillwell, *Classical topology and combinatorial group theory*, second edition,
Graduate Texts in Mathematics **72**, Springer, 1993 [first edition 1980]
{V.5, V.27, V.46}.

Still–82. J. Stillwell, *The word problem and the isomorphism problem for groups*, Bull.
Amer. Math. Soc. **6** (1982), 33–56				{III.14, VIII.49}.

Stoll–92. M. Stoll, *Galois groups over $\mathbb{Q}$ of some iterated polynomials*, Arch. Math. **59**
(1992), 239–244							{VIII.8}.

Stoll–96. M. Stoll, *Rational and transcendental growth series for the higher Heisenberg
groups*, Inventiones Math. **126** (1996), 85–109			{VI.8}.

Stoll–98. M. Stoll, *On the asymptotics of the growth of 2-step nilpotent groups*, J. London
Math. Soc. **58** (1998), 38–48					{VII.33}.

Streb–84. R. Strebel, *Finitely presented soluble groups*, in "Group theory, essays for Philip
Hall", K.W. Gruenberg and J.E. Roseblade, Editors, Academic Press (1984),
257–314								{V.19}.

Stroj–99. A. Strojnowski, *On residually finite groups and their generalizations*, Collo-
quium Math. **79** (1999), 25–35					{III.18}.

Sulli–81. D. Sullivan, *Travaux de Thurston sur les groupes quasi-fuchsiens et les variétés
hyperboliques de dimension 3 fibrées sur $\mathbb{S}^1$*, Séminaire Bourbaki, vol. 1979/80,
Lecture Notes in Mathematics **842**, Springer (1981), 196–214		{IV.49}.

Swan–69. R. Swan, *Groups of cohomological dimension one*, J. of Algebra **12** (1969),
585–601								{II.19}.

Swan–71. R. Swan, *Generators and relations for certain special linear groups*, Adv. in
Math. **6** (1971), 1–77						{IV.23}.

Synge–36. J.L. Synge, *On the connectivity of spaces of positive curvature*, Quarterly J.
Math. **7** (1936), 316–320						{V.51}.

Tauge–91. O.I. Taugen, *Bounded generators of Chevalley groups over rings of algebraic
S-integers*, Math. USSR Izvestiya **36:1** (1991), 101–128		{III.2}.

TaWG–94. M.C. Tamburini, J.S. Wilson, and N. Gavioli, *On the (2,3)-generation of some
classical groups. I*, J. of Algebra **168** (1994), 353–370		{III.39}.

Thomp–80. R.J. Thompson, *Embeddings into finitely generated groups which preserve the word problem*, in "Word Problem II", S.D. Adian, W.W. Boone, and G. Higman, Editors, North Holland (1980), 401–441   {V.26}.

ThoWo–93. C. Thomassen and W. Woess, *Vertex-transitive graphs and accessibility*, J. Combinatorial Theory, Series B **56** (1993), 248–268   {III.15, IV.50}.

Thurs–82. W.P. Thurston, *Three dimensional manifolds, Kleinian groups and hyperbolic geometry*, Bull. Amer. Math. Soc. **6** (1982), 357–381
{Introduction, III.18, IV.29}.

Tits–72. J. Tits, *Free subgroups in linear groups*, J. of Algebra **20** (1972), 250–270
{II.B, II.41–42, VII.27}.

Tits–81a. J. Tits, *Appendix* (to our [Gromo–81]), Publ. Math. I.H.E.S. **53** (1981), 74–78
{VII.27}.

Tits–81b. J. Tits, *Groupes à croissance polynomiale*, Séminaire Bourbaki, vol. 1980/81, Lecture Notes in Mathematics **901**, Springer (1981), 176–188
{VII.4–5, VII.29}.

Trofi–89. V.I. Trofimov, *Certain asymptotic characteristics of groups*, Math. Notes **46** (1989), 945–951   {IV.3.ix}.

Trofi–92. V.I. Trofimov, *On the action of a group on a graph*, Acta Applicandae Math. **29** (1992), 161–170   {IV.3.ix}.

Tyrer–74. J.M. Tyrer Jones, *Direct products and the Hopf property*, J. Austral Math. Soc. **17** (74), 174–196   {VIII.2}.

Ulam–74. S. Ulam, *Sets, numbers, and universes; selected works*, W.A. Beyer, J. Mycielski, and G.-C. Rota, Editors, M.I.T. Press, 1974   {III.7}.

UlavN–46. S.M. Ulam and J. von Neumann, *Random ergodic theorems*, Bull. Amer. Math. Soc. (Abstract of a paper submitted to the Society) **51** (1946), 660   {III.46}.

VaLWi–92. J.H. van Lint and R.M. Wilson, *A course in combinatorics*, Cambridge Univ. Press, 1992   {IV.3.ii, VI.55}.

VandW–48. B.L. van der Waerden, *Free products of groups*, Amer. J. Math. **70** (1948), 527–528   {II.1}.

Varop–99. N.Th. Varopoulos, *Distance distortion on Lie groups*, in "Random walks and discrete potential theory, Cortona 1997 ", M. Picardella and W. Woess Editors, Symposia Math. **XXXIX**, Cambridge Univ. Press (1999), 320–357   {VI.43}.

VarSC–92. N.Th. Varopoulos, L. Saloff-Coste, and T. Coulhon, *Analysis and geometry on groups*, Cambridge Univ. Pres, 1992   {I.4}.

VauZe–99. M. Vaughan-Lee and E.I. Zel'manov, *Bounds in the restricted Burnside problem*, J. Austral. Math. Soc. (Series A) **67** (1999), 261–271.   {VIII.19}.

VdDW–84a. L. van den Dries and A.J. Wilkie, *On Gromov's theorem concerning groups of polynomial growth and elementary logic*, J. of Algebra **89** (1984), 349–374
{Introduction, VII.29}.

VdDW–84b. L. van den Dries and A.J. Wilkie, *An effective bound for groups of linear growth*, Arch. Math. **42** (1984), 391–396   {VII.29–30, VIII.60}.

vonNe–29. J. von Neumann, *Zur allgemeinen Theorie des Masses*, Fund. Math. **13** (1929), 73–116 [Collected Works, vol. I, 599–642 & 643]   {II.31}.

Wagon–85. S. Wagon, *The Banach-Tarski paradox*, Cambridge Univ. Press, 1985 {II.37}.

Wagre–82. P. Wagreich, *The growth series of discrete groups*, in "Proc. Conference on algebraic varieties with group actions", Lecture Notes in Mathematics **956**, Springer (1982), 125–144   {VI.15}.

Walte–82. P. Walters, *An introduction to ergodic theory*, Graduate Texts in Mathematics **79**, Springer, 1982   {VI.55}.

Wang–63. H.C. Wang, *On the deformations of lattices in a Lie group*, Amer. J. Math. **85** (1963), 189–212   {V.20.vi}.

Wang–72. H.C. Wang, *Topics on totally discontinuous groups*, in "Symmetric spaces, short courses presented at Washington University ", W.M. Boothby and G.L. Weiss, Editors, M. Dekker (1972), 459–487   {V.20.vi}.

Wang–81. S.P. Wang, *A note on free subgroups in linear groups*, J. of Algebra **71** (1981), 232–234   {II.42}.

Weber–89. B. Weber, *Zur Rationalität polynomialer Wachstumsfunktionen*, Bonner Mathematische Schriften **197**, 1989   {VI.6}.

Weiss–81. B. Weiss, *Orbit equivalence of nonsingular actions*, in "Théorie ergodique, Les Plans-sur-Bex, 23–29 mars 1980", l'Enseignement math., monographie **29** (1981), 77–107   {VIII.7}.

Weyl–39. H. Weyl, *The classical groups, their invariants and representations*, Princeton Univ. Press, 1939   {VIII.87}.

White–88. S. White, *The group generated by $x \mapsto x + 1$ and $x \mapsto x^p$ is free*, J. of Algebra **118** (1988), 408–422   {II.40}.

Whyte–99. K. Whyte, *Amenability, bilipschitz equivalence, and the von Neumann conjecture*, Duke Math. J. **99** (1999), 93–112   {IV.46–47}.

Whyte–a. K. Whyte, *Quasi-isometry types of graphs of $\mathbb{Z}$s*, preprint, University of Utah, 1998   {IV.43}.

Wilf–90. H.S. Wilf, *Generatingfunctionology*, Academic Press, 1990   {VI.14}.

Wilso–76. J.S. Wilson, *On characteristically simple groups*, Math. Proc. Camb. Phil. Soc. **80** (1976), 19–35   {III.2}.

Wilso. J.S. Wilson, *On just infinite abstract and profinite groups*, in "Horizons in profinite groups", D. Segal, Editor, Birkhäuser, to appear   {VIII.31}.

Wink–97a. J. Winkelmann, *On discrete Zariski-dense subgroups of algebraic groups*, Math. Nachr. **186** (1997), 285–302   {II.41}.

Wink–97b. J.Winkelmann, *Only countably many simply-connected Lie groups admit lattices*, in "Complex analysis and geometry (Trento 1995)", Pitman Res. Notes Math. Ser. **366** (1997), 182–185   {IV.25.viii}.

Wise–96. D.T. Wise, *A non-Hopfian automatic group*, J. of Algebra **180** (1996), 845–847   {III.19}.

Wise–98. D.T. Wise, *Incoherent negatively curved groups*, Proc. Amer. Math. Soc. **126** (1998), 957–964   {V.31, V.55}.

Wise–99. D.T. Wise, *A continually descending endomorphism of a finitely generated residually finite group*, Bull. London Math. Soc. **31** (1999), 45–49   {III.19}.

Witte–95. D. Witte, *Superrigidity of lattices in solvable Lie groups*, Inventiones Math. **122** (1995), 147–193   {IV.25.viii, IV.27, VII.26}.

Woess–89. W. Woess, *Graphs and groups with tree-like properties*, J. Combinatorial Theory, Series B **47** (1989), 361–371   {IV.50}.

Woess–94. W. Woess, *Random walks on infinite graphs and groups — a survey on selected topics*, Bull. London Math. Soc. **26** (1994), 1–60   {I.4, VII.37}.

Woess–00. W. Woess, *Random walks on infinite graphs and groups*, Cambridge Tracts in Mathematics, to appear   {Introduction, I.11, VII.37}.

Wolf–64. J.A. Wolf, *Curvature in nilpotent Lie groups*, Proc. Amer. Math. Soc. **15** (1964), 271–274   {VII.7}.

Wolf–68. J.A. Wolf, *Growth of finitely generated solvable groups and curvature of Riemannian manifolds*, J. Diff. Geom. **2** (1968), 421–446   {VII.27}.

Wrigh–75. P. Wright, *Group presentations and formal deformations*, Trans. Amer. Math. Soc. **208** (1975), 161–169   {II.45}.

Wrigh–92. D. Wright, *Two-dimensional Cremona groups acting on simplicial complexes*, Trans. Amer. Math. Soc. **331** (1992), 281–300   {V.2}.

Wu–88. T.W. Wu, *A note on a theorem on lattices in Lie groups*, Canadian Math. Bull. **31** (1988), 190–193   {V.20.vi}.

Zelma–91. E.I. Zel'manov, *A solution of the restricted Burnside problem for groups of odd exponent*, Math. USSR Izvestiya **36:1** (1991), 41–60   {VIII.19, VIII.33}.

Zelma–92. E.I. Zel'manov, *A solution of the restricted Burnside problem for 2-groups*, Math. USSR Sbornik **72:2** (1992), 543–565   {VIII.19, VIII.33}.

Zelma–95. E.I. Zel'manov, *More on Burnside's problem*, in "Combinatorial and geometric group theory, Edinburgh 1993", A.J. Duncan, N.D. Gilbert, and J. Howie, Editors, London Math. Soc. Lecture Note Series **204** (1995), 314–321 {VIII.19}.

Zelma–99. E. Zel'manov, *On some open problems related to the restricted Burnside prob-
           lem*, in "Recent progress in algebra", S.G. Hahn, H.C. Myung and E. Zel'manov,
           Editors, Contemporary Mathematics **224** (1999), 237–243   {VIII.19}.
Zimme–84. R.J. Zimmer, *Ergodic theory and semi-simple groups*, Birkhäuser, 1984
                                                          {II.28, IV.29, V.20.viii}.
Zimme–87. R.J. Zimmer, *Amenable actions and dense subgroups of Lie groups*, J. Func-
           tional Analysis **72** (1987), 58–64   {V.20.vi}.

INDEX OF RESEARCH PROBLEMS

I.A. Is it true that $\sharp\left\{\,(a,b)\in\mathbb{Z}^2 \mid a^2+b^2\le t\,\right\} - \pi t = O\left(t^{\frac{1}{4}+\epsilon}\right)$ for all $\epsilon > 0$?

II.40. Use the Table-Tennis Lemma to show that $t \longmapsto t+1$ and $t \longmapsto t^3$ generate a free group.

II.41. Do dense free subgroups of $PSL(2,\mathbb{R})$ give rise to equidistributed orbits in the hyperbolic plane ? (Similar problems for other Lie groups of isometries).

II.43. Does $SL(n,\mathbb{Z})$ have Property $P_{\text{naï}}$ about free subgroups for $n \ge 3$?

II.45. On homomorphisms of F_{2g} onto $F_g \times F_g$, and the Andrews-Curtis conjecture.

II.46. The Kervaire conjecture: $(\Gamma * \mathbb{Z})/(r) = 1 \overset{???}{\Longrightarrow} \Gamma = 1$.

II.47. The 3-dimensional Poincaré conjecture.

III.5. Is any countable abelian group a subgroup of the centre of some finitely-presented group ?

III.6. Can any finitely-generated group be realized as the fundamental group of a compact metric space which is path-connected and locally path-connected ?

III.17. Find "nice and explicit" embeddings of $\mathbb{Q}$ in finitely-generated and finitely-presented groups.

III.35.v. Compute the exponential growth rates of the groups $G_{\mathcal{U}}$ of B.H. Neumann.

III.45. Existence of uncountably many finitely-generated groups with pairwise non-isomorphic II_1 factors.

III.46. What is the minimal number of elements which generate a dense subgroup in the unitary group of a given factor of type II_1, with the strong topology ?

IV.9. On properties of a group which are reflected in one of its Cayley graphs; see also VI.2.viii.

IV.25.vii. Classify torsion-free finitely-generated nilpotent groups up to quasi-isometry, and classify connected simply-connected nilpotent real Lie groups up to quasi-isometry.

IV.25.viii. Does there exist a finitely-generated group quasi-isometric to $\mathbb{R}^2 \rtimes SL(2, \mathbb{R})$?

IV.29.iii. Existence of pairs (M_1, M_2) of closed hyperbolic 3-manifolds with incommensurable Riemannian volumes.

IV.36. On torsion of large order in a finitely-generated group which is quasi-isometric to a torsion-free group.

IV.44.A. When are two lamplighter groups Γ_F and $\Gamma_{F'}$ quasi-isometric ? (An important case to understand is that with $|F| = 2$ and $|F'| = 3$.)

IV.44.B. Find two *finitely-presented* quasi-isometric groups, one solvable and the other not almost solvable.

IV.46.vii. For Γ an amenable finitely-generated infinite group and F a finite group, when are Γ and $\Gamma \times F$ Lipschitz equivalent ?
Do there exist two infinite finitely-generated groups which are quasi-isometric and not Lipschitz equivalent ?

IV.48. Let G be a non-linear connected real Lie group and let Γ be a lattice in G. Can one show that Γ is linear if and only if it is residually finite ?

IV.50. Are the following properties geometric ? having a polycyclic subgroup of finite index, having Kazhdan Property (T); see also VII.20.

V.20.iv. Extending to orbifolds the proof showing that appropriate lattices are finitely presented.

V.32. Does there exist a smooth projective complex variety of which the fundamental group is infinite and simple ?

V.33.A. Characterize the fundamental groups of the complete Riemannian manifolds with curvature $\kappa \leq 0$.

V.33.B. Characterize the finitely-generated groups which admit a discrete cocompact action on some metric geodesic space of strictly negative curvature.

V.60. Does every hyperbolic group contain a subgroup of finite index without torsion ?
Is every hyperbolic group residually finite ?

VI.2.v. Find more examples of pairs (Γ, S), with growth function $\beta(\Gamma, S; k)$ rational in k, for which $\beta(\Gamma, S; -k)$ has an interpretation, at least for some $k \geq 1$.

VI.2.viii. On properties of a group Γ which are reflected in growth series of the form $\Sigma(\Gamma, S; z)$; see also IV.9.

VI.14. Investigate exponential series and Dirichlet series for the growth of balls in finitely-generated groups.

VI.19. In a finitely-generated group which is not almost cyclic, does the size of spheres tend to infinity ?

VI.20. What can be said about the growth series of braid groups, knot groups, Thompson's group, Baumslag-Solitar groups, Burnside groups, and the Grigorchuk group ?

VI.40. Which functions are growth functions of finitely-generated groups ?

VI.53. Can a finitely-generated group of exponential growth be generated by a finitely-generated sub-semi-group of intermediate growth ?

VI.61. What are the properties of the set of positive numbers which are exponential growth rates of finitely-generated groups ?

VI.62. Does large growth imply non-amenability ?

VI.63. Does there exist a finitely-presented group of intermediate growth ?

VI.64. Is it true that a finitely-generated group of subexponential growth is residually finite ?

VI.65. How do spherical growth functions depend on the choice of generating sets for groups of subexponential growth ?

VI.66. Further study of admissible groups.

VII.11. On the regularity of the growth functions for groups of exponential growth; see also VII.34.B.

VII.16. What is the minimal growth rate of the fundamental group of a closed orientable surface of genus $g \geq 2$?

VII.19.A. Does there exist a finitely-generated group of exponential growth which is not of uniformly exponential growth ?

VII.19.B. Is it true that a soluble group of exponential growth has uniformly exponential growth ?

VII.19.C. Is it true that a lattice in a simple Lie group of higher rank has uniformly exponential growth ?

VII.19.D. Does there exist an infinite group with Kazhdan Property (T) which is not of uniformly exponential growth ?

VII.19.E. What are the finitely-generated groups Γ for which there exist finite generating sets S such that $\omega(\Gamma, S) = \omega(\Gamma)$?

VII.20. Is uniformly exponential growth a geometric property ?

VII.29. Is it true that a finitely-generated group of subradical growth is virtually nilpotent ? (see also Remark VIII.66).

Does there exist a finitely-generated group Γ of which the growth function β_Γ satisfes $\beta_\Gamma(k) \sim e^{\sqrt{k}}$?

VII.30. Is there a direct proof that $\limsup_{k\to\infty} \frac{\log \beta_\Gamma(k)}{\log k}$ is an integer when Γ is a group of polynomial growth ?

VII.34.A. A question about Følner sequences of linear growth.

VII.34.B. A question on groups of intermediate growth and on pairs (Γ, S) with $\lim_{k\to\infty} \frac{\beta(\Gamma,S;k+1)}{\beta(\Gamma,S;k)} = 1$.

VII.36. Extending to hyperbolic groups a formula of Grigorchuk on relative growth and spectral radius.

VII.39. Growth of coset spaces (Schreier graphs) and of double coset spaces.

VIII.19. Burnside problems (see also III.23).

VIII.85. On the Grigorchuk group as a quotient of a surface group.

VIII.86. On the automorphism group of the Grigorchuk group.

VIII.87. Investigate finite-dimensional representations of the Grigorchuk group in a way analogous to that of Passman and Temple for p-groups with p an odd prime.

We would like to mention the lists of *Open problems in combinatorial and geometric group theory* on the MAGNUS web site

http://zebra.sci.ccny.cuny.edu/web/

(see also [BauMS–99]) as well as all *Unsolved problems in group theory* of the Kourovka notebook [Kouro–95].

APPENDIX

CORRECTIONS AND UPDATES – 1st JANUARY, 2003

The book by K. Ohshika has been translated in english [Ohshi–02]. The main chapters are on Gromov's hyperbolic groups, on automatic groups, and on Kleinian groups.

I.B, and random walks on groups.

There is a nice introduction to random walks and diffusion on groups in [Salof–01], starting with a discussion on shuffling cards. A short exposition of Pólya's recurrence theorem can be found in [DymMc–72].

II.21 and VII.38, subgroup growth, and normal subgroup growth.

For further work concerning numbers of subgroups and normal subgroups of finite index in various groups, see among others [LiSMe–00] and [LarLu].

II.24, and a strong Schottky Lemma.

The classical Table-Tennis Lemma, or Schottky Lemma, is often used to show that a pair of isometries g, h of some hyperbolic space have *powers* g^n, h^n which generate a free group. There is a criterion for g, h to generate a free group in [AlFaN].

On free subgroups of isometry groups, see also [Woess–93] and [Karls].

II.25, II.33, and Möbius groups generated by two parabolics which are not free.

Let $\tilde{\Gamma}_z$ denote the subgroup of $SL(2, \mathbb{C})$ generated by $\begin{pmatrix} 1 & z \\ 0 & 1 \end{pmatrix}$ and $\begin{pmatrix} 1 & 0 \\ z & 1 \end{pmatrix}$, so that $\tilde{\Gamma}_z$ is free if $|z| \geq 2$ or if z is transcendental.

Grytczuk and Wójtowicz have shown that $\tilde{\Gamma}_{p/q}$ is *not* free for a set of rational values $z = p/q$ of the parameters which contains infinitely many accumulation points [GryWó–99].

II.28, and arithmeticity of lattices.

In $PSL(2, \mathbb{C})$, all arithmetic lattices which are generated by two elliptic elements and which are not co-compact have been determined [MacMa–01].

II.29$\frac{1}{3}$, more flowers for the herbarium of free groups.

Margulis has discovered a remarkable example of a free subgroup of the affine group of $\mathbb{R}^3$ acting *properly* on $\mathbb{R}^3$ [Margu–83]; an exposition appears in [Drumm–92].

II.29$\frac{2}{3}$, complement on groups with free subgroups.

We reproduce (most of) Problem 12.24 from the Kourovka Notebook.

Given a ring R with identity, the automorphisms of $R[[x]]$ sending x to $x(1 + \sum_{i=1}^{\infty} a_i x^i)$ form a group $N(R)$. We know that $N(\mathbb{Z})$ contains a copy of the free group F_2 of rank 2 (...). Does $N(\mathbb{Z}/p\mathbb{Z})$ contain a copy of F_2 ?

The answer is "yes": see [Camin–97]; it could be a challenge to find a table-tennis proof of this fact. For generalities on these "Nottingham groups" $N(R)$, see [Camin–00].

II.41, a misprint.

There is a misprint in the reference to [Bourb–75], which should be to Chapter VIII, § 2, Exercise 10.

II.41, and dense free subgroups of Lie groups.

The following result [BreGe] answers a question raised by A. Lubotzky and R. Zimmer: *if Γ is a dense subgroup of a connected semisimple real Lie group G, then Γ contains two elements which generate a dense free subgroup of G.* Also: in a connected non-solvable real Lie group of dimension d, any finitely generated dense subgroup contains a dense free subgroup of rank $2d$.

II.42, on Tits' alternative.

Let Γ be a subgroup of the group of homeomorphisms of the circle such that the action of Γ on the circle is minimal. Then, either the action is a conjugate of an isometric action, and therefore Γ contains a commutative subgroup of index at most 2, or Γ contains a so-called quasi-Schottky subgroup, which is in particular a non-abelian free subgroup [Margu–00]. A variation (possibly a simplification ?) of Margulis' original ideas appear in Section 5.2 of [Ghys–01].

For a group Γ of orientation preserving $\mathcal{C}^2$-diffeomorphisms of the circle, it is also known that the existence of an exceptional minimal set implies that Γ has non-abelian free subgroups [Navas].

On $Out(F_n)$, see also [BesFe–00].

Tits' alternative holds for automorphism groups of free soluble groups [Licht–95] and for linear groups over rings of fractions of polycyclic group rings [Licht–93], [Licht–99]. It also holds in a strong sense for subgroups of Coxeter groups [NosVi–02].

If Γ is a Bieberbach group, either both its automorphism group and its outer automorphism group are polycyclic, or both contain non-abelian free subgroups. See [MalSz] for precise criteria to decide which situation holds for a given Bieberbach group, in terms of the associated holonomy representation.

III.4, and examples of non-uniform tree lattices.

For the existence of such non-uniform lattices on uniform trees, see the work of L. Carbone [Carbo–01]. For tree lattices in general, see [BasLu–01].

III.6$\frac{1}{2}$, and further examples of finitely-genrated groups.

Let A be a commutative ring which is a finitely-generated $\mathbb{Z}$-module. Then the group A^* of invertible elements in A is a finitely-generated abelian group.

There is a proof in Section 4.7 of [Samue–67]; its main ingredient is Dirichlet's theorem, according to which the group of units in the ring of integers of a number field $\mathbb{K}$ is a direct product $F \times \mathbb{Z}^{r_1+r_2-1}$, where F is a finite group and r_1 [respectively $2r_2$] is the number of real [respectively complex] embeddings of $\mathbb{K}$ in $\mathbb{C}$.

More generally, if B is a commutative ring which is reduced (this means that 0 is the *only* nilpotent element) and finitely generated over $\mathbb{Z}$, then B^* is finitely generated [Samue–66].

III.18.iv, III.20, and residual finiteness.

On residual finiteness and topological dynamics: see also [Egoro–00].

A proof that finitely-generated linear groups are residually finite appears as Proposition III.7.11 in [LynSc–77].

III.21, on Baumslag-Solitar groups which are Hopfian.

For the equivalence between "$\Gamma_{p,q}$ Hopfian" and "p, q meshed" to hold, the definition should be

two integers $p, q \geq 1$ are *meshed* if they have precisely the same prime divisors

and *not* the definition as it reads in [BauSo–62], or on page 57. I am grateful to E. Souche who pointed out this correction to me.

III.21, on actions of Baumslag-Solitar groups on the line.

For any p, q with $p > q \geq 1$, there exists a faithful action of the group $BS(p,q)$ on the line by orientation preserving real-analytic diffeomorphisms. In particular, $Diff_+^\omega(\mathbb{R})$ contains Baumslag-Solitar groups which are not residually finite [FarFr].

III.24, on maximal subgroups.

In "familiar" uncountable groups, maximal subgroups cannot be countable. More precisely, Pettis [Petti–52] has shown that, if G is a second category[1] nondiscrete Hausdorff group containing a countable everywhere dense subset, then any proper subgroup H of G lies in an uncountable proper sugroup H_+ of G; if H is countable, H_+ can be taken to be everywhere dense as well.

In their work on maximal subgroups of infinite index in finitely generated linear groups (exluding extensions of solvable groups by finite kernels), Margulis and Soifer have shown that such a group Γ contains a free (*infinitely* generated) subgroup F_∞ which maps *onto* any finite quotient of Γ; they deduce from this that any maximal subgroup of Γ which contains F_∞ is necessarily of infinite index. Soifer and Venkataramana have shown the following result: if Γ is an arithmetic subgroup of a non-compact linear semi-simple group G such that the associated simply connected algebraic group over $\mathbb{Q}$ has the so-called congruence subgroup property, for example if $\Gamma = SL(n, \mathbb{Z})$ with $n \geq 3$, then Γ contains

---

[1]Recall that a topological space X is "second category" (= non-meager) if it is *not* the union of countably many subsets whose closures have empty interiors ("ensembles rares"). Baire's theorem shows that locally compact spaces and complete metric spaces are second category, indeed are Baire spaces (= spaces in which countable unions of closed subspaces with empty interiors have empty interiors).

a *finitely generated* free subgroup which maps onto any finite quotient of Γ [SeiVe–00].

III.24 and VIII.39. The Grigorchuk group has the following property: any maximal subgroup in it is of finite index [Pervo–00]. The same property holds for any group commensurable with Γ [GriWi].

III.B, an additional problem: does $SO(3)$ act non-trivially on $\mathbb{Z}$? (Ulam's problem).

I do not know which uncountable groups can act faithfully on a countable set. Of course, the group $Sym(\mathbb{N})$ of all permutations of $\mathbb{N}$ is itself uncountable, and it has received attention at least since [SchUl–33]. Here is a sketch to show that $\mathbb{R}$, viewed as a discrete group, acts faithfully on $\mathbb{N}$; in other and somehow biased words, this produces "a continuous flow on a discrete space". I am most grateful to Tim Steger for several helpful conversations on this material.

Choose a basis (e_t) of $\mathbb{R}$ as a vector space over $\mathbb{Q}$ which is indexed by the open interval $]0, 1[$ of the line. Let C denote the countable set of pairs (a, b) of rational numbers such that $0 < a < b < 1$. For each $(a, b) \in C$, the map

$$\phi_{a,b} : \mathbb{R} \ni \sum_{t \in (a,b)} x_t e_t \longmapsto \sum_{t \in]a,b[} x_t \in \mathbb{Q}$$

is well-defined, $\mathbb{Q}$-linear and onto. Observe that, for any $x \neq 0$ in $\mathbb{R}$, there exists $(a, b) \in C$ such that $\phi_{a,b}(x) \neq 0$. Now $\mathbb{N}$ is in bijection with the disjoint union $\bigsqcup_{(a,b) \in C} \mathbb{Q}_{a,b}$ of copies of $\mathbb{Q}$ indexed by C. Define an action ϕ of $\mathbb{R}$ on this union which leaves each $\mathbb{Q}_{a,b}$ invariant and for which $x \in \mathbb{R}$ transforms $q \in \mathbb{Q}_{a,b}$ to $q + \phi_{a,b}(x)$. This ϕ is a faithful action. [Even if it is not important for our argument, observe that the product over $(a, b) \in C$ of the $\phi_{a,b}$ is a $\mathbb{Q}$-linear bijection from $\mathbb{R}$ onto a subspace of the vector space which is a direct product over C of copies of $\mathbb{Q}$.]

The group $\mathbb{R}/\mathbb{Z}$ is a direct sum of the torsion group $\mathbb{Q}/\mathbb{Z}$, which is countable, and a group isomorphic to $\mathbb{R}$ (a $\mathbb{Q}$-vector space of dimension the power of the continuum). It follows from the previous construction that there exists an injective homomorphism from $\mathbb{R}/\mathbb{Z}$ into $Sym(\mathbb{N})$.

In 1960, Ulam asked if the compact group $SO(3)$ of rotations of the usual space, viewed as a discrete group, can act on a countable set (see Section V.2 in [Ulam–60]). As far as I know, this is still open. Previous observations are possibly near what Ulam had in mind when writing his comments in Section II.7 of [Ulam–60].

III.38 and III.D, on finite quotients of the modular group.

For more on which finite simple groups are quotients of $PSL(2, \mathbb{Z})$, see the exposition of [Shale–01].

III.45, uncountably many finitely-generated groups with pairwise non-isomorphic von Neumann algebras.

Let Γ be a torsion-free Gromov-hyperbolic group which is not cyclic. Building up on results of Gromov, Ol'shanskii has shown that Γ has an uncountable family $(\Gamma_\iota)_{\iota \in I}$ of pairwise non-isomorphic quotient groups, all of which are simple and icc [Ol's–93]. N. Ozawa [Ozawa] has shown that, for any given separable factor M of type II_1, the set of those $\iota \in I$ for which the unitary group $\mathcal{U}(M)$ has a subgroup isomorphic to Γ_ι is a countable set. In particular, the set of von Neumann algebras of the groups Γ_ι (which are factors of type II_1) contains uncountably many isomorphism classes.

III.46, on groups with two generators.

It has been shown that two randomly chosen elements of a finite simple group G generate G with probability 1 as $|G| \to \infty$ (work of Dixon, Kantor-Lubotzky, Liebeck-Shalev, see [Shale–01]).

IV.1 and VI.1, on infinite generating sets and related word lengths.

Consider an integer $n \geq 2$, the group $\Gamma = SL(n, \mathbb{Z})$, and the infinite subset S of Γ consisting of those matrices of the form $I + kE_{i,j}$, with $k \in \mathbb{Z}$, $i, j \in \{1, \dots, n\}$, $i \neq j$, and $E_{i,j}$ the matrix with all entries 0 except one 1 at the intersection of the ith row and the jth column.

As stated in Item III.2, the diameter of Γ with respect to the corresponding S-word length is finite as soon as $n \geq 3$.

IV.3.viii, on stable length: a correction.

The subadditivity

$$\tau(\gamma\gamma') \leq \tau(\gamma) + \tau(\gamma')$$

holds for *commuting* elements $\gamma, \gamma' \in \Gamma$ (as correctly stated by Gersten and Short).

For example, if γ, γ' are the two standard generators of the infinite dihedral group, then $\tau(\gamma\gamma') > 0$ and $\tau(\gamma) = \tau(\gamma') = 0$.

IV.24.i, and values of the indices for subgroups: a question.

Consider the following property of a group Γ: whenever two subgroups Γ_1, Γ_2 of finite indices are abstractly isomorphic, the indices $[\Gamma : \Gamma_1]$ and $[\Gamma : \Gamma_2]$ are equal.

Finitely generated free groups and fundamental groups of closed surfaces have this property, by an easy argument using Euler characteristics.

More generally, it would be interesting to know which groups have this property and which groups don't.

IV.25.vii, a quasi-isometry criterion for existence of lattices.

B. Chaluleau and C. Pittet [ChaPi–01] have answered one of the questions there and have shown:

Let N be a graded simply connected nilpotent real Lie group. If there exists a finitely-generated group which is quasi-isometric to N, then N has lattices.

IV.25, and examples of quasi-isometries.

(x) Say that a metric space X is *quasi-isometrically incompressible* if any quasi-isometric embedding from X into itself is a quasi-isometry. E. Souche [Souch] has shown that finitely generated nilpotent groups and uniform lattices in simple connected real Lie groups are quasi-isometrically incompressible, but that finitely-generated free groups and Baumslag-Solitar groups are not.

(xi) A finitely-generated group cannot be quasi-isometric to an infinite dimensional Hilbert space. Indeed, such a space has the following quasi-isometric-invariant property: for any positive number r, there exists a positive number R such that a ball of radius R contains infinitely many pairwise disjoint balls of radius r; and a finitely-generated group does not have this property.

IV.27, groups which are commensurable up to finite kernels.

Another terminology for commensurable up to finite kernels is *weakly commensurable* subgroups. See § 5.5 in [GorAn–93]; these authors also point out the following fact.

If M is a manifold on which some Lie group act transitively, then $\pi_1(M)$ contains a subgroup of finite index which is isomorphic to a discrete subgroup of a connected Lie group; if M is also compact, then $\pi_1(M)$ contains a subgroup of finite index which is isomorphic to a uniform lattice in some connected Lie group.

IV.29.v and VII.26, and the classification of lattices up to commensurability in some nilpotent Lie groups.

Y. Semenov has classified $\mathbb{Q}$-forms of some real nilpotent Lie algebras, and thus the commensurability classes of lattices in the corresponding nilpotent Lie groups [Semen]. It seems that the following question is open:

does there exist a finite dimensional real nilpotent Lie algebra of which the number k of $\mathbb{Q}$-forms (up to isomorphism) is such that $1 < k < \infty$?

IV.34 & 35, and commensurability. The following exercise is taken from [Gabor–02] and is clearly missing just before IV.34.

Exercise. (i) Show that two groups Γ_1, Γ_2 are commensurable if and only if they have commuting free actions on a set X such that both quotients $\Gamma_1 \backslash X, \Gamma_2 \backslash X$ are finite.

[Hint for one direction. Let Γ'_j be a subgroup of finite index in Γ_j, $j = 1, 2$, such that there exists an isomorphism $\varphi : \Gamma'_1 \longrightarrow \Gamma'_2$. Set $\Delta = \{ (\gamma_1, \gamma_2) \in \Gamma_1 \times \Gamma_2 \mid \gamma_1 \in \Gamma'_1, \gamma_2 = \varphi(\gamma_1) \}$. Consider the natural actions of Γ_1 and Γ_2 on $(\Gamma_1 \times \Gamma_2)/\Delta$.

Hint for the other direction. Choose $x_0 \in X$. Consider the natural action of Γ_1 on $\Gamma_2 \backslash X$ and the canonical projection $[x_0]_2$ of x_0 in $\Gamma_2 \backslash X$. Let Γ'_1 be the isotropy subgroup of Γ_1 defined by $[x_0]_2$ and set $\gamma_1 x_0 = \varphi(\gamma_1) x_0$. Check that φ is a well-defined group homomorphism $\Gamma'_1 \longrightarrow \Gamma_2$ which is injective and whose image is of finite index in Γ_2.]

(ii) Assume that Γ_1, Γ_2 have commuting free actions on X such that both $\Gamma_1 \backslash X, \Gamma_2 \backslash X$ are finite, and let Γ'_1, Γ'_2 be as in the previous hints. Check that

$$\frac{[\Gamma_1 : \Gamma'_1]}{[\Gamma_2 : \Gamma'_2]} = \frac{|\Gamma_2 \backslash X|}{|\Gamma_1 \backslash X|} .$$

IV.36, on commensurability and torsion.

G. Levitt has observed that a group Γ with infinitely many torsion conjugacy classes can have a subgroup of finite index Γ_0 which is torsion-free.

Indeed, let first Γ_0 be the wreath product $\mathbb{Z} \wr \mathbb{Z} = (\oplus_{i \in \mathbb{Z}} \mathbb{Z} a_i) \rtimes \mathbb{Z}$, where the generator t of $\mathbb{Z}$ acts on the direct sum by a shift; this group is torsion-free. Then let Γ be the semi-direct product of Γ_0 with the automorphism ϕ of Γ_0 of order 2 defined by $\phi(a_i) = -a_i$ for all $i \in \mathbb{Z}$ and $\phi(t) = t$; and let $s \in \Gamma$ denote the element of order 2 which implements ϕ on the subgroup Γ_0. For $v, v' \in \oplus_{i \in \mathbb{Z}} \mathbb{Z} a_i$, the elements sv, sv' are on the one hand of order 2; on the other hand, they are conjugate in Γ if and only if there exist $\epsilon \in \{\pm 1\}$, $k \in \mathbb{Z}$, and $w \in \oplus_{i \in \mathbb{Z}} \mathbb{Z} a_i$ such that $v' = \epsilon t^k v t^{-k} + 2w$; it follows that the conjugacy classes in Γ of $s(a_1 + \cdots + a_n)$ are pairwise distinct $(n \geq 0)$.

A. Erschler has shown that a torsion-free group can be quasi-isometric to a group having torsion of unbounded order [Ersch–b].

The main ingredient of the proof is the construction, for any finitely-generated group A, of another finitely generated group $W^{\infty}(A)$, using an iterated wreath product construction and an HNN-extension. On the one hand, if A, B are Lipschitz equivalent groups, then $W^{\infty}(A), W^{\infty}(B)$ are Lipschitz equivalent; on the other hand, if A is torsion-free and if B has torsion, then $W^{\infty}(A)$ is torsion-free and $W^{\infty}(B)$ has torsion of unbounded order. One example is provided by $A = \mathbb{Z}$ and $B = \mathbb{Z} \oplus (\mathbb{Z}/p\mathbb{Z})$.

IV.40, and groups quasi-isometric to abelian groups.

Some of Shalom's ideas are now available in [Shalo].

IV.41, on groups of classes of quasi-isometries.

J. Taback has studied the quasi-isometry groups of $PSL_2(\mathbb{Z}[1/p])$, for p prime. These quasi-isometry groups are all isomorphic to $PSL_2(\mathbb{Q})$, even though the groups are not quasi-isometric for different values of the prime p. For this and other results, see [Tabac–00].

IV.43, and quasi-isometries of Baumslag-Solitar groups.

For the results of K. Whyte quoted from [Whyte–a], see now [Whyte–01].

IV.46, and Lipschitz equivalence.

Here is a question of B. Bowditch. (Private communication, March 2000. See also Item 1.A' in [Gromo–93].) Consider a Penrose tiling of the plane with two prototiles D and K (dart and kite), more precisely a tiling $\mathbb{R}^2 = \bigsqcup_{j \in J} T_j$ with each T_j given together with an isometry onto either D or K. This defines a cell decomposition X of the plane, of which the 0-skeleton $X^{(0)}$ is a discrete subset of the plane.

Is $X^{(0)}$ Lipschitz equivalent to a lattice in $\mathbb{R}^2$?

IV.47.vi, on costs and ℓ^2–Betti numbers.

For a group Γ with cost $\mathcal{C}(\Gamma)$ and ℓ^2–Betti numbers $\beta_j^{(2)}(\Gamma)$, we have always

$$\mathcal{C}(\Gamma) - 1 \geq \beta_1^{(2)}(\Gamma) - \beta_0^{(2)}(\Gamma).$$

Moreover, for a large class of groups (including all groups for which both terms are known), the two terms are indeed equal. See [Gabor], in particular Corollary 3.22.

IV.50, geometric properties and weakly geometric properties.

Following [Ersch–b], it can be useful to be more precise in the terminology concerning a property $(\mathcal{P})$ of finitely generated groups. She suggests the following definitions.

Say $(\mathcal{P})$ is *geometric* if, for a pair (Γ_1, Γ_2) of finitely-generated groups which are quasi-isometric, Γ_1 has Property $(\mathcal{P})$ if and only if Γ_2 is commensurable to a group which has Property $(\mathcal{P})$.

Say $(\mathcal{P})$ is *weakly geometric* if, for a pair (Γ_1, Γ_2) of finitely-generated groups which are quasi-isometric, Γ_1 has Property $(\mathcal{P})$ if and only if Γ_2 is commensurable up to finite kernels to a group which has Property $(\mathcal{P})$.

An example of a property which is weakly geometric and which is not geometric is "being a lattice in $Spin(2,5)$"; see III.18.vi, III.18.x, and IV.42.

V.18, on the group of a remarkable simple closed curve.

It has been shown by Anna Erschler Dyubina that the group of V.18 is not finitely generated. Finding a proof is proposed as Problem 10 835 in the American Mathematical Monthly [DyuHa–00].

PROBLEM. *Let Γ be the group defined by the presentation which has an infinite sequence $b_0, b_1, b_2, \ldots$ of generators and an infinite sequence $b_1 b_0 b_1^{-1} = b_2 b_1 b_2^{-1} = b_3 b_2 b_3^{-1} = \ldots$ of relations. Show that Γ is not finitely generated.*

We would like to add a comment and our solution. The nice solution of S.M. Gagola has appeared in the *Monthly*, November 2002.

COMMENT. In a short paper on wild knots, R.H. Fox discovered *A remarkable simple closed curve* (Annals of Math. **50**, 1949, pages 264–265) which is almost unknotted, a fact that Fox thinks "should be obvious to anyone who has ever dropped a stitch". The fundamental group Γ of the complement of this curve in 3-space has the presentation described above.

For other fundamental groups of complements of wild knots, see [Myers–00].

OUR SOLUTION. Observe first that there is a homomorphism $\Gamma \longrightarrow \mathbb{Z}$ mapping b_k onto 1 for each $k \geq 0$; hence b_k is of infinite order in Γ for each $k \geq 0$. Observe also that there is a homomorphism σ from Γ onto the symmetric group $\langle x, y \mid x^2 = y^2 = (xy)^3 = 1 \rangle$ such that $\sigma(b_{2j}) = x$ and $\sigma(b_{2j+1}) = y$ for all $j \geq 0$; hence $b_k b_{k+1} \neq b_{k+1} b_k$ for all $k \geq 0$.

For each $n \geq 0$, there is a homomorphism $\phi_n : \Gamma \longrightarrow \Gamma$ such that $\phi_n(b_k) = b_{k+n}$ for all $k \geq 0$. Since the first relation of the presentation defining Γ can be written as $b_0 = b_1^{-1} b_2 b_1 b_2^{-1} b_1$ and since the other relations do not involve b_0, the group Γ has another presentation with generators b_k and relations $b_{k+1} b_k b_{k+1}^{-1} = b_{k+2} b_{k+1} b_{k+2}^{-1}$ for $k \geq 1$. Similarly, for each $n \geq 0$, the group Γ has a presentation with generators b_k and relations $b_{k+1} b_k b_{k+1}^{-1} = b_{k+2} b_{k+1} b_{k+2}^{-1}$ for $k \geq n$, so that ϕ_n is an automorphism of Γ.

Assume now by contradiction that Γ is finitely generated, and therefore generated by $b_0, b_1, \ldots, b_{n+1}$ for some $n \geq 0$. Using again the relations $b_{k+1} b_k b_{k+1}^{-1} = b_{k+2} b_{k+1} b_{k+2}^{-1}$, this time for $0 \leq k \leq n - 1$, we see that Γ is generated by $\{b_n, b_{n+1}\}$. Thus Γ is also generated by $\{b_0, b_1\} = \phi_n^{-1}(\{b_n, b_{n+1}\})$, as well as by $\{b_1, b_2\} = \phi_1(\{b_0, b_1\})$.

For each $k \geq 0$, let $\tilde{\Gamma}_{k+1}$ the group abstractly defined by $k + 2$ generators $b_0, \ldots, b_{k+1}$ and k relations $b_1 b_0 b_1^{-1} = \ldots = b_{k+1} b_k b_{k+1}^{-1}$. The same argument as above shows that $\tilde{\Gamma}_{k+1}$ has another presentation with 2 generators b_k, b_{k+1} and no relation, hence that $\tilde{\Gamma}_{k+1}$ is free of rank two. As b_0, b_1 do not commute in $\tilde{\Gamma}_{k+1}$, they generate a subgroup of $\tilde{\Gamma}_{k+1}$ which is free of rank two. As this holds for any $k \geq 0$, it follows that the group Γ, generated by b_0 and b_1, is itself free of rank two.

As Γ is free on $\{b_1, b_2\}$, there is a homomorphism $\psi : \Gamma \longrightarrow \mathbb{Z}$ such that $\psi(b_1) = 0$ and $\psi(b_2) = 1$, which is *onto*. On the other hand, as Γ is generated by b_0 and b_1, and as $\psi(b_0) = \psi(b_1^{-1} b_2 b_1 b_2^{-1} b_1) = 0 = \psi(b_1)$, we have $\psi(\Gamma) = \{0\}$. This is a contradiction and ends the proof. $\square$

The group Γ has other straightforward non-finitenes properties. (i) It is not Hopfian, since it is isomorphic to its quotient by the relation $b_0 = 1$. (ii) It maps onto the Baumslag-Solitar group $\langle t, z \mid tzt^{-1} = z^2 \rangle$ by $b_{2n} \longmapsto zt^{-1}$ and $b_{2n+1} \longmapsto t^{-1}$.

V.20, and lattices in Lie groups.

Information on lattices in *complex* Lie groups can be found in [Winke–98].

V.21, and finiteness homological properties of $SL(n, \mathbb{F}_q[T])$.

The finiteness result according to which $SL(n, \mathbb{F}_q[T])$ is of type (F_{n-2}) and not of type (F_{n-1}) for $q \geq 2^{n-2}$ is due independently to H. Abels (as recorded in V.21) and P. Abramenko [Abram–96].

V.22, on commensurability and groups of automorphisms.

G. Levitt has drawn my attention to the fact that, given a group Γ and a subgroup Γ_0 of finite index, there can exist an infinity of automorphisms of Γ which coincide with the identity on Γ_0.

Indeed, let Γ be the infinite dihedral group and let Γ_0 be its infinite cyclic subgroup of index 2. Then the conjugations of Γ by elements of Γ_0 are pairwise distinct.

V.22, on large groups of automorphisms.

The automorphism group of a finitely-generated group is clearly countable. The automorphism group of a countable group need not be; an easy example is provided by an infinite direct sum of copies of any countable group not reduced to one element.

Here is another example, inspired from Ulam and using the notation of the addendum to III.B above. For each $(a, b) \in C$, let $\phi_{a,b} : \mathbb{R} \longrightarrow \mathbb{Q}$ be the homomorphism defined there and let $\mathbb{Q}^2_{a,b}$ be a copy of $\mathbb{Q}$. The mapping

$$\psi_{a,b} \; : \; \begin{cases} \mathbb{R} & \longmapsto \; \mathrm{Aut}\,(\mathbb{Q}^2_{a,b}) \approx GL_2(\mathbb{Q}) \\[2mm] x & \longmapsto \; \begin{pmatrix} 1 & \phi_{a,b}(x) \\ 0 & 1 \end{pmatrix} \end{cases}$$

is a homomorphism of groups. Define Γ to be the direct sum, over $(a, b) \in C$, of the groups $\mathbb{Q}^2_{a,b}$; then the direct sum of the homomorphisms $\psi_{a,b}$ is an injection of $\mathbb{R}$ into $\mathrm{Aut}(\Gamma)$.

V.26, and some groups of Richard Thompson.

There are three groups, acting respectively on an interval, the circle, and the Cantor set, denoted by F, T, and V in [CanFP–96], and which appear in many different contexts. For T in the context of Teichmüller theory, see several articles by R.C. Penner, including [Penne–97]; for the isomorphism of Penner's group with T, see [Imber–97]. One interesting byproduct of this circle of ideas is that T can be generated by two elements α, β satisfying $\alpha^4 = \beta^3 = 1$, and other relations, such that the subgroup of T generated by α^2, β is the free product $\mathbb{Z}/2\mathbb{Z} * (\mathbb{Z}/3\mathbb{Z}) \approx PSL_2(\mathbb{Z})$; see [LocSc–97].

V.31, and efficiency.

A. Çevik gives in [Çevik–00] a sufficient condition for the efficiency of wreath products of efficient finite groups.

VI.9, an example of spherical growth series which is not monotonic.

On page 161, the last display, the coefficient of z^2 should be 8 not 6. This was pointed out to me by N.J.A. Sloane. Several growth series which appear in the the book appear also in his database of integer sequences: see

http://www.research.att.com/ njas/sequences/

on the web.

VI.19, on groups with the size of spheres not tending to infinity.

Groups in which the size of spheres does not tend to infinity are virtually cyclic (communicated by Anna Erschler Dyubina). More precisely:

PROPOSITION. *If $\sigma(\Gamma, S; k) \leq C$ for infinitely many values of k, then Γ is virtually cyclic.*

PROOF. Consider an arbitrary infinite finitely generated group, and let Φ be its inverse growth function, as in VII.32. First, it follows from the definition and from the obvious inequality $\beta(4k) > 2\beta(k)$ that $\Phi(2\beta(k)) \leq 4k$. Then, it follows from the first result quoted in VII.32 that, for an appropriate constant K, we have

$$\frac{\sigma(n)}{\beta(n-1)} \geq \frac{1}{8|S|\Phi(2\beta(n-1))} \geq \frac{1}{8|S|4(n-1)} \geq \frac{1}{Kn},$$

whence

$$\beta(n-1) \leq K\sigma(n)n$$

for all $n \geq 1$.

Assume now that $\sigma(n_j) \leq C$ for some constant C and a strictly increasing infinite sequence $(n_j)_{j \geq 1}$. Thus $\beta(n_j - 1) \leq KCn_j$ for any $j \geq 1$. By the strong form of Gromov's theorem (VII.29) on groups of polynomial growth, which is elementary for linear growth and which is due to Van den Dries and Wilkie [VdDW–84b], this implies that Γ is a group of linear growth and therefore a virtually cyclic group. $\square$

VI.20, and the growth of braid groups for Artin generators.

For any integer $n \geq 2$, Artin's *braid group* on n strings has presentation

$$B_n = \left\langle \sigma_1, \ldots, \sigma_{n-1} \middle| \begin{array}{ll} \sigma_i\sigma_{i+1}\sigma_i = \sigma_{i+1}\sigma_i\sigma_{i+1} & (1 \leq i \leq n-2) \\ \sigma_i\sigma_j = \sigma_j\sigma_i & (1 \leq i,j \leq n-1, \, |i-j| \geq 2) \end{array} \right\rangle$$

[Magnu–73] and is obviously a quotient of the *locally free group of depth* 1 with $n - 1$ generators

$$LF_n = \langle f_1, \ldots, f_{n-1} \mid f_if_j = f_jf_i \quad (1 \leq i,j \leq n-1, \, |i-j| \geq 2) \rangle$$

[Versh–90], [Versh–00], [VeNeB–00]. The value of the exponential growth rate of B_n for the generators σ_i is still unknown; however, Vershik and his co-authors have obtained partial results by comparing B_n with LF_n, more precisely by using the fact that LF_n appears both as a group of which B_n is a quotient and as a subgroup of B_n, the image of the injective homomorphism which maps f_i onto σ_i^2 for $i \in \{1, \ldots, n-1\}$.

For example, if ω_n^B, ω_n^{LF} denote respectively the exponential growth rates of B_n, LF_n for the generators discussed here, then

$$\lim_{n \to \infty} \omega_n^{LF} = 7 \quad \text{and} \quad \sqrt{7} \leq \omega_n^B \leq 7 \quad \text{for } n \text{ large enough.}$$

VI.B, early papers on growth of groups, and Dye's theorem on orbit equivalence for groups of polynomial growth.

Growth occurs in a paper by Margulis [Margu–67] published one year before those of Milnor ([Miln–68a], [Miln–68b]), where Margulis shows that if a compact three-dimensional manifold admits an Anosov flow, then its fundamental group has exponential growth. For a generalization to higher dimensions, see [PlaTh–72].

Also, between the mid fifties and 1968, some mathematicians in France were aware of the notion of growth of groups. Besides Dixmier (quoted on page 187), Avez had learned this from Arnold in 1965 [Avez–76].

We should also mention the following results of H. Dye. On the one hand, consider the compact abelian group $\prod_{j=0}^{\infty} C_j$, where each C_j is a copy of the group $\{0, 1\}$ of order 2, with its normalised Haar measure μ. Let $T : G \longrightarrow G$ be the adding machine, defined by

$$T(x_0, x_1, x_2, \ldots) = (0, 0, \ldots, 1, x_{j+1}, x_{j+2}, \ldots)$$

where j is the smallest index such that $x_j = 0$, and

$$T(1,1,1,1,\ldots) = (0,0,0,0,\ldots).$$

Then T defines an ergodic action of $\mathbb{Z}$ by measure preserving transformations of the probability space (G, μ). On the other hand, consider any finitely generated group Γ acting by measure preserving transformations on a standard Borel space furnished with a non-atomic probability measure, the action being ergodic.

One of Dye's theorems is that, if Γ is of polynomial growth, then the action of Γ is orbit-equivalent to the odometer action of $\mathbb{Z}$ [Dye–63]; if $\Gamma \approx \mathbb{Z}$, this is already in [Dye–59]. See [Weiss–81] for an exposition, and [OrnWe–80], [CoFeW–81] for related results; in particular, Dye's theorem carries over to *amenable* countable groups.

VI.40, and the functions which are growth functions of semigroups.

Let M be a monoid generated by a finite set S and let $\beta(k) = \beta(M, S; k)$ denote the corresponding growth function (see VI.12). It is obvious that if $\beta(k)$ is unbounded, then $k \prec \beta(k)$; moreover,

$$k \prec \beta(k) \quad \text{and} \quad k \approx \beta(k) \qquad \text{imply} \qquad k^2 \prec \beta(k)$$

as has been shown[2] by V.V. Beljaev (reported in [Trofi–80]).

Let $f, g : \mathbb{N} \longrightarrow \mathbb{N}$ be two functions such that $k^2 \prec f(k)$ and $g(k) \prec 2^k$. Then there exists a monoid M generated by a finite set S such that the sets

$$\{k \in \mathbb{N} \,|\, \beta(M, S; k) \le f(k)\} \qquad \text{and} \qquad \{k \in \mathbb{N} \,|\, \beta(M, S; k) \ge g(k)\}$$

are both infinite.

VI.40, and the growth functions of Riemannian manifolds.

For further work after the paper of Grimaldi and Pansu quoted in VI.40, see [GriPa–01] and its bibliography.

VI.42, and growth of groups with respect to weights.

Growth with respect to generating sets and given weights are older than suggested by the references of Chapters VI and VII. In particular, in [PlaTh–76], Plante and Thurston define the growth of a countable group (*not* necessarily of finite type) with respect to a generating set (*not* necessarily finite) and a proper weight on it.

VI.42–43 and VII.35, on relative length functions and relative growth.

In the last line of page 176, read "relative length function" instead of "relative growth function". For relative growth of subgroups of solvable and linear groups, see [Osin–00].

VI.45, on word and Riemannian metrics.

See also [LubMR–00].

---

[2]This has been shown independently by several other mathematicians.

VI.56, on asymptotics of subadditive functions.

The correct conclusion of (i) should be that the sequence $\left(\frac{\alpha(k)}{k}\right)_{k \geq 1}$ either converges to $\inf_{k \geq 1} \frac{\alpha(k)}{k}$ or *diverges properly to* $-\infty$. (Since sequences appearing in the book are bounded below, the second case does not occur.)

VI.64, and groups of intermediate growth which are not residually finite.

Anna Erschler has shown that there exist uncountably many groups of intermediate growth which are commensurable up to finite kernel with the first Grigorchuk group, but which are not residually finite. She has also shown that there exist groups of intermediate growth which are not commensurable up to finite kernels with any residually finite group. See [Ersch–b].

VII.2, and a version of the Table-Tennis Lemma due to Margulis.

PROPOSITION. *Let* Γ *be a group acting on a set* X *and let* $a, b \in \Gamma$. *Assume that there exists a non-empty subset* U *of* X *such that* $b(U) \cap U = \emptyset$ *and* $ab(U) \cup a^2 b(U) \subset U$. *Then the semi-group generated in* Γ *by* ab *and* $a^2 b$ *is free; in particular, it is of exponential growth if* Γ *is finitely generated.*

PROOF. Inside $U_\emptyset \doteq U$, the sets $U_1 = ab(U)$ and $U_2 = a^2 b(U)$ are disjoint, since

$$ab(U) \cap a^2 b(U) = a\Big(b(U) \cap U_1\Big) \subset a\Big(b(U) \cap U\Big) = \emptyset.$$

More generally, for each $n \geq 0$, let J_n denote the set of sequences of length n with elements in $\{1, 2\}$; for each $\underline{j} = (j_1, \ldots, j_n) \in J_n$, define a subset $U_{\underline{j}} = a^{j_1} b a^{j_2} b \ldots a^{j_n} b(U)$ of U. For any $n \geq 1$ and $\underline{j}' \in J_{n-1}$, observe that the sets $U_{(1,\underline{j}')}$ and $U_{(2,\underline{j}')}$ are disjoint, since

$$U_{(1,\underline{j}')} \cap U_{(2,\underline{j}')} = a\Big(b(U_{\underline{j}'}) \cap U_{(1,\underline{j}')}\Big) \subset a\Big(b(U) \cap U\Big) = \emptyset,$$

and that both are inside $U_{\underline{j}'}$. Thus, for two sequences $\underline{j}, \underline{j}' \in \bigcup_{n=0}^{\infty} J_n$, either the corresponding subsets $U_{\underline{j}}, U_{\underline{j}'}$ are disjoint, or one is strictly contained in the other; in other words, their inclusion order is that of the infinite rooted 2-ary tree (see Item VIII.1). The proposition follows. □

This version of the Table-Tennis Lemma was communicated by G.A. Margulis to the authors of [EsMoO–02], see VII.19 below.

VII.13, on tight growth of free groups and hyperbolic groups.

It is easy to show that, for any normal subgroup $N \neq 1$ of F_k and the canonical image $\underline{S}_k$ of S_k in Γ/N, the corresponding exponential growth rates satisfy the strict inequality $\omega(F_k/N, \underline{S}_k) < 2k - 1$. G. Arzhantseva and I.G. Lysenok have shown the following generalization, which answers a question of [GrHa–97]. Let Γ be a non-elementary hyperbolic group, S a finite generating set and N an infinite normal subgroup of Γ; denote by $\underline{S}$ the canonical image of S in the quotient group Γ/N; then $\omega(\Gamma/N, \underline{S}) < \omega(\Gamma, S)$ [ArjLy].

VII.19, on uniformly exponential growth of solvable groups.

D. Osin has shown that any solvable group of exponential growth has uniformly exponential growth [Osin–a], thus solving Problem VII.19.B (see page 297); this has also been shown independently and shortly afterwards by J. Wilson (unpublished). More generally, Osin has shown that any elementary amenable group of exponential growth has uniformly exponential growth [Osin–b].

Also, D. Osin has shown that the uniform Kazhdan constant of an infinite Gromov hyperbolic groups is zero [Osin–c]

John Wilson has discovered *examples of groups which answer the main problem of Item VII.19* [Wilso]. More precisely, there exist groups which are isomorphic to their permutational wreath product with the alternating group on 31 letters. Let $\Gamma \approx \Gamma \wr A_{31}$ be any group of this kind; on the one hand, there exists a sequence $(S_n = \{x_n, y_n\})_{n \geq 1}$ of generating sets of Γ, with $x_n^2 = y_n^3 = 1$ for all $n \geq 1$, such that the limit of the corresponding exponential growth rates is 1, in formula $\lim_{n \to \infty} \omega(\Gamma, S_n) = 1$; on the other hand, for an appropriate choice of Γ, there exist non-abelian free subgroups in Γ, so that in particular Γ is of exponential growth.

VII.19, on uniformly exponential growth of linear groups.

It is a theorem of A. Eskin, S. Mozes and Hee Oh that, given an integer $N \geq 1$ and a field $\mathbb{K}$ of characteristic 0, a finitely generated subgroup of $GL(N, \mathbb{K})$ is of uniformly exponential growth if and only if it is not virtually nilpotent, namely if and only if it is of exponential growth (result of [EsMoO], announced in [EsMoO–02]).

In particular, this *solves Research Problem VII.19.C* (see page 297).

For other progress on uniformly exponential growth, see [BucHa-00], [GrHa–01a], and [GrHa–01b]. For an exposition on uniformly exponential growth, see [Harpe].

If constants measuring exponential growth often have uniform bounds in terms of the generating sets, other constants exhibit the opposite behaviour. For example, T. Gelander and A. Zuk have shown that, in many cases, Kazhdan constants depend in a crucial way on the chosen generating set [GelZu–02].

VII.29, group growth, and Gromov's theorem.

There is a brief survey on group growth and Gromov's theorem by D.L. Johnson [Johns–00].

Concerning polynomial growth for locally compact groups, V. Losert has published a second part to [Loser–87]: see [Loser–01].

VII.29, and growth of double coset classes.

Consider a *Hecke pair* (G, H), namely a group G and a subgroup H such that all orbits of the natural action of H on G/H are finite, or equivalently such that, for each $g \in G$, the indices of $H \cap gHg^{-1}$ in both H and gHg^{-1} are finite. It is a natural counting problem to estimate for each $g \in G$ the cardinality of the

H-orbit of gH in G/H, or equivalently the number of one-sided classes $g_j H$ in the double class HgH.

The specific case of the pair $(SL(2, \mathbb{Z}[1/p]), SL(2, \mathbb{Z}))$, p a prime, appears in [BeCuH–02].

VII.34, and the growth of Følner sequences.

A question related to our Problem VII.34.A appears as Problem 14.27 in the Kourovka Notebook [Kouro–95], and has been answered in [Barda–01].

VII.38, and the growth of normal subgroups of finite index.

See [LarLu].

VII.39, growth of conjugacy classes, and growth of pseudogroups.

For growth of conjugacy classes in hyperbolic groups, see [CooKn–02] and [CooKn–b].

Growth of pseudogroups appears in connection with foliations in [Plant–75].

VII.40, and growth of infinitely generated groups.

See [PlaTh–76], and the above comment on Item VI.42.

VII.61, on the set of exponential growth rates.

Part of the problem was solved by Anna Erschler Dyubina, who has shown that *the set Ω_2 of exponential growth rates of 2-generated groups has the power of the continuum* (see [Ersch–02], [Ersch–a]).

VIII.7, and the adding machine.

The adding machine on the infinite 2-ary tree $T^{(2)}$ can be economically (and recursively, compare VIII.9) defined as the element $\tau \in Aut(T^{(2)})$ such that

$$\tau = a(1, \tau).$$

Observe that $\tau \neq 1$ since τ exchanges 0 and 1, and that τ is of infinite order since

$$\tau^2 = a(1, \tau)a(1, \tau) = (\tau, 1)(1, \tau) = (\tau, \tau).$$

The simple and clever Proposition 20 of [Sidki–00] shows that an element $g \in Aut(T^{(2)})$ is conjugate to τ if and only if it acts transitively on the set of 2^k vertices of the level $L^{(k)}$ for each $k \geq 0$.

Later, Sidki has shown that a solvable subgroup K of $Aut(T^{(2)})$ which contains an element such as τ above is an extension of a torsion-free metabelian group by a finite 2-group. If furthermore K is nilpotent then it is torsion-free abelian [Sidki].

VIII.10.ii on automata and finitely generated groups.

This connexion is a very active subject of research; see among others [GriNS–00], [GriZu–a], [GriZu–b], and [Sidki–00].

VIII.31, a result of John Wilson.

At the end of this item, the "recent result" which is quoted was in fact essentially in John Wilson's Ph.D. thesis of 1971, as well as in [Wilso–72]. ("Essentially" in the sense that he did not use the words "branch groups".)

For these, [Grigo] contains comments and a sketchy proof, whereas details can be found in [Wilso].

VIII.32 and VIII.71, and elements of small lengths and large orders in the Grigorchuk group.

PROPOSITION. *For any* $n \in \mathbb{N}$, *there exists* $\gamma \in \Gamma$ *such that*

$$\gamma^{2^n} \neq 1 \qquad and \qquad \ell(\gamma) \leq 2^n.$$

PROOF (FOLLOWING A SKETCH OF L. BARTHOLDI). Let $K = \langle abab \rangle^\Gamma$ be the normal subgroup of Γ of index 16 defined in VIII.30; recall that K is generated by

$$t = (ab)^2 , \quad v = (bada)^2 , \quad w = (abad)^2$$

and that $\psi^{-1}(K \times K)$ is a subgroup of K (the index is 4 by Exercise VIII.81, but we do not use this here). Let σ be the endomorphism of Γ defined in VIII.57. Since $\sigma(a) = aca$, $\sigma(b) = d$, $\sigma(d) = c$, and $\sigma(c) = b$, we have $\psi\sigma(a) = (d, a)$, $\psi\sigma(b) = (1, b)$, $\psi\sigma(c) = (a, c)$, $\psi\sigma(d) = (a, d)$. It follows that $\psi\sigma(x) = (1, x)$ for $x \in \{t, v, w\}$, and therefore for all $x \in K$.

Define inductively a sequence $(x_i)_{i \geq 0}$ by $x_0 = abab$ and $x_{i+1} = a\sigma(x_i)$. Since

$$\psi \left(a\sigma(x_i)a\sigma(x_i) \right) = (x_i, x_i),$$

the order of x_{i+1} is twice that of x_i. As x_0 is of order 8 by Proposition VIII.16, it follows that the order of x_i is 2^{i+3} for all $i \geq 0$.

On the other hand, denote by w_0 the word $abab$ representing x_0; for each $i \geq 0$, let w_{i+1} the word obtained from w_i by

· substitution of aca, d, b, c in place of a, b, c, d respectively,
· deletion of a if it appears as the first letter and addition of a as a prefix letter otherwise,

so that w_{i+1} represents x_{i+1}. Thus $w_1 = cadacad$ and, for each $j \geq 0$,

· w_{2j+1} is a word of length $2\ell(w_{2j}) - 1$ which begins with c and ends with a letter from $\{b, c, d\}$,
· w_{2j+2} is a word of length $2\ell(w_{2j+1})$ which begins with a and ends with a letter from $\{b, c, d\}$;

in particular, $\ell(x_i) \leq \ell(w_i) < 2^{i+2}$ for all $i > 0$. The proposition follows (with $x = x_{n-2}$ for $n \geq 2$). $\square$

VIII.67, and power series with finitely many different coefficients.

Here is a result of Szegö: a power series with finitely many different coefficients that converges inside the unit disk is either a rational function, or has the unit circle as natural boundary [Szegö–22].

VIII.88, complement on commensurability of finitely-generated subgroups.

It is a remarkable result of Grigorchuk and Wilson that any infinite finitely-generated subgroup of the Grigorchuk group Γ is commensurable to Γ [GriWi]. In other words, Γ has exactly two commensurability classes of finitely-generated subgroups: itself and $\{1\}$.

Here are a few examples of other groups for which all commensurability classes of finitely-generated subgroups are known; in case of torsion-free groups, we do not list the class of $\{1\}$.

(i) Free abelian groups $\mathbb{Z}^n$, with $\mathbb{Z}^j$ for $j \in \{1, \ldots, n\}$.

(iii) The Heisenberg group $\begin{pmatrix} 1 & \mathbb{Z} & \mathbb{Z} \\ 0 & 1 & \mathbb{Z} \\ 0 & 0 & 1 \end{pmatrix}$, with $\mathbb{Z}$, $\mathbb{Z}^2$ and the group itself.

(iii) Non-abelian free groups F_n, with $\mathbb{Z}$ and F_2.

(iv) Virtually free groups, for example $PSL(2, \mathbb{Z})$, with finite subgroups, $\mathbb{Z}$ and F_2.

(v) The fundamental group Γ_g of a closed surface of genus $g \geq 2$, with $\mathbb{Z}$, F_2 and the group itself.

(vi) Olshanskii's "monsters" (see the reference in III.5, as well as [AdyLy–92]), in which any proper subgroup is cyclic.

VIII.87, on complex linear representations of the Grigorchuk group.

For each $k \geq 0$, let Γ_k denote as in VIII.35 the finite quotient of the Grigorchuk group which acts naturally on the level $L(k)$ of the binary tree. Choose some point in $L(k)$ and denote by P_k the corresponding isotropy subgroup of Γ_k. Then (Γ_k, P_k) is a *Gelfand pair*, and the natural linear representation of Γ_k on the space $\mathbb{C}^{L(k)}$ splits as a direct sum of $k + 1$ pairwise inequivalent irreducible representations, of dimensions $1, 1, 2, 4, \ldots, 2^{k-1}$ [BeHaG].

References

Abram–96. P. Abramenko, *Twin buildings and applications to S-arithmetic groups*, Lecture Notes in Math. **1641**, Springer, 1996.

AdyLy–92. S.I. Adyan and I.G. Lysënok, *Groups, all of whose proper subgroups are finite cyclic*, Math. USSR Izvestiya **39** (1992), 905–957.

AlFaN. R.C. Alperin, B. Farb, and G.A. Noskov, *A strong Schottky Lemma for non-positively curved singular spaces*, Preprint (January, 2001).

AnoSi–67. D.V. Anosov and Ya.G. Sinai, *Some smooth ergodic systems*, Russian Math. Surveys **22:5** (1967), 103–167.

ArzLy–02. G. Arzhantseva and I.G. Lysenok, *Growth tightness for word hyperbolic groups*, Math. Z. **241** (2002), 597–611.

Barda–01. V.G. Bardakov, *Construction of a regularly exhausting sequence for groups with subexponential growth*, Algebra i Logica **40** (2001), 22–29.

BarGr–00. L. Bartholdi and R. Grigorchuk, *Spectra of non-commutative dynamical systems and graphs related to fractal groups*, C.R. Acad. Sci. Paris, Série I **331** (2000), 429–434.

BarGr–01. L. Bartholdi and R. Grigorchuk, *Sous-groupes paraboliques et représentations de groupes branchés*, C.R. Acad. Sci. Paris, Série I **332** (2001), 789–794.

BasLu–01. H. Bass and A. Lubotzky, with appendices by H. Bass, L. Carbone, A. Lubotzky, G. Rosenberg, and J. Tits, *Tree lattices*, Birkhäuser, 2001.

BeCuH–02. M.B. Bekka, R. Curtis, and P. de la Harpe, *Familles de graphes expanseurs et paires de Hecke*, C.R. Acad. Sci. Paris, Série I **335** (2002), 463–468.

BeHaG. M.B. Bekka, P. de la Harpe, *Irreducibility of unitary group representations and reproducing kernels Hilbert spaces*, Appendix on *Two point homogeneous compact ultrametric spaces* in collaboration with Rostislav Grigorchuk, Expositiones Math. (to appear).

BesFe–00. M. Bestvina and M. Feighn, *The topology at infinity of $Out(F_n)$*, Inventiones Math. **146** (2000), 651–692.

BreGe. E. Breuillard and T. Gelander, *On dense free subgroups of Lie groups*, Preprint (2002).

Camin–97. R. Camina, *Subgroups of the Nottingham group*, J. of Algebra **196** (1997), 101–113.

Camin–00. R. Camina, *The Nottingham group*, in "New horizons in pro-p groups", M. du Sautoy, D. Segal, and A. Shalev Editors, Birkhäuser (2000), 205–221.

Carbo–01. L. Carbone, *Non-uniform lattices on uniform trees*, Memoir Amer. Math. Soc. **724**, 2001.

Çevik–00. A. S. Çevik, *The efficiency of standard wreath product*, Proc. Edinburgh Math. Soc. **43** (2000), 415–423.

ChaPi–01. B. Chaluleau and C. Pittet, *Exemples de variétés riemanniennes homogènes qui ne sont pas quasi isométriques à un groupe de type fini*, C.R. Acad. Sci. Paris, Sér. I **332** (2001), 593–595.

CoFeW–81. A. Connes, J. Feldman, and B. Weiss, *An amenable equivalence relation is generated by a single transformation*, Ergod. Th. & Dynam. Sys. **1** (1981), 431–450.

CooKn–02. M. Coornaert and G. Knieper, *Growth of conjugacy classes in Gromov hyperbolic groups*, Geometric and Functional Analysis **12** (2002), 464–478.

CooKn–b. M. Coornaert and G. Knieper, *An upper bound for the growth of conjugacy classes in torsionfree word hyperbolic groups*, to appear.

Drumm–92. T.A. Drumm, *Fundamental polyhedra for Margulis space-times*, Topology **31** (1992), 677–683.

Dye–59. H. Dye, *On groups of measure preserving transformations I*, Amer. J. Math. **81** (1959), 119–159.

Dye–63. H. Dye, *On groups of measure preserving transformations II*, Amer. J. Math. **85** (1963), 551–576.

DymMc–72. H. Dym and H.P. McKean, *Fourier series and integrals*, Academic Press, 1972.

Dyubi–00. A. Dyubina, *Instability of the virtual solvability and the property of being virtually torsion-free for quasi-isometric groups*, International Math. Res. Notices **21** (2000), 1098–1101.

DyuHa–00. A. Dyubina Erschler and P. de la Harpe, *Problem 108 35*, Amer. Math. Monthly **107^9** (2000), 864.

Egoro–00. A.V. Egorov, *Residual finiteness of groups and topological dynamics*, Sbornik Math. **191^4** (2000), 529–541.

Ersch–02. A. Erschler, *On growth rates of small cancellation groups*, Funct. Analy and its Appl. **36** (2002), 93–95.

Ersch–a. A. Erschler, *Growth rates of small cancellation groups*, Proceedings of the workshop Random Walks and Geometry, Vienna, ESI (to appear).

Ersch–b. A. Erschler, *Not residually finite groups of intermediate growth, commensurability and non geometricity*, Preprint (2002).

EsMoO–02. A. Eskin, S. Mozes and Hee Oh, *Uniform exponential growth for linear groups*, International Math. Res. Notices **2002:31** (2002), 1675–1683.

EsMoO. A. Eskin, S. Mozes and Hee Oh, *On uniform exponential growth for linear groups*, Preprint (2002).

FarFr. B. Farb and J. Franks, *Groups of homeomorphisms of one-manifolds I: actions of nonlinear groups*, Preprint (2001).

Gabor–02. D. Gaboriau, *Arbres, groupes, quotients*, Thèse d'habilitation, ENS-Lyon, 8 avril 2002.

Gabor. D. Gaboriau, *Invariants ℓ^2 de relations d'équivalence et de groupes*, Publ. Math. I.H.E.S. (to appear).

GelZu–02. T. Gelander and A. Zuk, *Dependence of Kazhdan constants on generating subsets*, Israel J. Math. **129** (2002), 93–98.

Ghys–01. E. Ghys, *Groups acting on the circle*, l'Enseignement math. (2) **47** (2001), 329–407.

GorOn–93. V.V. Gorbatsevich and A.L. Onishchik, *Lie transformation groups*, in "Lie groups and Lie algebras I", Encycl. Math. Sciences **20**, Springer (1993), 95–229.

GrHa–97. R.I. Grigorchuk and P. de la Harpe, *On problems related to growth, entropy and spectrum in group theory*, J. of Dynamical and Control Systems **3:1** (1997), 51–89.

GrHa–01a. R.I. Grigorchuk and P. de la Harpe, *One-relator groups of exponential growth have uniformly exponential growth*, Math. Notes **69** (2001), 575–577.

GrHa–01b. R.I. Grigorchuk and P. de la Harpe, *Limit behaviour of exponential growth rates for finitely generated groups*, l'Enseignement math., monographie **38²** (2001), 351–370.

GriNS–00. R.I. Grigorchuk, V.V. Nekrashevich, and V.I. Sushchanskii, *Automata, dynamical systems, and groups*, Proc. Steklov Inst. Math. **231** (2000), 128–203.

GriPa–01. R. Grimaldi and P. Pansu, *Nombre de singularités de la fonction croissance en dimension 2*, Bull. Belgian Math. Soc. **8** (2001), 395–404.

GriWi. R.I. Grigorchuk and J. Wilson, *A rigidity property concerning abstract commensurability of subgroups*, Preprint, 2001.

GriZu–a. R. Grigorchuk and A. Zuk, *The lamplighter group as a group generated by a 2-state automaton and its spectrum*, Geometriae Dedicata, to appear.

GriZu–b. R. Grigorchuk and A. Zuk, *A free group generated by a three state automaton*, Internat. J. Algebra Comput., to appear.

GryWó–99. A. Grytczuk and M. Wójtowicz, *Beardon's diophantine equations and non-free Möbius groups*, Bull. London Math. Soc. **32** (1999), 305–310.

Harpe. P. de la Harpe, *Uniform growth in groups of exponential growth*, Geometriae Dedicata, to appear.

Imber–97. M. Imbert, *Sur l'isomorphisme du groupe de Richard Thompson avec le groupe de Ptolémée*, in "Geometric Galois Actions, 2", L. Schneps and P. Lochak Editors, London Math. Soc. Lecture Notes Series **243** (Cambridge Univ. Press 1997), 313–324.

Johns–00. D.L. Johnson, *Growth of groups*, The Arabian Journal for Science and Engineering **25–2C** (2000), 53–68.

Karls. A. Karlsson, *Free subgroups of groups with non-trivial Floyd boundary*, Preprint (January 2002).

LarLu. M. Larsen and A. Lubotzky, *Normal subgroup growth of linear groups: the (G_2, F_4, E_8)-theorem*, Prepublication (2001).

Licht–93. A.I. Lichtman, *The soluble subgroups and the Tits alternative in linear groups over rings of fractions of polycyclic group, I*, J. of Pure and Appl. Algebra **86** (1993), 231–287.

Licht–95. A.I. Lichtman, *Automorphism groups of free soluble groups*, J. Algebra **174** (1995), 132–149.

Licht–99. A.I. Lichtman, *The soluble subgroups and the Tits alternative in linear groups over rings of fractions of polycyclic group, II*, J. Group Theory **2** (1999), 173–189.

LisMe–00. V. Liskovets and A. Mednykh, *Enumeration of subgroups in the fundamental groups of orientable circle bundles over surfaces*, Comm. in Algebra **28⁴** (2000), 1717–1738.

LocSc–97. P. Lochak et L. Schneps, *The universal Ptolemy-Teichmüller groupoid*, in "Geometric Galois Actions, 2", L. Schneps and P. Lochak Editors, London Math. Soc. Lecture Notes Series **243** (Cambridge Univ. Press 1997), 325–347.

Loser–01. V. Losert, *On the structure of groups with polynomial growth II*, Journal London Math. Soc. **63** (2001), 640–654.

LubMR–00. A. Lubotzky, S. Mozes, and M.S. Raghunathan, *The word and Riemannian metrics on lattices of semisimple groups*, Publ. Math. I.H.E.S. **91** (2000), 5–53.

MacMa–01. C. Maclachlan and G.J. Martin, *The non-compact arithmetic generalized triangle groups*, Topology **40** (2001), 927-944.

Magnu–74. W. Magnus, *Braid groups: a survey*, in "Proceedings of the Second International Conference on the Theory of Groups (Australian Nat. Univ., Canberra, 1973)", Lecture Notes in Math. **372** (Springer, 1974), 463–487.

MalSz. W. Malfait and A. Szczepański, *The structure of the (outer) automorphism group of a Bieberbach group*, Preprint (2002).

Margu–67. G.A. Margulis, *Y-flows and three-dimensional manifolds (Appendix to [AnoSi–67])*, Russian Math. Surveys **22:5** (1967), 164–166.

Margu–83. G.A. Margulis, *Free completely discontinuous groups of affine transformations*, Dokl. Akad. Nauk SSSR **272** (1983), 785–788.

Margu–00. G.A. Margulis, *Free subgroups of the homeomorphism group of the circle*, C.R. Acad. Sci. Paris, Série I **331** (2000), 669–674.

Myers–00. R. Myers, *Uncountably many arcs in $\mathbb{S}^3$ whose complements have non-isomorphic, indecomposable fundamental groups*, J. Knot Theory Ramifications **9** (2000), 505–521.

Navas. A. Navas, *Sur les groupes de difféomorphismes du cercle engendrés par des éléments proches des rotations*, Preprint (2002).

NosVi–02. G.A. Noskov and E.B. Vinberg, *Strong Tits alternative for subgroups of Coxeter groups*, J. Lie Theory **12** (2002), 259–264.

Ohshi–02. K. Ohshika, *Discrete groups*, Translations of mathematical monographs **207**, Amer. Math. Soc., 2002.

OrnWe–80. D.S. Ornstein and B. Weiss, *Ergodic theory of amenable group actions. I: the Rohlin lemma*, Bull. Amer. Math. Soc. **2** (1980), 161–164.

Osin–00. D. Osin, *Problem of intermediate relative growth of subgroups in solvable and linear groups*, Proc. Steklov Inst. Math. **231** (2000), 316–338.

Osin–01. D. Osin, *subgroup distortions in nilpotent groups*, Comm. in Alg. **29:12** (2001), 5439–5464.

Osin–a. D. Osin, *The entropy of solvable groups*, Ergod. Th. & Dynam. Sys., to appear.

Osin–b. D. Osin, *Algebraic entropy and amenability of groups*, Preprint (June, 2001).

Osin–c. D. Osin, *Kazhdan constants of hyperbolic groups*, Preprint (November, 2001).

Ozawa. N. Ozawa, *There is no separable universal II_1-factor*, Preprint (November 2002).

Penne–97. R. Penner, *The universal Ptolemy group and its completions*, in "Geometric Galois Actions, 2", L. Schneps and P. Lochak Editors, London Math. Soc. Lecture Notes Series **243** (Cambridge Univ. Press 1997), 293–312.

Pervo–00. E.L. Pervova, *Everywhere dense subgroups of one group of tree automorphisms*, Proc. Steklov Inst. Math. **231** (2000), 339–350.

Petti–52. B.J. Pettis, *A note on everywhere dense subgroups*, Proc. Amer. Math. Soc. **3** (1952), 322–326.

Plant–75. J.P. Plante, *Foliations with measure preserving holonomy*, Annals of Math. (2) **102** (1975), 327–361.

PlaTh–72. J.P. Plante and W.P. Thurston, *Anosov flows and the fundamental group*, Topology **11** (1972), 147–150.

PlaTh–76. J.P. Plante and W.P. Thurston, *Polynomial growth in holonomy groups of foliations*, Comment. Math. Helv. **51** (1976), 567–584.

Salof–01. L. Saloff-Coste, *Probability on groups: random walks and invariant diffusion*, Notices of the AMS **48:9** (October 2001), 968–977.

Samue–66. P. Samuel, *A propos du théorème des unités*, Bull. Sci. math. **90** (1966), 89–96.

Samue–67. P. Samuel, *Théorie algébrique des nombres*, Hermann, 1967.

SchUl–33. J. Schreier and S. Ulam, *Über die Permutationsgruppe der natürlichen Zahlenfolge*, Studia Math. **4** (1933), 134–141.

Semen. Y. Semenov, *On the rational forms of nilpotent Lie algebras and lattices in nilpotent Lie groups*, l'Enseignement math. (to appear).

Shale–01. A. Shalev, *Asymptotic group theory*, Notices of the Amer. Math. Soc. **48**[4] (April 2001), 383–389.

Shalo. Y. Shalom, *Harmonic analysis, cohomology, and the large scale geometry of amenable groups*, Preprint (2002).

Sidki–00. S. Sidki, *Automorphisms of one-rooted trees: growth, circuit structure, and acyclicity*, J. Math. Sci. (New York) **100** (2000), 1925–1943.

Sidki. S. Sidki, *The binary adding machine and solvable groups*, Preprint (2001).

SoiVe–00. G.A. Soifer and T.N. Venkataramana, *Finitely generated profinitely dense free groups in higher rank semi-simple groups*, Transf. Groups **5** (2000), 93–100.

Souch. E. Souche, *Quasi-isométries et quasi-plans dans l'étude des groupes discrets*, Ph.D. Thesis, Marseille (2001).

Szegö–22. G. Szegö, *Über Potenzreihen mit endlich vielen verschiedenen Koeffizienten*, Sitzungberichte der Preussischen Akademie der Wissenschaften, Phys.-Math. Klasse (1922), 88–91 [Collected Papers, Vol. 1, pages 667–561].

Tabac–00. J. Taback, *Quasi-isometric rigidity for $PSL_2(\mathbb{Z}[1/p])$*, Duke Math. J. **101** (2001), 335–357.

Trofi–80. V.I. Trofimov, *The growth functions of finitely generated semigroups*, Semigroup Forum **21** (1980), 351–360.

Ulam–60. S.M. Ulam, *A Collection of mathematical problems*, Interscience, 1960 [See also S. Ulam, *Sets, numbers, and universes – Selected works*, W.A. Beyer, J. Mycielski and G.-C. Rota Editors, MIT Pres, 1974, pages 503–670].

VdDW–84b. L. van den Dries and A.J. Wilkie, *An effective bound for groups of linear growth*, Arch. Math. **42** (1984), 391–396.

VeNeB–00. A.M. Vershik, S. Nechaev, and R. Bikbov, *Statistical properties of locally free groups with applications to braid groups and growth of random heaps*, Commun. Math. Phys. **212** (2000), 469–501.

Versh–90. A.M. Vershik, *Local algebras and a new version of Young's orthogonal form*, in "Topics in algebra", Banach Center Publications **26, 2** (PWN, 1990), 467–473.

Versh–00. A.M. Vershik, *Dynamic theory of growth in groups: entropy, boundaries, examples*, Russian Math. Surveys **55:4** (2000), 667–753.

Weiss–81. B. Weiss, *Orbit equivalence of nonsingular actions*, Monographie de l'Enseignement mathématique **29** (1981), 77–107.

Whyte–01. K. Whyte, *The large scale geometry of the higher Baumslag-Solitar groups*, GAFA Geom. Funct. Anal. **11** (2001), 1327–1343.

Wilso–72. J. Wilson, *Groups with every proper quotient finite*, Math. Proc. Camb. Phil. Soc. **69** (1972), 373–391.

Wilso. J.S. Wilson, *On exponential growth and uniformly exponential growth of groups*, Preprint (2002).

Winke–98. J. Winkelmann, *Complex analytic geometry of complex parallelizable manifolds*, Mémoire **72–73** , Soc. Math. France, 1998.

Woess–93. W. Woess, *Fixed sets and free subgroups of groups acting on metric spaces*, Math. Zeit. **214** (1993), 425–440.

The following references, firstly quoted as preprints, have now appeared.

BacVd. R. Bacher and A. Vdovina, *Counting 1-vertex triangulations of oriented surfaces*, Discrete Math. **246** (2002), 13–27.

Bambe. J. Bamberg, *Non-free points for groups generated by a pair of 2×2 matrices*, J. London Math. Soc. (2) **62** (2000), 795–801.

BarCe–b. L. Bartholdi and T.G. Ceccherini-Silberstein, *Salem numbers and growth series of some hyperbolic graphs*, Geometriae Dedicata **90** (2002), 107–114.

BarGr–a. L. Bartholdi and R. Grigorchuk, *Lie methods in growth of groups and groups of finite width*, in "Computational and geometric aspects of modern algebra (Edinburgh, 1998)", N. Gilbert, Editor, London Math. Soc. Lecture Note Ser. **275**, Cambridge Univ. Press (2000), 1–27.

BarGr–c. L. Bartholdi and R. Grigorchuk, *On the spectrum of Hecke type operators related to some fractal groups*, Proc. Steklov Inst. Math. **231** (2000), 1–41.

Barth. L. Bartholdi, *Lower bounds on the growth of a group acting on the binary rooted tree*, Internat. J. Algebra Comput. **11** (2001), 73–88.

Bavar. C. Bavard, *Classes minimales de réseaux et rétractions géométriques équivariantes dans les espaces symétriques*, J. London Math. Soc. **64** (2001), 275–286.

BekMa. B. Bekka and M. Mayer, *Ergodic theory and topological dynamics of group actions on homogeneous spaces*, London Math. Soc. Lecture Note Ser. **269**, Cambridge University Press, 2000.

BesFH–a. M. Bestvina, M. Feighn and M. Handel, *The Tits alternative for $Out(F_n)$ I: Dynamics of exponentially growing automorphisms*, Annals of Math. (2) **151** (2000), 517–623.

Bigel. S. Bigelow, *Braid groups are linear*, J. Amer. Math. Soc. **14** (2001), 471–486.

BonSc. M. Bonk and O. Schramm, *Embeddings of Gromov hyperbolic spaces*, GAFA Geom. Funct. Anal. **10** (2000), 266-306.

BruSi. A.M. Brunner and S. Sidki, *The generation of $GL(n, \mathbb{Z})$ by finite state automata*, Internat. J. Algebra Comput. **8** (1998), 127–139.

BucHa. M. Bucher and P. de la Harpe, *Free products with amalgamation and HNN-extensions of uniformly exponential growth*, Mathematical Notes **67** (2000), 686–689.

BuxGo. K-U. Bux and C. Gonzalez, *The Bestvina-Brady construction revisited — geometric computation of Σ-invariants for right angled Artin groups*, Journal London Math. Soc. **60** (1999), 793–801.

CanCo. J.W. Cannon and G.R. Conner, *The combinatorial structure of the Hawaiian earring group*, Topology and its appl. **106** (2000), 225–271.

CecMS. T. Ceccherini-Silberstein, A. Machì and F. Scarabotti, *Il gruppo di Grigorchuk di crescita intermedia*, Rend. Circ. Mat. Palermo (2) **50** (2001), 67–102.

Champ. C. Champetier, *L'espace des groupes de type fini*, Topology **39** (2000), 657–680.

FarMo. B. Farb and L. Mosher, *On the asymptotic geometry of abelian-by-cyclic groups*, Acta Math. **184** (2000), 145–202.

Grigo. R.I. Grigorchuk, *Just infinite branch groups*, in "New horizons in pro-p groups", M. du Sautoy, D. Segal, and A. Shalev Editors, Birkhäuser (2000), 121–179.

Jones. V.F.R. Jones, *Ten problems*, in "Mathematics: frontiers and perspectives", V. Arnold, M. Atiyah, P. Lax, and B. Mazur Editors, Amer. Math. Soc. (2000), 79–91.

Kramm–a. D. Krammer, *The braid group B_4 is linear*, Inventiones Math. **142** (2000), 451–486.

Lamy. S. Lamy, *L'alternative de Tits pour Aut $[\mathbb{C}^2]$*, J. of Algebra **239** (2001), 413–437.

Ledra. F. Ledrappier, *Some asymptotic properties of random walks on free groups*, in "Topics in probability and Lie groups: boundary theory", J.C. Taylor, Editor, CRM Proceedings and Lecture Notes **28** (Amer. Math. Soc. 2001), 117–152.

Leono–01. Yu.G. Leonov, *A lower bound for the growth of a 3-generator 2-group*, Sbornik Math. **192:11** (2001), 1661–1676.

LucTW. A. Lucchini, M.C. Tamburini, and J.S. Wilson, *Hurwitz groups of large rank*, J. London Math. Soc. **61** (2000), 81–92.

MarVi. G.A. Margulis and E.B. Vinberg, *Some linear groups virtually having a free quotient*, J. Lie Theory **10** (2000), 171–180.

Nekra. V. Nekrashevych, *On equivalence of nets in hyperbolic spaces*, Dopov. Nats. Akad. Nauk Ukr. Mat. Prirodozn. Tekh. Nauki **11** (1997), 18–21.

Osin. D. Osin, *Subgroup distortions in nilpotent groups*, Comm. in Algebra **29**[12] (2001), 5439–5463.

PapWh. P. Papasoglu and K. Whyte, *Quasi-isometries between groups with infinitely many ends*, Comment. Math. Helvetici **77** (2002), 1343–144.

Pauli. F. Paulin, *Un groupe hyperbolique est déterminé par son bord*, J. London Math. Soc., to appear.

Pitte. C. Pittet, *The isoperimetric profile of homogeneous Riemannian manifolds*, J. Differential Geom. **54** (2000), 255–302.

Shalo–a. Y. Shalom, *Explicit Kazhdan constants for representations of semisimple and arithmetic groups*, Ann. Inst. Fourier **50** (2000), 833–863.

Shalo–b. Y. Shalom, *Bounded generation and Kazhdan's property (T)*, Publ. Math. I.H.E.S. **90** (1999), 145–168.

Wilso. J.S. Wilson, *On just infinite abstract and profinite groups*, in "New horizons in pro-*p* groups", M. du Sautoy, D. Segal, and A. Shalev Editors, Birkhäuser (2000), 181–203.

Woess. W. Woess, *Random walks on infinite graphs and groups*, Cambridge Tracts in Mathematics **138**, Cambridge University Press, 2000.

Later updates may appear at
http://www.unige.ch/math/biblio/?preprint/liste.html

INDEX

$\mathbb{Z}/2\mathbb{Z}$ denotes the group of order 2 (additive *or multiplicative* notation) . . .
. . . . {VIII.5, footnote}